LAKE VILLA DISTRICT LIBRARY
3 1981 00604 5375

GANGES

Yale UNIVERSITY PRESS

NEW HAVEN AND LONDON

SUDIPTA SEN

GANGES

The Many Pasts of an Indian River

Lake Villa District Library
Lake Villa, Illinois 60046
(847) 356-7711

Copyright © 2019 by Yale University. All rights reserved.
This book may not be reproduced, in whole or in part, including illustrations, in any form (beyond that copying permitted by Sections 107 and 108 of the U.S. Copyright Law and except by reviewers for the public press), without written permission from the publishers.

Yale University Press books may be purchased in quantity for educational, business, or promotional use. For information, please e-mail sales.press@yale.edu (U.S. office) or sales@yaleup.co.uk (U.K. office).

Designed by Nancy Ovedovitz and set in Meridien LT and Brandon Grotesque type by Newgen.
Printed in the United States of America.

Library of Congress Control Number: 2018953983
ISBN 978-0-300-11916-9 (hardcover : alk. paper)

A catalogue record for this book is available from the British Library.

This paper meets the requirements of ANSI/NISO Z39.48-1992 (Permanence of Paper).

10 9 8 7 6 5 4 3 2 1

Frontispiece: Gulls alighting on the Ganges at Varanasi.
Photograph by Debal Sen.

In memory of my maternal grandfather, Sisir Mallik, who taught us always to think of others first

CONTENTS

PREFACE

I visited the upper reaches of the Ganges for the first time as a child in 1968. I distinctly remember its vivid aquatint blue-green ripples near the swinging suspension bridge at Lakshman Jhula. I was told that the two valiant brothers from the epic *Ramayana* had once crossed the river at that place. I remember being afraid of the beggars and lepers along its sides who solicited alms from travelers and pilgrims with their lusty cries and dramatic gesticulations. I also remember crossing the river on a crowded and unsteady ferry boat near Hrishikesh, where I could see the fish swimming in dark emerald shoals and swarming to the sides of the boat without fear as pilgrims scattered puffed rice over the water. I was familiar with the industrial waterscape of the Hugli River in Calcutta, now Kolkata, the city where I was born and brought up. The Ganges that I saw near the foothills of the Himalayas was different. Magical. Now those quiet and reclusive pilgrim towns are gone, changed beyond recognition. The river is not the same as I remember it.

Such recollection is hardly a reliable starting point for a history as large and overwhelming as this. But it is the nature of the Ganges and its presence in the minds and imagination of the people of my subcontinent that make it difficult not to think about the past without a sense of loss, no matter how indistinct or disarticulated. There is no getting away from the fact that India's great river is in trouble, suffocated by dams, overcrowded, and polluted.

This book is not about global warming or the debates surrounding the Anthropocene, but the discerning reader will detect the weight of such concerns and anxieties throughout these pages. I started writing this book twelve years ago when disputes about the future of our planet had not quite acquired the same stridency as they have today. It would be remiss not to acknowledge the fact that the fear of climate change has occupied my thoughts about the connection of the past and the future of the river as I have traveled to various points along its course to get a better sense of the terrain that I was writing about. I have followed the course of the river over the last two decades at Rajmahal, Bhagalpur, and Munger in the corridor between Bengal and Bihar; between Sasaram, Banaras, and Allahabad along the vestiges of the old Grand Trunk Road; and between the pilgrimages of Rishikesh and Gangotri in the mountains. I have crossed over the glacial cave of Gaumukh, where the river emerges, to the high-Himalayan glades of Tapovan at the base of the mighty Shivling peak. I have visited the eerie mangrove forests and the salty flats of its extensive delta.

A book like this would not have been possible without a multitude of friends, well-wishers, and extended family who have inquired, encouraged, cajoled, and needled me to get on with the project, urging me to see the end while it became larger, more daunting, and more unruly with each passing year. I will not be able to thank them all in the confines of this printed space, but I would like to remember friends who braved both instinct and logic to accompany me in many of my impulsive expeditions along the Ganges: Pradeep Gooptu for some memorable trips to Banaras, Chunar, and Triveni; Tapati and Shrimoy Roychaudhuri for their company at Munger, Bhagalpur, Rajmahal, and Gaumukh; and Sharmadip Basu for making it possible for us to trek to the magnificent glacial valley of Tapovan. I thank my friend Litan Dhar for taking me to my first tentative dip at the confluence of Prayag back in 1991. I am also grateful to Shikha Mukherjee for arranging trips to the mangroves and islands of the Sunderbans. Finally, I thank Debal Sen, a mentor in matters of aesthetic and spiritual abandon, for his nonpareil camerawork that graces the frontispiece and many of the following pages, and especially for his company at the ghats of Banaras and for one unforgettable afternoon spent with the ancient trees near the ruins of the Mulagandhakuti Vihara that are the sentinels of Sarnath. Among people who have had to listen to my long discourses about this project over the years, I must mention colleagues that I worked with at the Davis Humanities Institute, James Smith, Beth Freeman, Simon Sadler, Julia Simon, and Tobias Menely; and my departmental colleagues Ali Anooshahr, Baki Tezcan, Omnia El Shakry, Louis Warren, and Alan Taylor. To this list I must add Faisal Devji, Ramendra Sarkar, Gautam Bhadra, Daud Ali, Nita Kumar, Kevin Grant, Lisa Trivedi, Nadine Berardi, Kathryn Babayan, Jayanta Sengupta, Panchali Sen, Osmund Bopearachchi, Bishnupriya Basak, Carola Erika Lorea, Jo Guldi, Shakeel Hossain, Martha Selby, Padma Kaimal, Rick and Cathy Asher, Vandana Sinha, and my ever-supportive colleague Nicole Ranganath.

I have received many helpful comments and feedback at the conferences and talks where I have presented chapters or sections of this book. Of these I would like to mention in particular the addresses I delivered for the conferences "Land and Water: A Long-Term Perspective" (September 2015) organized by the Watson Institute, Brown University, and "Sharpening the Edges: Instating State and Power in Indian Ocean History" (August 2015), a workshop funded by the Andrew W. Mellon Foundation and organized by the International Institute for Asian Studies, Leiden. Other venues where I have presented parts of this book include Sacramento State University; Victoria Memorial Hall, Kolkata; the Instituto Italiano di Studi Orientali, La Sapienza University, Rome; the American Folklore Society; the American Historical Association; the Marian Miner Cook Athenaeum Speakers Series, Claremont McKenna College; and Hamilton College, New York.

Thanks are also due to Neel Amin for tracking down the sources and copyrights of images, to Molly Roy for her terrific work on the maps that are an integral part of this book, to Jaya Chatterjee for her patience and advice through the last and the most difficult stages of bringing the book to a close, and to my compadre and partner-in-crime for all sorts of intellectual and aesthetic transgressions, novelist Stephen Barnett, whose editorial wizardry is hidden in plain sight throughout the book. Lastly, I must acknowledge and atone for the endless frustration I must have put my family—L, D, and M—through in my years of writing "the Ganges book."

GANGES

INTRODUCTION

Our story begins with the basin of the Kali Gandaki River, which flows from Nepal into India through the desolate Ladakh Range defile over a bed that is 5,000 feet high, joining at last the River Ganges above Saran in the northern Indian plains. Along the course of this river lie some of the oldest trails frequented by yak caravans, Buddhist monks, and salt traders since antiquity. In a distant geological epoch the Kali Gandaki used to flow all the way from Tibet. The river is much older than the Himalayas, which make up one of the youngest mountain systems in the world. Geologists have been fascinated by the Kali Gandaki and its tributaries because it is the only river that has retained its path through the Himalayan massif. In a geological process known as antecedence,[1] it has cut its path at a pace faster than the rise of the mountains, resulting in one of the world's deepest and most stunning gorges, lying between the Dhaulagiri and Annapurna peaks.

Rivers are tireless dynamos of nature that erode, transport, and deposit rocks, continually retracing the surface of the earth. The Kali Gandaki has dissected the continental crust from time immemorial, exposing layer upon

Ammonite fossil, worshipped as the insignia of Vishnu. Courtesy the Gooptu family, Kolkata. Photograph by Pradeep Gooptu.

layer of sedimentary rock that once formed the bed of the great Sea of Tethys, a relatively shallow body of water that covered much of what later became Eurasia more than 190 million years ago. The mass of seashells and marine organisms deposited on the ocean floor laid much of the limestone grid of the Asian and European continents. Tectonic activity during the early Cretaceous created a massive upheaval that squeezed the Sea of Tethys and eventually raised its bed, creating the Himalayas by pushing Cretaceous and Eocene limestone along with Paleozoic shale skyward over 10,000 feet. An abundance of marine fossils are still embedded in the high Himalayas and strewn along the gorge of the Kali Gandaki River.

During the formation of the Himalayan massif, sulfate-reducing bacteria residing in shale created a myriad of fossil layers rich in deep marine belemnites such as cuttlefish, gastropods such as slugs and snails, bottom-feeding brachiopods, and some of the most beautiful ammonites in the world. Ammonites were mollusks that scoured the floors of the Sea of Tethys a hundred million years ago. Their dark polished fossils, typically found in the Kali Gandaki gorge, are known as shaligrams (*śālagrāma* in Bengali) and have been worshipped throughout the Indian subcontinent for thousands of years as the preeminent sign of the god Vishnu. The ancient settlers of the Indian subcontinent were entranced by the presence of creaturely stones that spoke of time outside ordinary human dimensions. The proto-Hindu scriptures pit human finitude against eternal creation (*kāla* and *ākāla*); stories of how the world came to be have always straddled the divide between god and nature, cosmology and ecology.[2] Vishnu, the keeper of all creation, took the form of the fish (*matsya*) avatar during the great deluge to save the world.[3] Many of his insignia are aquatic: ammonite, conch shell, lotus. The River Ganges in Hindu mythology is the purest emanation of Vishnu in fluid, natural form. This book begins with the simple and puzzling question: How did the Ganges come to assume such a central place in the civilization and culture of the Indian subcontinent?

River of Afterlife

The seventh-century Chinese traveler Xuanzang visited the confluence of the Ganges and the Yamuna at Prayag, a spot frequented by Hindu pilgrims during the early part of the seventh century C.E., when it was surrounded by forests full of wild elephants.[4] Here he saw hundreds of men fasting for days before immersing themselves in the river for a last ritual bath before committing suicide by drowning. Xuanzang describes this in a matter-of-fact way, noting also that Hindus believed that the confluence was enchanted, a spot where monkeys and deer from the nearby hills also came to die in large numbers for salvation. He saw that the Brahmins had raised a high column in the middle of the river where they prayed by holding on to the pillar with one arm and foot while stretching outward with the rest of their bodies, eyes transfixed on the setting sun. Such grueling homage went on for years and even decades. The devout Brahmins of Prayag believed that their exertions would release them from the endless karmic cycle of life and death.

It is remarkable how long the Ganges has been a comfort for the dying. An untold number of Hindus have spent the last moments of their mortal lives with a dribble of sacred Ganges water on their lips. In the fifteenth century it was common practice for penitent Hindus from all over India to end their lives at Prayag by jumping from a great banyan tree either into a well or directly into the Ganges—a custom that was finally ended by the Mughal emperor Akbar. Early Europeans, especially missionaries, were morbidly fascinated by the sight of the sick and aged crowding the banks of the river, half immersed in the water, praying and counting the moments to the final call. Such scenes of a last "journey to the Ganges" (*gaṅgā-yātrā*) were common in early British Calcutta; they were painted by artists like Balthazar Solvyns in the late eighteenth century and transmitted to European viewers as evidence of the bizarre customs of the natives of India. Many devout Bengali Hindus less than two hundred years ago still believed that a death that did not occur by the side of the river was a dire misfortune. The best way to die was to have spent at least three nights by the river, immersed up to one's navel in the water, chanting "*gaṅgānārāyaṇabrahma*" ("Ganga is Vishnu and Brahma"). A wealthy Calcutta native, Chudamani Datta, keen to outdo his social rivals, ordered a crowd of drummers for a procession the moment he was declared gravely ill. Seated in a silver carriage draped in cloth printed with Sanskrit hymns and strewn with holy basil, he asked the bearers to carry him to the Ganges. As the carriage wound its way through the city, a throng of singers sang about how, having conquered the world with his money, Chudamani was now going to conquer Yama, the Lord of Death.[5]

Such unflinching conviction in the face of death reveals the extent to which a river can become synonymous with the threshold of the afterlife, integral to the continuum between the worlds of the living and the dead.

Ritual ablution or immersion in this context is not only an outward symbol of purification but very much a visceral act of rebirth and renewal, an essential element, for example, in the consecration of kings. An extraordinary act credited to King Dindiga (Prithvipati I) of the Ganga Dynasty of Talkad, in present-day Mysore in the state of Andhra Pradesh in southern India, was recorded in a copperplate inscription commissioned by his son. During the late ninth century C.E., at the battle of Vaimbalguri, fighting vassals of the formidable Rashtrakuta kingdom of the north, he was badly wounded.[6] Anticipating death but unwilling to lay down arms, he cut off a piece of his bone with his sword and sent off the fragment with a messenger so that it could "enter the water of the Ganges."[7] His dynasty had taken its name from the River Ganges, claiming genealogical descent from the Solar Dynasty of northern kings. Dindiga's self-infliction not only prefigured a warrior's honorable death and final oblation but was also meant to secure the blessing of his ancestral lineage through the consignment of his bone shard to the expiatory waters of the river. Panditaraja Jagannatha, Mughal court poet extraordinaire, a scholar of linguistics, poetics, and philosophy hounded by the Brahmin orthodoxy led by Hara Dikshita for marrying a Muslim woman, sought refuge on the steps of Banaras by the side of the Ganges.[8] Forbidden to step into the water lest he pollute the river with his transgression, he was moved to compose his famous devotional eulogy of the Ganges known as the *Piyushalahari*. As he composed each verse, legend has it, the river rose step by step, and at the end of his recitation swept him and his devoted wife away.[9]

Along with such stories of voluntary sacrifice, various kinds of ritualistic living death were also common among the many orders of Hindu monks in India, whose rite of passage into a state that is free of social ties and obligations (*jīvanamukta*, freed from life itself) was marked by the symbolic act of cremation. In Varanasi, the haunt of some of the oldest monastic orders in the world, countless such virtual funerals have taken place on the steps leading to the waters of the Ganges. Viennese anthropologist and scholar of Sanskrit Leopold Fischer, known as the monk Swami Agehananda Bharati, has left an eloquent account of his initiation into the Dasanami ascetic order in Banaras, performed by his guru Swami Visvananda.[10] The ceremony took place at the cremation grounds of the Manikarnika ghat, where a platform was set up next to the blazing funeral pyre of a young woman and her stillborn child. Agehananda recalls the muted chants of his guru offering black and white sesame seeds to the heavens as tokens of the removal of all earthly blemishes, propitiations for the goddess of fortune proclaiming the initiate free from all ties of the world, liberated from the pangs of grief, hunger, and death. The oblation was followed by the symbolic act of cremation, where he recited his own funerary rites and his body was touched by a smoldering charcoal from the pyre in seven places, commencing his death and integrating his consciousness with the ambient unity of the cosmos. He was then asked to disrobe, walk down the steps, and immerse his newly anointed body in the water. As soon as he "emerged from Gaṅgā's womb," Agehananda writes, he was given an ocher robe and the name *agehānanda* ("homeless bliss"). The guru handed him a wooden staff to be flung into the middle of the holy river, a reminder of the order of human society for which he now had no use. The sanctity of Ganges water in the mind of the dying soldier facing death in battle and the ritual funeral of a sannyasi are affective examples of ancient life-cycle rituals that have defied the passage of time and history, where the river, fashioned from the primordial bodies of earth and water, survives as a visible reminder of human transience.

Thinking with a River

The river as a clearly defined object—with a beginning, a middle, and an end—is, after all, a human fabrication. As a natural phenomenon it is part of the earth's water cycle, the endless succession of clouds, rain, snow, and glacial melt that merges into other rivers, lakes, or the ocean. This kind of reckoning was known to Indian philosophers of antiquity. A verse in the *Prashna Upanishad,* for instance, explaining personhood in terms of the specific and the universal, gives an analogy of the river and the ocean: when the river reaches the ocean, it *becomes* the ocean, and its name and form are dissolved at that instant.[11] The Buddhist text of the *Anguttara Nikaya* speaks of raindrops gathering on mountainsides and pouring down clefts, gullies, and creeks, filling lakes and streams and replenishing the rivers that fill the oceans in the ceaseless play of the infinitesimal and the infinite.[12]

At the same time, the Ganges is also a river incarnate, indispensable to thinking about the history and culture of the Indian subcontinent. In this sense it is not only a natural entity outside the frame of ordinary human experience but also a reflexive extension of something akin to a uniquely Indian consciousness. The relationship between anthropomorphic and naturalistic conceptions of the Ganges thus has deep roots in Indic culture. This connection is also manifest in various ritual and bodily practices, especially in the Hindu tantric traditions. The river Ganges and her two sisters, Yamuna and Saraswati, are identified, for example, with the three subtle conduits (*nāḍīs*) of the life force in Yogic conceptions of the human body. The solar *piṅgalā* is the Yamuna, the lunar *iḍā* is the Ganges and the medial, fiery *suṣumṇā* is the hidden Saraswati. When yogis perform breath control (*prāṇāyāma*) they are supposed to inflate and clear the two peripheral channels so that the third is filled with the subtle breath of life, forcing it upward through the body, leading them toward blissful liberation. As David Gordon White points out, the identification of the body with the triad of the sacred Indian rivers is fundamental to the learning and practice of traditional yoga.[13]

India, of course, is not the only place where a river has placed such weight on the definition of civilization. The Yangtze and the Yellow Rivers (Huang He) have equally shaped the long-term ecological and political orientation of Chinese culture. They have both determined to a large extent the intensity of agricultural sustenance and population density in China and the fate of successive regimes that attempted to harness water as an essential resource. The Yellow River, as David Pietz shows, has always tested the resolve and ingenuity of the Chinese state in mitigating the devastating floods and seasons of want since the time of the Warring Kingdoms of the fifth century B.C.E.[14] While the hydraulic models of top-down Oriental despotism advanced by theorists from Karl Marx to Karl Wittfogel have now largely been put to rest, it is still true that many foundational ideas of statecraft and moral legitimacy in China emerged from the struggle to find a balance between natural phenomena and the organization of society. The sorrows of the Yellow River taught Chinese imperial regimes the value of long-term stability in the distribution of natural resources, and while perennial floods took a terrible toll on the lives and effects of their subjects across the northern plains, they were also prized as a gift from above that made the dispersal of sediments and the distribution of rich alluvium possible.[15] As Philip Ball points out in a recent book, the Chinese language has had a historic obsession with water as a fundamental element in the natural world, and Chinese rivers have "shaped the way the country has been governed" across centuries.[16] Rivers have dictated the coordinates of the Middle Kingdom, and by association the essential ideas of Chinese nationhood as a geographical expression; river-myths have played a significant role in shaping the ideals of Chinese civilization, in which legends of deluge were often meant to justify political rules and social norms, especially in the interests of the state. Seen as punishment for the wrongdoing of subjects, the occurrence of floods added further credence to the belief that responsible management of water was a benchmark of legitimate and enlightened authority.

The Ganges offers a different window into the culture and civilization of the greater Indian subcontinent. The rich and boisterous mythology surrounding the river does not reveal such a straightforward relationship to the dominant social and political order. If anything, the practice of mass pilgrimage to sacred sites along the river has defied the everyday strictures of status and caste. While kings and emperors fought endless battles to lay claim to the valley that is India's own "middle kingdom," along with its confluences and sacred sites, and attempted to seal their authority with the blessings of the river goddess and her waters, no regime could afford to disregard the deeper imprints of cosmology, myth, and metaphysics preserved and nurtured, although unevenly, in scriptures, art, and oral traditions. Even during the periods of the last great empires such as the Mughal and the British, which were built on surplus generated from one of the largest peasant economies in the world, when irrigation and the redistribution of water became a significant priority of statecraft, the virtuous river and its pilgrim landscape remained vivid in the popular imagination. Unlike the distant and forgotten gods of the Nile such as Hapi and Khnum, the Ganga as a deity is very much alive to this day. The only other example of such a venerated body of water with similar universal salience and mythic geography is the River Jordan, enshrined in both Judaic and biblical traditions. It is a reminder of what Rachel Havrelock describes as a "cosmic boundary" between two overlapping templates of civilization.[17]

The mythical sacredness of the Ganges is not simply a veiled acknowledgment of its ecology or the environment. Myths, to echo the great anthropologist Bronislaw Malinowski, are much more than the musings of primitive naturalists.[18] Apart from the river's ritual significance, its history includes intimate forms of knowledge of the natural environment and its bearing on forms of social organization: tribes, villages, cities, kingdoms, and empires. To prize out such facets of this past, following the lead of geographer Nigel Thrift, this book focuses not only on the linguistic and textual representations of natural objects but also on the experiences and material practices that may not fall neatly into the categories of nature and culture.[19] This notion of the river as a living presence in the history and culture of a people is also a response to the recent provocation of anthropologists such as Eduardo Kohn, who in his study of the forests and peoples of the Upper

Amazon has argued for a deliberate unsettling of the conceptual divide between human and nonhuman aspects of everyday life.[20]

The history of the Ganges since antiquity has also been a history of divine manifestations: gods, goddesses, ancestors, spirits of the forest, trees, water bodies, and the underworld. It is also tied to the history of changing conceptions of birth, death, and rebirth and forms of social organization and caste hierarchy based on such distinctions. As the political philosopher Eric Voegelin once pointed out, every society must create and maintain an order that imparts meaning to ordinary human lives, one that fulfills the purpose of their existence toward divine and human ends.[21] Elements of this duality appear across the mosaic of Indian religious traditions, which includes various forms of Buddhist, Jain, and Hindu beliefs and practices, often dominated by deterministic views of the purpose of human existence. The ulterior ends of human life are carefully laid out in their core doctrines, which stress prescribed forms of gnosis and ritual practice over immediate and intuitive experiences of divinity and nature. These tendencies have helped carve a cultural order in India distinguished by what Mircea Eliade saw as the stubborn persistence of archaic forms of ontology intended to discredit alternative or discordant visions of the world: sensory, hedonistic, or utopian.[22] The weight of such moral and social sanction—most evident in outward forms of Brahminical orthodoxy—make it challenging to find in the historical record experiences of ordinary people who have lived and died alongside the Ganges for generations untold, or indeed a view of the river from the perspectives of boaters, fishers, or poor cultivators whose livelihoods and survival depended on different and intimate forms of knowledge about the river.

Take, for instance, the seasonal and transient settlers of the sandbanks (*diyārā*s) of northern Bihar and eastern Uttar Pradesh, or the more extensive *car*s of the Bengal delta overgrown by reeds and catkin grass. These remote habitats have their own microhistories of subsistence and struggle for survival. A recent study by geographers Kuntala Lahiri-Dutt and Gopa Samanta show that along the flood-prone lower reaches of the Ganges and its tributaries, the predominant suffix for place names is the word *bhāṅgā* ("broken"), denoting the continuous acts of erosion and subsidence of an unstable and dynamic delta.[23] In the extended deltaic fans of the river, now shared by the coastlines of India and Bangladesh, one can find even more compelling stories of survival among communities that have lived by the turbulent waterscape of salty, serpentine creeks, tidal lowlands, and brooding mangrove forests, rendered timeless in Samaresh Basu's 1957 novel *Ganga*. Basu's story is based on the lives of the *mālo* fishers of the Sundarban delta, "fish killers" (*māchamārā*), boat makers, net weavers of the estuary, and deep-sea fishers, whose primal ancestor appeared from the marine depths, dark and sinewy, with fishing spear in hand.[24] Recounting the life story of an aging fisherman, Basu's novel captures the terrifying, murderous lure of the river, the forest, and the sea, and the lives of fisherfolk living at their mercy, braving hidden undertows and the retreating tides that snatch away boats to the lethal depths of the outer ocean. Basu writes about the spirits of the impenetrable forest, of the fish god with unblinking eyes who watches the hunter and the hunted at sea, and of Ganga herself—a mother-goddess who holds the lives and deaths of the fishers in the palm of her hand. This book cannot pretend to do justice to such subaltern figures that crowd the margins of this long-unfolding history, just as it cannot bring back to life the textured pasts of peasants and tribes of antiquity who have battled the forests, swamps, and waterways of the Ganges basin with plow and ax, or the faceless multitudes of pilgrims and devotees, or the forgotten guilds of traders, artisans, and temple sculptors who made the face of Ganga the goddess appear on rock and stone. I am convinced, nevertheless, that the cultural

Boaters of the Sunderbans, West Bengal. Photograph by author.

gravitas of the river is not simply a function of myth and religion but also of this richly layered human landscape.

Such a story can only be narrated along an extended timescale. I have taken the liberty of foreshortening passages of history that have unfolded over centuries to try and capture the broader sweep of change across time. These questions inevitably dredge up old controversies among environmentalists about the distinction between the natural and the human, especially the conflict—as geographer Carl Sauer saw it—between evolving human mastery over the environment and the "revenge of outraged nature."[25] This book also examines the assertion of historian William Cronon that not only is the natural environment a historical actor in its own right, but plants, animals, soils, and climates and a host of nonhuman entities have shaped the history of the planet both with and without humans.[26] From such a perspective, the river is neither an independent nor a dependent variable; it is both cause and context for ecological change. In this regard, the contrasting pasts of the Ganges presented in the book recall Jamie Linton's perceptive analysis of water as a subject of history in its own right, and his suggestion that such history must also be moved by everything else that water acts on, sustains, or is part of—atmosphere, climate, soil, elevation, microbes, flora, and fauna—the whole array of factors that affect patterns of survival and sustenance and test the limits of human instinct and ingenuity.[27] In relation to the use and exploitation of water, rather than a linear ascendancy of human industry over forces of nature, the history of the Ganges shows both the vulnerability and resolve of humans who have survived the onslaught of monsoons, drought, and floods and their roles as agents and victims in the exploitation of resources, the depletion of habitats, industrial pollution, the precipitous fall of the water tables, and groundwater contamination.

The plains of the Ganges stretch from the valley of the Brahmaputra River in the east to the Indus River system in the west, a combined distance of more than 1,600 miles, skirting the southern flank of the Himalayas, occupying an area of roughly 300,000 square miles. The river measured from the source to the Bhagirathi channel in western Bengal is over 1,500 miles across, an immense, active, and turbulent body of water. Measurements of its flood discharge taken during the late nineteenth century from Rajmahal, a point before its entry into the Bengal plains after it has received the water from all its major tributaries, were recorded at 1.35 million cubic feet of water per second.[28] Charles Lyell in his 1840 textbook on geology estimated that the Ganges pours 355,361,464 tons of mud into the ocean every year, roughly the weight of sixty Egyptian pyramids.[29] The Ganges basin lies over the remains of many ancient and abandoned floodplains. Hidden in its clay, sand, and gravel are pockets of coarse peat with remains of the shells of aquatic creatures and the bones of long-extinct fauna. At Lucknow, in the state of Uttar Pradesh, artesian wells 1,000 feet below sea level seldom reach the bottom, revealing layers of ancient silt, peat, gravel, and tree stumps—evidence of continuous subsidence.[30] Such records of habitation are particularly evident in the delta, where peat beds of centuries past lie buried underneath low-lying mangrove flats.[31] They remind us of the tremendous power such a body of water can wield on human subsistence. Generations of British officials noted the continuous shifts and meanders of the main bed of the Ganges in Bengal, Bihar, and the United Provinces of British India, which led to the proliferation of abandoned channels. They saw the rise and subsidence of the great sandbanks in places like Munger in Bihar, where the river in spate could fan out six to eight miles across.[32] British irrigation engineers during the early nineteenth century tried to alter the channel of the Bhagirathi River in Bengal, but within a year after they began dredging, the entrance of the river at Murshidabad shifted five miles westward, the force of the current changing its breadth from 250 feet to half a mile. In 1825 the river changed its course again; the same point moved eight miles to the southeast, a deviation that severely disrupted navigation from 1826 to 1831. These examples show how unruly and capricious the Ganges has been at certain points in its history. In the 1820s, during the attempted dredging of the main channel of the Ganges at Nadia, a great number of timber logs of the sal tree (*Shorea robusta*) were found buried in the riverbed along with hundreds of sunken boats, many at a depth of 12 feet in the sand.[33] During the digging of irrigation canals, remains of abandoned villages, ancient cities, and lateral aqueducts (*rajbuha*s) turned up routinely. Such evidence suggests that the course of a river and the lives of the people along its banks can change abruptly over time. In this, the river does not act alone. A convergence of unrelenting forces, including the lift and subsidence of soil and rocks as well as changes in climate and atmosphere, shape the contours of a river basin, and, in turn, the human dimensions of its ecology.

The cultural memory of the River Ganges dates much further back than the recorded history of the Indian subcontinent, mired in genealogy, legend, and fable. This ancient body of myths attending the revered figure of the river, a goddess who descended from heaven to absolve sinners with her purifying waters, are still vital, not just to Hindus but to a vast majority of the people across southern Asia and beyond. Without her one could hardly imagine the destinies of the people who have lived by the slopes of her descent, her fertile basin, or her expansive delta, over 2,000 years. This book explores the evolution of this image of a cosmic river at the intersection of myth, history, and ecology. It asks why the Ganges was held in such reverence across a multitude of religious traditions and how it became a trophy fought over by tribes, kingdoms, and empires that sought to command its valley to proclaim their territorial prowess. It is an account of the changing ecological profile of the river, whose burden of population and contaminants, along with axes, plows, and animals, has transformed the contour of its plains and forests. It shows how the river embodies the collective past of an entire subcontinent as a bounteous, maternal figure; a nurturer of villages, cities, and kingdoms; a vessel for the concourse of traders, pilgrims, and invaders; and, in more recent times, a beacon for the Indian nation-state torn between promises of development and risks of environmental degradation.

Two threads tie this book together. One is the mythical and historic significance of the sacred river, worshipped in human form and fought over as a political icon for centuries by various regimes. The other is the long-term convergence of climate and ecology and the cumulative consequences of human activity—the expansion of irrigation, agriculture, industry, and population—that has contributed to the transformation of the Ganges valley in the long Anthropocene, setting up the paradox that the purest of all rivers has also become one of the most polluted in the world. These two strands of history are plotted along discrepant and mutually incompatible scales of history and time. This book argues that these uneven contours of the past are very much at work today, especially at the heart of contemporary debates on its future, its valley, and its people.

CHAPTER 1

THE WORLD OF PILGRIMS

Most Himalayan pilgrim trails are designed to remind those who have taken the pledge that the journey will test their stamina and resolve. My students and I observed this firsthand during an expedition in the fall of 2007. We boarded the train to the city of Rishikesh in the Garhwal Himalayas. From there we hopped onto a crowded bus and then a commuter van that lumbered all the way up to Gangotri, where the mighty Ganges begins its long descent through the rugged Bhairon Ghati gorge. The trail lay across the last foothills of the Himalayan massif—one of the oldest pilgrim routes in India. As we looked at the Miocene and Pleistocene molassic sediments accumulated over millennia, rising skyward from the quaternary alluvium deposited by the tireless river, the mud, silt, and sand grains of the plains down below seemed like a distant memory. We were now in the kingdom of pebbles, rocks, and boulders through which the pilgrim path meandered over unforgiving crags, leading us toward the glacial source of the Ganges.

Our van had disgorged us at Gangotri, the little pilgrim town that had once been much closer to the glacial melt that becomes the Bhagirathi River,

The pilgrim trail toward the Gaumukh glacier.
Photograph by author.

which is what the Ganges is called in the hills. During a different part of the Holocene, before the earth's climate became so volatile, the Gaumukh glacier must have reached all the way down here. The ice cave has now moved at least eleven miles upstream. I remember a tangle of deodar and pine trees conspiring in the wind at the head of the gorge overlooking the trail through granite cliffs and glacial moraines. As we began to climb the stairs of the temple built by Amar Singh Thapa, the Nepali conqueror of Garwhal, we heard the rushing water of the Ganges, cascading over beige and gray granite polished to a mirror finish. Sometimes with a murmur, sometimes

The cave of Gaumukh. Photograph by author.

with a roar, the Ganges was with us for the entire journey. The pilgrim path began after ninety unbelievably steep, paved steps that were likely to give pause to the infirm and the faint of heart. Incredibly, they did not seem to deter the fragile old men clothed in faded saffron, some sporting raggedy old sneakers, who strode merrily ahead of us. As we labored upward through slopes of grass, scrub, and thorn, we were greeted by moru oaks, rhododendrons, firs, and forests of deodar, birch, and the long-needled chir pine (*Pinte longifolia*). We were on our way to the Gaumukh glacier at the base of the Bhagirathi peaks at about 12,000 feet. That night we camped in Chirbasa amid the pines, under a starlit sky by the gurgling river.

The River Ganges near its source. Photograph by Debal Sen.

The next evening, by sundown, famished and exhausted, we reached the dark frozen cave from which the main channel of the Ganges emerges. The nose of the glacier was visibly active, with unstable lateral moraines. We were told by Indian army *jawan*s who were on the way to their base camp not to venture too close to the cave. Two days before our arrival, a massive chunk of ice had fallen away from the roof of the cave and splashed near the edge of the precipice, sweeping away a German photographer and his guide who were trying to set their tripod at the edge of the floe. Their bodies, the *jawan*s said, would never be found again. Next morning we were on our way again, heading across the glacier toward the remote and rugged alpine meadow of Tapovan at more than 15,000 feet. This was a serious, precipitous climb over jagged rocks and shifting debris, over slow-moving glacial formations. Our feet ached. We had finally reached the base of Shivling, Lord Shiva's crested mountain—an expansive green and red meadow of succulents and glacial rivulets that shone briefly in the afternoon sun before the clouds came down again.

I offer readers a brief glimpse of this journey made by millions before us to emphasize the daunting prospect that this kind of climb poses for the elderly and the sick. Yet at the core of the Indian cultural mosaic, great value is still placed on such arduous journeys. They are the tribute of a lifetime

Glacial melts at the base of the Shivling peak. Photograph by author.

paid in voluntary suffering to witness the descent of the purest river in the world, whose waters have the power to cleanse—as most Hindus believe—every kind of sin known to humankind. This fundamental idea of ablution and redemption can be traced to the dawn of Indian civilization; it not only runs through Hindu beliefs and practices but animates the imagination of people from all walks of spiritual life. In this regard, the Ganges is much more than a river; it is a metaphysical threshold. This chapter explores the ways in which the Ganges has been held sacred through the lens of pilgrimage—especially as embodied in the ancient idea of *tīrtha*: travel as penance and redemption—as well as the ways in which the river enters and exits the world of rituals, and the ways in which the water itself becomes an object of worship, a cleanser of sins, a healer of afflictions, and even a destroyer of microbial pathogens.

Praise to the River

Of the many odes written to the Ganges in various Indian languages, these lines are perhaps the most poignant:

O River, daughter of Sage Janhu, you redeem the virtuous
But they are redeemed by their own good deeds—
where's your marvel there?
If you can give me salvation—I, a hopeless sinner—then I would say
That is your greatness, your true greatness
Those who have been abandoned by their own mothers,
Those that friends and relatives will not even touch
Those whose very sight makes a passerby gasp and take the name of the Lord
You take such living dead in your own arms
O Bhagirathi, you are the most compassionate mother of all[1]

These Sanskrit *śloka*s, taken from an eight-stanza ode to the Ganges, have been a part of the oral tradition in Bengal for centuries, and many people knew them by heart just a generation ago. They were composed—surprisingly—not by a Brahmin, not even by a Hindu, but by a thirteenth-century author who went by the popular name of Darap Khan Gaji. The noted Bengali linguist Suniti Kumar Chatterji identified "'Darap Khan'" as Zafar Khan Ghazi, who is credited with daring military exploits during the first major phase of Islamic expansion in Bengal toward the end of the thirteenth century, after the Turkish Sultanate had been established in northern India around Delhi as the new capital. To find what remains of the memory of Zafar Khan, a self-proclaimed virtuous warrior of Islam in Bengal, you have to travel to Tribeni, a small town in Hugli in West Bengal on the banks of the Bhagirathi, which is the name of the Ganges there. This place, which was considered very sacred in antiquity, is where the Ganges once branched off into three streams: the Saraswati River flowed southwest beyond the port of Saptagram, the Jamuna River[2] flowed southeast, and the Bhagirathi proper flowed through the present Hugli channel all the way to the location where English traders much later erected a city, Calcutta.

Zafar Khan Ghazi was said to have struck terror among the local Hindus, attacking their temples and idols during the late thirteenth and early fourteenth centuries. He conquered the pilgrimage of Tribeni and the port of Saptagram, destroyed a large and ancient temple there, and allegedly used the spoils to build an imposing mosque. He took the title of Ghazi, warrior of Islam, and established a school for Arabic learning and a charity (*dar-ul-khairat*). One of the oldest Bengali Shia texts has a curious tribute to the Ghazi:

On the quays of Tribeni pay respect to Daraf Khan
Whose water for *wazu* [ritual ablutions] came from the River Ganges[3]

The little we can surmise about Zafar Khan's life and death reminds us that the sacredness and the value of a river and the landscape that it flows through are entwined with the practice of everyday life. He seemed to have realized this later in life. He became a friend of the poor, donned the robes of a Sufi mystic, learned to write beautiful Sanskrit, and eventually won the hearts of the local people he had tried to convert forcibly to Islam in his youth. For generations to come, Zafar Khan became an emblem of the composite culture of Hindus and Muslims in Bengal, which shared a sense of enchantment with the landscape of the delta. The noted Bengali critic, novelist, and historian Bhudev Mukhopadhyay, writing in the latter half of the nineteenth century in his utopian history of India, imagined a country where people of all faiths paid obeisance to the river by singing the hymns of Darap Khan.[4] Mirza Ghalib, the foremost Urdu poet of his time, echoed a similar sentiment when he visited Varanasi (Banaras) in the early spring of 1828 and fell in love with the city, composing a poem in Persian called "Chiragh-e-Dair" [The Lamp of the Temple], a memorable tribute in which he named the city the "Kaba of Hindustan":[5]

> May Heaven keep
> The Grandeur of Banaras,
> Arbour of bliss, meadow of joy,
> For oft-returning souls
> Their journey's end.[6]

He almost wished that he could have left his own religion to pass his life on the bank of the Ganges with prayer beads, a sacred thread, and a mark on his forehead.[7]

One of the oldest explanations for this abiding faith in the purity of the waters of the Ganges has to do with the practice of pilgrimage that has for centuries provided a stage on which to reenact the difficult inner journey of reconciliation and atonement, often imagined through the pristine Himalayan landscape of mountains and glacial melts—a terrain that the great Sanskrit poet Kalidasa describes as "*diśī devatātmā*" [the country of divine beings]. Ritual baths and offerings at sacred spots along the river are tied to this sense of geography, which is steeped in ideas and images drawn from history, myth, and nature as shared forms of reckoning, an experience (*tīrthabhāva*) difficult to capture in words. When Guru Nanak, the founder of Sikhism, traveled in search of wisdom on the Hindu pilgrimages of the east, as recorded in the *Janam Sakhi*, it was not the usual places of ritual obeisance that impressed him. He was entranced instead by a flock of migratory swans alighting nearby. They appeared much closer to heaven than did the throngs of pious Hindus. With their shining silver-white plumage and burnished eyes, the swans were messengers who flew across the Himalayas, from India to Central Asia and back, year after year.[8] It is the journey, the story conveys, and not the destination, that defines the purpose of pilgrimage. Such convictions and practices, along with other aspects of Hindu practice, have been misunderstood by an array of observers and critics from the West, including missionaries, colonial administrators, authors, and travel writers.

Pilgrim Trails

There is no single explanation of why the ritual significance of the Ganges and its water has permeated so deeply into the realms of everyday culture across the Indian subcontinent, but an important clue lies in the historical evolution of certain practices of pilgrimage. In most northern Indian languages derived from Sanskrit, the main purpose of the pilgrim's journey is encapsulated in the term *tīrtha*. Jain texts contain some of the earliest references to this term, also tied to the term *tīrthaṅkara* (literally, "ford-makers"), denoting the founders of the faith who had overcome worldly attachments and led their followers toward eternal life. The places associated with such ascetics eventually became part of the Jain repertoire of holy sites.[9] The standard Sanskrit definition of the term denotes a passage or road, and also typically a ford across a body of water, or a bathing place or stairs leading down to a river, and thus a sacred place on the banks of a stream or river. The pilgrim's journey is referred to as *tīrthayātrā*, where *yātrā* denotes journey or travel, physical and spiritual. The distance between one side of the river and the other is a common metaphor for the journey from this life to the next—an act of atonement and redemption from time immemorial among Hindus of every persuasion, ascetic or householder. Obligatory and meritorious, such difficult journeys are reserved for the latter part of one's life on earth when the ties of family, social duties, and pleasure become less important and thoughts of the afterlife take over. A pilgrim's journey is also measured by the difficulty of reaching the site. A *tīrtha* in this sense is a physical and metaphorical reenactment of the struggles of a lifetime. This penultimate journey for the elderly, the sick, or the dying is considered part of the penance performed to earn a place with ancestors and gods in the afterlife. In this eternal quest, the mountainous terrain of the Garhwal Hi-

Gori Ganga (also known locally as Bhagirathi) descending through the Garhwal Himalayas. Photograph by Debal Sen.

malayas, through which the Ganges descends to the Indian plains, embodies the archetypal pilgrim landscape for millions of merit seekers whose dusty feet and walking staves have marked these trails for centuries.

The confluence of rivers (*saṅgama*, which literally means "to go together"), especially of the Ganges and its tributaries, is one of the most significant geographical spaces for the pilgrim, just as the geomancy of certain rites is determined by the migration and conjunction of stars and planets (*saṅkrānti*), the phases of the moon (*pakṣa*), the equinox and solstice, and eclipses (*grahaṇa*).

A common name for such a place in Sanskrit and Sanskrit-derived languages is *prayāga*. Prayag is also one of the most celebrated pilgrimages at the meeting of the Yamuna and Ganges Rivers near the city of Allahabad, but there are others, such as Rudraprayag, situated at the meeting of two rivers: the Mandakini River, coming down from the steep glaciers beyond Kedarnath, and the Alakananda River, making its way from Badrinath. Jim Corbett, in one of his memorable hunting yarns, *The Man-Eating Leopard of Rudraprayag*, gives us an endearing and intimate account of the pilgrimage to Kedarnath that starts from the cities of Haridwar and Rishikesh on the banks of the Ganges, through the mountains of Garhwal, across the gorge of the Alakananda River, and up the left bank of the Mandakini River to the sacred abode of Kedarnath. A native of the hills, born to an Irish family settled in Dehradun, Corbett mused over the obligatory penance of the pilgrim from the plains of Hindustan who must trudge resolutely upward through an unfamiliar and forbidding landscape battling altitude, gravity, and fear:

> The road in front of you, which has been trodden by the feet of millions of pilgrims like you, is excessively steep and incredibly rough, and you, whose lungs have never breathed air above sea level, who have never climbed anything higher than the roof of your house, and whose feet have never trodden anything harder than yielding sand, will suffer greatly. Times there will be, a-many, when gasping for breath, you toil up the face of steep mountains on feet torn and bleeding by passage over rough rocks, sharp shale, and frozen ground, when you will question whether the prospective reward you seek is worth the present price you pay in suffering; but being a good Hindu you will toil on, comforting yourself with the thought that merit is not gained without suffering, and the greater the suffering in this world, the greater the reward in the next.[10]

Corbett's view is remarkably unencumbered compared with those of many of his contemporaries and predecessors who thought poorly of such masses of pilgrims. Other European observers who wrote about meeting Hindu pilgrims on these trails were both fascinated and appalled by the simple faith of poor Hindus. Abject and absolute renunciation, even in old age, had always baffled Englishmen in India. As Rudyard Kipling put it in one of his *Jungle Book* stories, "The Miracle of Purun Bhagat," when the highly decorated Indian servant of the British Empire, Sir Purun Dass, at the end of his career took up the begging bowl and ocher robe of a Hindu holy man, letting wealth, power, and even honor go "as a man drops the cloak he longer needs," he was doing something "no Englishman would have dreamed of doing."[11]

Of all the passages that have been written to capture the visceral experience of the Western observer in colonial India recoiling at the sight of Hindu pilgrimage, Aldous Huxley's notes on Varanasi in *The Jesting Pilate* are perhaps the most ingenious and disarming. Huxley's words are frank and merciless. Of devout Hindus he says:

> They have been taught that this present world is more or less illusory, that the aim of every man should be to break out of the cycle of recurrent birth, that the "soul" is everything and that the highest values are purely "spiritual." Owing to their early inculcation, such beliefs have tended to become almost instinctive, even in the minds of those whose consciously formulated philosophy of life is of an entirely different character.[12]

For Huxley, all such piety is bunk. Moral obligation is nothing but the "theory of pre-existing social habits" expressed in mindless, slavish behavior.[13] Spirituality then is "the primal curse of India."[14] There is much to be celebrated in a materialist view of the world instead, Huxley urges. Why should we not be preoccupied with "the actual world in which we live"? Why isn't an attachment to the tangible world "something wholly admirable"?

On January 14, 1926, Huxley witnessed the gathering of holy bathers at the steps by the river in Varanasi—"acres upon sloping acres of humanity"—waiting for a solar eclipse.[15] Huxley was bemused by the fact that people actually believed the story of Rahu the serpent eating and spewing out the sun.

> There were, at the lowest estimate, a million of them on the bathing ghats that morning. A million. All the previous night and day they had been streaming into the town. We had met them on every road, trudging with bare feet through the dust, an endless and silent procession. In bundles balanced on their heads they carried provisions and cooking utensils and dried dung for fuel, with the new clothes which it is incumbent on pious Hindus to put on after their bath in honour of the eclipsed sun. Many had come far. The old men leaned wearily on their bamboo staves. Their children astride of their hips, the burdens on their heads automatically balanced, the women walked in a trance of fatigue. The serpent went on nibbling imperceptibly at the sun. The Hindus counted their beads and prayed, made ritual gestures, ducked under the sacred slime, drank, and were moved on by the police to make room for another installment of the patient million.[16]

To Huxley such superstition was simply inexcusable.

> To save the sun (which might, one feels, very safely be left to look after itself) a million of Hindus will assemble on the banks of the Ganges. How many, I wonder, would assemble to save India?[17]

James Prinsep, *Eve of the Eclipse of the Moon* (ca. 1823).
From James Prinsep, *Benares Illustrated in a Series of Drawings*, Calcutta: Baptist Mission Press, 1831.

Huxley's passages have the effect of discomfiture, because they foreshadow the way in which Hindu pilgrimage is written about even today from a modernist, secular, or liberal perspective.

Such an attitude, prickly and unbelieving, captures in retrospect the views of an entire genre of European observers who were incredulous at the common religious practices of millions of Hindus in India. Another well-known account is that of British adventurer Frank Smythe in his popular travelogue *The Valley of Flowers*. Smythe, a mountaineer, author, and photographer, described in 1931 a similar trail of pilgrims on the road above Vishnu-Prayag by the gorge of the Alakananda River. His description is even less charitable:

> They were an amazing assortment of men and women: fakirs, with wild, haggard, sunken faces and unkempt beards, clad only in a loin-cloth, their bodies smeared with ashes, and fat bunnias [*banias*, traders and merchants], squatting like bloated bull-frogs on charpoys [sisal beds] borne by sweating coolies, their women-folk plodding dutifully in the rear, carrying the family bedding, cooking pots and food . . . little old men and women, so old it seemed impossible that life could persist within their

> fragile shrunken bodies, hunched uncomfortably in wicker baskets on the backs of coolies.[18]

Smythe was struck and repulsed by the sheer number of pilgrims and their wretched conditions. In the same account, his animadversion becomes clear when he describes another group of pilgrims traveling to Joshimath, 1,500 feet above the meeting point of the Alakananda and Dhauli Rivers. According to Smythe, Joshimath was

> a halting-place of pilgrims who journey to Badrinath during the summer months to pay their respects to Mother Ganges, and worship at the shrines of deities associated with the sacred snows of Himachal (Himalaya). Unhappily, they bring with them many diseases from the plains, and cholera, dysentery, typhoid and malaria exact their toll from the devout, whose notions as to sanitation, cleanliness and hygiene are at constant variance with the well-intentioned preventive and remedial efforts of the Indian Government.[19]

And yet Smythe could not but be moved by the sheer resoluteness of such pitiable folks.

> Something sustains these pilgrims; few seem to enjoy their pilgrimage, yet their faces are intent, their minds set on their goal. They are over-awed, too, by their stupendous environment; you can see this in their faces. Europeans who have read and travelled cannot conceive what goes on in the minds of these simple folk.[20]

It is not that the authors of these passages did not know Hindus and their sacred sites in India well enough or grasp what lay behind the compulsion of seeking out some of the roughest terrains high in the mountains. Why did the glaciers of the Himalayas, along with every stream and rivulet, become associated with the overwhelming mythos of the Ganges? Why this entrancement with landscape and geography? Why such undisputed sanctity? Clearly there are certain deep and abiding associations between the landscape of high altitude, flowing waters, sacredness, and ritual purification. Why is the Ganges sacred, and what constitutes its sacred status? What is *not* sacred, and belongs rightfully to the realm of the profane? Water itself is a substance of routine and everyday use. How has one river retained its mythical and ritual significance as the redeemer of all deeds and misdeeds, bathing in the waters of which, as the *Vishnu Purana* says, all sins are instantly destroyed (*snānasya salile yasyaḥ sadyaḥ pāpaṃ praṇaśyati*)?[21]

To answer these questions we briefly turn to the distinction drawn by the great scholar of mythology Mircea Eliade between the sacred and the profane. Eliade was interested not only in the distinction itself; he wanted to show how sacredness derives from profanity. To describe this process, he borrowed the term *hierophany* (Greek *hiero*, "sacred," and *phainein*, "to show"), a process through which the sacred becomes manifest. According to this formulation, sacredness was not randomly accorded by human choice, but something that "showed itself." Eliade writes,

> It could be said that the history of religions—from the most primitive to the most highly developed—is constituted by a great number of hierophanies, by manifestations of sacred realities. From the most elementary hierophany—e.g., manifestation of the sacred in some ordinary object, a stone or a tree—to the supreme hierophany (which, for a Christian, is the incarnation of God in Jesus Christ) there is no solution of continuity. In each case we are confronted by the same mysterious act—the manifestation of something, of a wholly different order, a reality that does not belong to our world, in objects that are integral part of our natural, "profane" world.[22]

In this sense human beings have always been surrounded by sacred objects and artifacts. Émile Durkheim, in his celebrated work *The Elementary Forms of the Religious Life*, suggested that a profound and abiding sense of the sacred, which is the most vital element in all forms of religious belief, also makes the foundation of society possible. According to Durkheim, certain objects and entities inspire individual or collective respect, and such elementary forms of obeisance draw people instinctually and emotively toward those invested with moral authority.[23] Sacredness, then, lies in this emotion of respect, which prepares the members of a society to accept certain forms of authority and inevitability.

And yet there is an inherent and stubborn claim of Hindus of all denominations and castes to pilgrimage sites along the River Ganges. Just as there is ardent faith in the performance of arduous journeys, great respect is also given to all who return to their family, friends, and community from such long-distance, merit-seeking travel. This is one reason why the young still touch the feet of the elderly in greeting, to literally gather the dust of uncovered feet that have been on the pilgrim path. In many ways the pilgrim's journey is reminiscent of the final exit from mortal life itself, and thus the sacredness and merit reserved for such a journey are not ordinary. Towering mountains and the steep gorges cut by the glacial melts in the high Himalayas represent a numinous realm far removed from the strife and toil of ordinary worldly existence. To such extent, the waters of the upper tracts of the Ganges and its tributaries are also coveted as pure and fit for special ritual ablutions. But here lies an intriguing puzzle. Ganges water collected from any part of the river is deemed sacred, fit to offer to the sick, the dead, and ancestors in afterlife. Yet Ganges water is also used for every other common purpose.

The River of Last Resort

Across the body of Indian myths known as the Puranas, the waters of the Ganges are described as indispensable for the deliverance of both the virtuous and the sinful. The *Kashi Khanda* of the *Skanda Purana* (attributed to Agastya, a sage from ancient times), a Sanskrit text that most likely acquired its present form during the thirteenth century, contains one of the earliest and most extensive descriptions of the spiritual value of the city of Varanasi on the Ganges.[24] The *Kashi Khanda* offers some striking passages on the powers of the Ganges to wash away the sins of the most wretched among humankind. According to this text, true pilgrimage resides not just in traveling to holy places such as Kurukshetra, Ayodhya, Mathura, or Dwarka but in the journey that lies within (*mānasatīrtha*) and guides a seeker toward the path of kindness, sacrifice, and purity.[25] Similarly, immersion in the Ganges does not constitute a real ablution unless one can truly cleanse one's mind of profane, sensory attachments. In this regard, a ritual bath at a designated spot on the river does not by itself redeem the lustful, the greedy, the immodest, and the covetous. The *Kashi Khanda* explains that funerary oblations (*śrāddha*), including the ritual of *tarpaṇa*—in which the next of kin make a sacred offering to the gods so that the *ātman* (or "soul") of the departed ascends to the heavens—are best performed at an established place of pilgrimage. During the first epoch of creation, *satyayuga*, or the age of truth, every place on earth was sacred. In the next *tretāyuga* it was the lake of Pushkar in Rajasthan, followed in the *dvāparayuga* by the great battlefield of the Kurukshetra, where the Kauravas and Pandavas fought in the *Mahabharata*. In the phase of *kaliyuga*, it is not a place but the River Ganges that is the most sacred body of water.[26] One may even pray at such a pilgrimage on behalf of another who cannot be there by constructing a grass (*kuśa*) effigy of a person and washing it in the Ganges for blessings.[27] By virtue of a once-in-a-lifetime visit to Varanasi or Prayag in Allahabad, one can claim to have visited all the major rivers of India, which are but many variations of the primal sacredness of the Ganges. In the mind's eye all pilgrimage can be replicated in a ritual bath here if one has the requisite faith.[28] The early sixteenth-century digest of pilgrimages, the *Tirthacintamani*, credited to Vacaspati Misra of Mithila, says that in the Kaliyuga the Ganges surpasses Pushkara or Kurukshetra in the west as the repository of all pilgrimages (*sarvvatīrthāni*).[29]

The *Kashi Khanda* makes an intriguing association between the physical and metaphysical, inner and outer, aspects of ablution. Just as some parts of the human body are cleaner than others, some places on the surface of the earth are cleaner than others.[30] Thus the cleaning of the body in the flowing Ganges water at a prescribed time and place, tonsure, nail trimming, and many other aspects of outward hygiene help rid the material human body (*jīvadeha*) of accumulated transgressions.[31] In the *Gangastakam*, the eight odes to the Ganges allegedly composed by the author of the *Ramayana* Valmiki himself, a similar sentiment is expressed about the waters of the Ganges as the final destiny for the desecrated and profane mortal body. The river is addressed here as a "resplendent necklace on the bosom of this earth" illuminating a path towards the heavens:

Mātaḥ śailasutasapatnī vasudhāsṛṅgārahārāvalī
Svargārohaṇavaijayantī bhavatiṃ bhāgīrathīṃ prārthaye
Tvattīre vāsastvadambu pivatastvadvīciṣu preṅkhata
Stvannāma smaratastvadarpitadṛśaḥ syānme śarīravyayaḥ[32]

The first two lines of this ode offer prayers to Mother Ganga, the daughter of the mountains, who dwells in the heavens. I translate the last two lines of the preceding quatrain as follows:

> I hope to live on your banks, to drink your water, to be lulled by your currents, to remember your name in prayer, and to gaze upon you until the day I die.

The suffering of birth and death is the lot of all humans, but this pilgrim cannot endure any more. He would rather be reborn as a bull, bird, owl, snake, or elephant near the kingdom of Varanasi and die in the forests by the banks of the River Ganges. He imagines that his body will be consigned after death to the river. He longs for his body to float on the bosom of the river, gradually turning into carrion for the crows, dogs, and jackals.[33]

This morbid image of the remains of a human being washed in the torrent of the river is not simply a figment of Hindu religious imagination. The practice of consigning bodies of the dead without cremation continues to this day. The view of bodies floating down the river became a staple vignette of travelogues and descriptions of the River Ganges and its continuation, the Hugli (Hooghly) in Calcutta when India was under British rule. Many were struck by the common sight of the adjutant bird (*Ardea argala*), a prime scavenger, feasting on the remains of a human body cast into the river without proper cremation. John Erskine Clarke, describing the Calcutta riverfront in the popular juvenile magazine *Chatterbox*, presented a sordid image of the river and its many denizens:

> The Hindoos often consign the body of their relatives, *sometimes before they are dead*, to the river Ganges, which is believed to be holy, and a safe passage to happiness. The body if it escapes the crocodiles, is always the prey of the "adjutant bird."[34]

These adjutant birds were as curious as the Hindu customs:

> The singular-looking stork-shaped "adjutant birds," which walk about quite tame, or float down the river upon a human corpse, their wings spread out to catch the wind like sails.[35]

As early as the 1830s, the *Library of Entertaining Knowledge* was distributing this image of the Hindu corpse on the Ganges to a host of armchair travelers. In the volume titled *The Hindoos*, the authors offered this view:

> Of all the rivers of India the Ganges is the most sacred. It is, in the estimation of the natives, a deity; and the most secure way to heaven is through its waters. Hence, whenever this is possible, the Hindoo comes to its banks to die, and piously carries thither his parents or relations, to ensure their eternal happiness. With the converse of the feelings of the Gheber, who would consider the eternal fire—the object of his worship—polluted by the touch of a corpse, the Hindoo casts the dead naked into the sacred stream; so that those who sail upon the Ganges frequently meet with corpses, floating down in various stages of corruption and decay towards the sea.[36]

Such descriptions abound among standard descriptions of India provided by European travelers during the period of British rule in India in the later eighteenth and nineteenth centuries. We can certainly regard them as part of an Orientalist repertoire of morbid characterizations that made the difference of customs in the East appealing reading for domestic audiences, but that is not the purpose of my illustration. For many generations, poor Hindu folk who could not or did not cremate their dead regarded the Ganges as a place for redemption. Unlike English observers, they did not find such a practice contrary to common sense or decency, because the body was already destined for dispersal into its constitutive natural elements. Even as rotting carrion visited by scavengers, it could not pollute or corrupt the sacred body of the river.

This paradox of purity and pollution led anthropologist Kelly Alley, conducting fieldwork in the city of Varanasi, to ask her Hindu informants the vexing question: If the Ganges River is considered sacred by Hindus across India and worshipped every day as purifier, goddess, and mother, why do the people allow it to become polluted by the wastes of cities, industries, and households?[37] How can they disregard acts that sully and contaminate the river and its waters every day? Searching for answers to these questions, Alley realized that even in the 1990s a vast gulf remained between scientific, official, and academic discourse about water pollution and the way Hindu practitioners—pilgrims, priests, and common people—understood the devotional aspects of ritual purification. These differences and their contradictions were not effaced by the progress of "modern" attitudes in medical knowledge regarding microbes and pathogens or notions of personal and public hygiene. Rather, there were complex and multilayered ways in which subjects negotiated the implications of dirt, ritual pollution, and cleanliness in particular contexts.[38]

In places like Varanasi, devout Hindus who bathe, pray, and worship by the murky waters of the Ganges do not distinguish in any simple way between organic dirt or environmental contamination and ritual impurity.[39] Faith in the purity and the cleansing powers of the waters of the Ganges thus has much more to do with inward aspects of being, encompassing morality, metaphysics, and cosmology rather than a narrow biological definition of the human body and its vulnerability. In considering the ritually purifying aspects of Ganges water, we must try to distinguish between different conceptions of the body itself as a metaphor.

Mary Douglas, in her celebrated treatise *Purity and Danger*, argued persuasively that seemingly undisputed modern ideas of dirt and hygiene were simply a product of social conventions.[40] Douglas argued that the word *dirt* itself was figuratively related to various anxieties of social order and hierarchy. She suggested that prescriptive aspects of hygiene arose typically out of normative Christian injunctions of behavior, ascribing propriety to a prescriptive set of habits. Dirt was thus to a degree evident to the beholder:

> As we know it, dirt is essentially disorder. There is no such thing as absolute dirt. If we shun dirt, it is not because of craven fear, still less dread of holy terror. Nor do our ideas about disease account for the range of our behaviour in cleaning or avoiding dirt. Dirt offends against order. Eliminating it is not a negative movement, but a positive effort to organise the environment.[41]

Douglas's suggestions are pertinent to an understanding of why the waters of the Ganges have had such an abiding association with notions of purity and cleanliness, especially in the domain of rituals and piety, where traditional and persistent elements of belief are still at work.

Water of Life and Death

The ritual purity of Ganges water (*gaṅgājala*) has persisted as a matter of common faith across the Indian subcontinent, despite its visible contamination through industrial effluence and organic waste.[42] How does a substance that is used every day retain its uncommon quality? Here the opposing duality of the sacred and the profane, after Eliade and Durkheim, while essential to an understanding of a religious imaginary, needs a further, quotidian context: the world of the ordinary and the everyday, where people have to negotiate constantly between sacred and secular pursuits, spiritual and material interests. The variety of sacred spaces and objects that are part of the daily life of people across the Indian subcontinent gives me pause on positing a singular relationship between the religious and social life of things, but Ganges water is sui generis. Vacaspati Misra, in the *Tirthacintamani*, cites the *Skanda Purana* to declare that the waters of the Ganges are secreted directly from the cosmic body of Vishnu (*viṣṇusṛtaswarūpī*), in effect, the divine in liquefied form (*dravarūpa*).[43] Not only are one's sins dissipated by immersion in this water, but just a sip can deliver the merits of a Vedic horse-sacrifice (*gaṇḍuṣamātrapānena aśvamedhaphalaṃ labhet*).[44] Similar blessings are acquired by touching Ganges water or by simply casting one's eyes upon the river (*gaṅgādarśanam*).[45] Misra sifts through a profuse body of Puranic material to establish the value of ritual ablution, tonsure (*muṇḍanam*), and the best methods of praying and making offerings to ancestors (*tarpaṇa*). Death cannot defile one who has been given Ganges water, or one who dies while immersed at the banks of the river (*gaṅgāntarjalamaraṇa*) or while besmeared with Ganges clay, all of which promise the ultimate release from the chain of endless rebirth and suffering.[46] One who remembers the river frequently in prayers is also freed from the ties (*bandhana*) of the temporal world. In such persistent usage, the human body is treated not as a self-contained biological entity but as analogous to an impermanent vessel of pure and polluted elements. The word *tīrtha* is also tied to the ritual dimensions of the human body in offering, especially parts of the right hand from which water is poured for ritual offerings.[47] Specific names are given to each part of the hand used. *Pitṛya,* for instance, is the interval between the right thumb and forefinger, *āgneya* the space of the middle of the palm, and *daiva* the tips of the fingers. Touching, sipping, and pouring water are all part of such extended bodily rites.

Given this cumulative weight of myth, history, and ritual practice, Ganges water is in many respects an object of extraordinary value, treated by its purveyors much like any other precious commodity. Of course this water is not produced by human artifice but is found in the "natural" state, consumed as a substance of ritual necessity, but not bought and sold like gold or silver in the open market. In many ways the social and ritual life of this water, to paraphrase Arjun Appadurai, straddles the space between commodity and sacra.[48] The taken-for-granted authenticity of the river as a natural entity and preternatural wonder lends value to its water as a means of purification. The more remote the reach of the river where the water is collected, the more value might be placed on its ritually cleansing properties. The arduous journeys undertaken by common folk, widows, the elderly, and end-of-life renouncers lend further sanctity to the snowy heights from which the Ganges originates and the landscape through which it flows. Most devout Hindus wish to see the source and confluences of the Ganges once in their lifetime. This desire is partly a historic recognition of the immense power, destructive and life-affirming, of the river and its various terrains. I suggest that the salience and the symbolic density of Ganges water can be grasped only in the larger and capacious context of its mythological, geographical, and environmental associations. These aspects of the history and culture of the subcontinent not only are fused; they have developed as fundamental orientations at the everyday level. As Alley found out during her ethnographic work, the Ganges is much more than a river with symbolic value and practical importance. Hindus in northern India in general do not think that Ganges water is simply a "natural resource"—it is part of a superordinate cosmic order. Here ecology and related aspects of material culture cannot be separated from the sacred configuration of the pilgrim's universe. As Richard White has eloquently stated in his study of the Columbia River, rivers are the most dynamic symbols of nature: "They absorb and emit energy, they rearrange the world."[49] By the same token they are also fundamental to the ideas of the cosmos and eschaton in the trans-Hindu belief systems of the Indian subcontinent. These ancient ideas of a sacred river and its landscape may have had elements in common with today's ecological and environmental concerns, a distillation of the reverence that has survived for powerful and life-affirming natural phenomena. How else can we explain the faith in the purity of the waters of Ganges, bereft of basic human curiosity about natural cause and effect? Or the enduring faith in the cleansing, votive, and healing properties of such a substance in the face of ominous indicators of environmental degradation: the corrosive onslaught of dams such as the Tehri, the chemical plants, the suffocating silt deposits along the lower delta, and the prodigious quantities of organic waste?

Indian mythology is replete with stories about the miraculous healing properties of Ganges water, which gave cripples and lepers glimpses of hope, if not in this life, then certainly the next. A report for the Epidemiological Society of London noted that during the late nineteenth century, lepers were thrown into the river in the hope that their bodies, even in death, would be scrupulously cleaned of the contagion.[50] In one of the many legends surrounding the famous weaver, songwriter, and mystic Kabir recorded in the Sikh sacred text of the Adi Granth, Kabir's son Kamal met a man with leprosy about to drown himself in the River Ganges. Kamal asked him to desist from this act. He then took some Ganges water into his palm, blew on it, and sprinkled it on the leper. Within an instant the disease disappeared. The cured man rewarded Kamal handsomely for this miracle. Kabir admonished his son for falling prey to the lure of wealth.[51] Faith in such miraculous powers of Ganges water spread across many distant parts of India. In the legend of King Rai Mandlik of Girnar, Gujarat, a man named Vijnal had leprosy.[52] He was a friend of the king, but he could not bear this state. He became a fugitive, hiding from the king, waiting to die. The king, meanwhile, set off in pursuit. On this journey, while encamped near a small rivulet, the king met a man who was carrying Ganges water from the east. The king had had no time to bathe. Seeing the water-bearer, he grabbed the pitcher and poured the waters over his clothes. When he finally caught up with his friend the leper, he embraced him. And again, without fail, the leprosy disappeared from his body in the blink of an eye. Such stories abound across India, the plot invoking a millennial familiarity.

Missionaries found it difficult to dissuade devout Hindus from flocking to crowded pilgrimages, which they saw as corrupt, heathen, and unenlightened. Harold Begbie, in his missionary tracts, expressed his staggering disbelief in the fact that Hindus "literally" took the Ganges to be clean. They mistakenly believed, he thought, that the Ganges "will wash away sin, cure lepers, and carry souls of the dead into Paradise."[53] Begbie argued that this Ganges had nothing to do with the river itself for the devout masses. It was therefore mistaken as "the laundry of souls, the hospital for disease, and the channel to Paradise."[54] Hindus regarded their pilgrimage "not as a sacrament but as a cure." Begbie's puzzlement reinforces my point, that is, that the true supplicants believed that Ganges water was endowed with a potency that could remove both their physical ills and moral failings.

The reputation of the purity of Ganges water made it a much-sought-after substance in all temple rituals of cleansing and oblation. It is said that at the fabled temple of Somnath constructed by the Solanki Dynasty of Gujarat, the idol of Shiva in the form of the Lord of the Moon—if Al Biruni's account in the *Tarikh al-Hind* is to be believed—was washed daily with urns of Ganges water brought in from hundreds of miles away.[55] The water of the Ganges was famous for its taste and lasting qualities. When Muhammad ibn-Tughluq decided to move his capital from Delhi to Deogir across the Vindhyas, Ganges water was procured for him every day; it took forty days to bring the water from the valley of the north.[56] Mughal emperor Akbar drank only Ganges water; he did not like the taste of well water.[57] Akbar extolled the purity and taste of Ganges water, calling it the source of life (*āb-i-hāyāt*). At home and during his many travels and campaigns, Akbar had Ganges water delivered sealed in jars from Saran, the city on the Ganges nearest Agra, his capital.[58] Only the most trustworthy retainers were sent to procure Ganges water from the banks of the river.

European trading companies that frequented the Indian Ocean littoral through the sixteenth and seventeenth centuries valued the taste and purity of Ganges water. Joannes de Laet, director of the Dutch East India Company, in his description of India during the reign of Jahangir extolled the virtue of Ganges water as a most pleasant staple.[59] English traders of the early East India Company found that Ganges water lasted for over a month during the transcontinental voyage, whereas water from near the docks in the home country spoiled in a matter of days.[60] Sir Edmund Hillary, the conqueror of Mount Everest and indefatigable traveler, recounting the long history of Ganges water used for overseas travel, marveled at the fact that it "keeps fresh and unspoiled indefinitely in its container, whereas water from any other river becomes tainted and unpleasant."[61] European inhabitants of Calcutta built reservoirs to store waters from the river all year around.[62]

In the late nineteenth century, British scientists and hydrologists became intrigued by the fact that Ganges water did not go bad, even after long periods of storage, contrary to the water of other rivers in which a mounting lack of oxygen quickly promoted the growth of anaerobic bacteria. In 1896, the British physician E. Hanbury Hankin wrote in the French journal *Annales de l'Institut Pasteur* that cholera microbes that had a life of forty-eight hours in distilled water died within three hours in Ganges water. Dr. Hankin, a government analyst of the United Provinces of British India posted in Agra, had become interested in the Hindu custom of depositing their dead in the river. Fearing reprisal from devout Hindus, the British government found it impracticable to interfere with such unsalutary practice, even during the frequent epidemics of cholera, plague, and other diseases. Hankin was able to secure corpses of cholera victims in the river and isolated samples of Ganges water with a large concentration of the bacillus *E. coli*. Much to his astonishment, he found that after six hours the microbes had completely disappeared.[63] Hankin concluded that the water of the Ganges and Yamuna Rivers in India was "energetically bactericidal" in general and particularly destructive of the cholera vibrio. Hankin attributed this to volatile acids present in the water, but the Frenchman Felix d'Herelle suggested that the disinfection was probably due to the presence of a bacteriophage.[64] Another British physician, C. E. Nelson, observed that Ganges water taken from the Hugli in Calcutta, one of its "filthiest" mouths, by ships traveling back to England remained fresh throughout the voyage.[65] D'Herelle was amazed to find that only a few feet below the corpses of people who had died of dysentery and cholera that were floating in the Ganges, the waters were free of any form of contagion.[66]

Water analysts, soil chemists, and environmental scientists are still debating whether Ganges water can destroy *E. coli* and other harmful bacteria. Many reasons are given for this potential. Some scientists have pointed out the effect of long exposure of the water surface to ultraviolet radiation. One plausible explanation, studied in great detail by the eminent Indian hydrologist Devendra Bhargava, is that the Ganges has chemical and biological properties that help the river rapidly absorb prodigious quantities of organic waste.[67] It seems to be able to reoxygenate itself almost miraculously. Author Julian Hollick, who has pursued this question with many Indian scientists, is convinced that the purifying qualities of the Ganges are due to the large concentration of the bacteriophage virus that is present in superabundance by the mud banks of the river and in shallow pools, and the virus is activated as soon as bacterial concentration reaches a certain threshold.[68] The bacteria-killing properties of the phage virus have been scientifically proven. However, exactly how the virus is related to the decontamination of Ganges water is a matter that has not been settled among researchers. There is still no clear explanation of the fact that festivals such as the Kumbh and Kartik Purnima, in which more than sixteen million people indulge in communal bathing on the Ganges, have seldom resulted in a pandemic or widespread outbreak. The Sankat Mochan Foundation laboratories in Varanasi report *E. coli* levels at a level exponentially higher than that considered safe by the World Health Organization (WHO). And yet again, bacteria seem to disappear after such major concentration in a matter of hours and not days.

In the 1930s, before the advent of penicillin, when bacteriophage therapy for the treatment of bacterial infections was coming into vogue, many claimed to have solved the mystery of the Ganges water cure for remedies in India. The magazine *Popular Science*, aimed at young audiences, proclaimed:

> The discovery of bacteriophage has solved the ancient riddle of the Ganges River in India. By the Brahman temples at Benares, thousands of natives bathe daily in the sacred stream. Yet a few miles upstream from the bathing *ghats* at the temple steps, the river is unbelievably filthy, since it receives the drainage from a densely populated land that for centuries has known no sewage system but the sluggish streams and rivers . . . It would be almost certain death to bathe in many parts of the Ganges. Still, for hundreds of years, millions of natives have washed in it, protected from infection as we now know, not by their own strange gods, but by bacteriophage that devour the disease germs.[69]

The assertion that a specific, isolable etiology must be behind the miracle work of Ganges water is a matter of outlook. Science and faith in this particular context permeate contrary views of the human body itself, perpetuating the stories of cures for diseases by the mere administration of Ganges water.

Ganges water is used for ritual ablution and offerings in millions of households across the world where some form of Hindu devotion is practiced. I have carried the hazy translucent liquid from Gangotri, Haridwar, and Varanasi in plastic bottles for family and relatives in Kolkata, London, New York, and California. In my hometown of Kolkata, where the murky waters of the Hugli laden with silt are readily available, clearer water from more distant pilgrimages of the upper or mountainous reaches of the river is much more highly prized. I have seen every second shop in the small town of Gangotri in the Garhwal Himalayas selling empty plastic bottles to pilgrims arriving from the plains, who duly fill them up during their journey and return with their prize in the crowded buses, cars, and trains. As I have argued before, material and devotional pursuits in India cannot be split in any simple analytic sense. As we shall see through the rest of this book, pilgrims over untold generations have had to parse out their meager resources, paying leprous beggars, innumerable fees and gratuities to guides, offerings to priests, and tolls to the various guardians of the road, rajas, zamindars, Mughal road keepers, and tax collectors of the East India Company. They have also jostled with crowds at the innumerable fairs and festivals along the banks of the Ganges.

The British sea captain Alexander Hamilton, visiting India during the early nineteenth century, observed a brisk sale of copper vessels full of Ganges water from upper India to the rest of the subcontinent:

> The priests sell brass and copper pots, made in the shape of short-necked bottles, with Ganges water, which they consecrate and seal up, and send those bottles, which contain about four English gallons, all over India, to their benefactors, who make them good returns, for whoever is washed with that water just before they expire, are washed as clean from their sins as a new-born babe.[70]

The travel writer Caleb Wright saw near a temple in Allahabad, where twenty to forty thousand pilgrims assembled every February, priests providing each pilgrim with two or three earthen bottles full of water from the Ganges. Equipped with these and some coins, they made offerings to Shiva by breaking the bottles against the temple. The next day, the Brahmins would pick up the money and, as trustees of the idol, secure it for their safekeeping. Wright saw that the broken bottles were removed to a short distance away, where they had formed a mound many times larger than the temple itself.[71] Accounts such as these reassert the limited exchange value of Ganges water. The social life of Ganges water resembles any other commodity on the market, and yet it has never been marketed or packaged for mass consumption. This self-possession among its adherents has some bearing on the historical and cultural gravitas of the river itself.

Pilgrim Maps

The worldview of the pilgrim has always been tied to the relationship between the ordinary and the extraordinary, the everyday and the afterlife, the immanent and the transcendent. The pilgrim's reverence for sacred spaces and sites is based on a mélange of functional geography and elements of common mythology derived from Puranic stories, regional sects, and deity-cults, what A. K. Ramanujan fittingly described as the demotic, "little" tradition of Indian civilization that lingers in the age-old veneration of stones, trees, and rivers.[72] In this ordinary, unremarkable sense, merit seekers have always followed the descent of the Ganges to the plains of India, as part of an ambient landscape configured in the unique geography of pilgrimage. As I found out during my visit to the base of the Shivling peak in the high Himayalas, every glacial melt that deserves to be called a river is termed *Ganga* in one form or the other: Akashganga, Amarganga, Kaliganga. *Ganga* here assumes the shape of a universal suffix, its fluvial cosmology fused with an unchanging and natural order of things. Akashganga, or the Sky Ganga in some northern Indian languages, is also a name for the Milky Way Galaxy.

Other unprinted, vernacular, and subjective maps seem to be at work here, governing the rudimentary spatial order of the pilgrim's landscape.[73] Underneath the random spectacle of ghats, temples, and markets lurks an emotive sense of place tied to cosmology and deep history that may have survived the encroachment of the present-day nation-state and global tourism. It is invoked each time a devotee visits a shrine. In the acts of travel and circumambulation, he or she is retracing the cognitive atlas of a much older proto-Indic, geographical space tied to the Puranic idea of Bharatavarsha: land of the descendants of King Bharata of the lunar clan, who defeated all the kings of the world and performed four-hundred-horse sacrifices on the banks of the Ganges. Puranic cosmology, places where deities are known to have come down to earth (Sanskrit *avataraṇa*, hence Avatar), the four principal abodes or *dhāms*, and the self-manifestation of divinity in graven images all converge in this imagined space.[74] Members of ascetic religious orders who frequent Varanasi or the holy bathing steps of the Allahabad confluence at Kumbh pitch their tents consciously according to rules derived from this sacred formula or device (*yantra*).[75] The Ganges in this ritual context reassumes its original threefold shape, *tripathagā*, charting its course from the original celestial abode to earth, but also to the subterranean realms of the known world (*pātāla*). During the time of the great fair of Kumbh, with the right concatenation of stars and planets at the vernal equinox, it transfixes its devotees once again as the primordial river, flowing from the foot of Vishnu; cascading on the matted locks of Shiva; splashing on to the roof of the world, which is the mountain of Meru; and gushing forth into the four cardinal directions.

Varanasi, which is an earthly form of Shiva's celestial body, is seen as an assemblage of concentric circles. Kashi is the largest unit, Varanasi is somewhat smaller, and the city between the two smaller rivers, Varuna and Asi in the south, is lesser. Still smaller is the Avimukta Khsetra, or "un-forsaken," zone of Shiva, which measures about two to three miles in all four directions from the temple of Krittivasa.[76] Diana Eck points out that the transcendent geography of Kashi lies astride the physical layout of its temples and ghats; such a space is traversed not by the physical body but only in the inner realms of the mind. In this form it encapsulates the logic of all sacred spaces of India. It is also a space of death and salvation. The *Kashi Khanda* stipulates that in Shiva's domain, people who leave their mortal bodies are instantly freed from the chain of birth and death.[77] The anthropologist Jonathan Parry describes how the Panchakosi road that marks the wider orbit of sanctity in Varanasi is for many a space coveted for an ideal death.[78] Many do not make it alive at the final hour to enter this charmed circle for cremation, where Shiva himself is supposed to recite the *tāraka* mantra, the seed-incantation of release. Varanasi is thus a place of ultimate destiny, the fires of its famed cremation grounds waiting to receive the fresh remains of supplicants; a city of the dead and the dying; a gateway to ancestors and other celestial beings. And thus history and legend abound with examples of those taken to the city to die who must not return to mundane life if they happen to live on. Many such houses that still stand by the banks of the Ganges were built for temporary occupants on their way to the afterlife.

Other texts also contain references to such cognitive maps of sacred spaces. Lakshmidhara Bhatta (1100–1160 C.E.), minister of the ruler Govindachandra of Kanauj and author of the well-known digest (*nibandha*) of rites and injunctions, the *Kritya-Kalpataru*, gives us an insight into how facets of this ritual geography persisted through the ages. Lakshmidhara says that human settlements outgrew the original abode of gods and Brahmins (*brahmāvarta*) as they expanded beyond the limits of Kurukshetra through the middle country (*madhyadeśa*) stretching from the Himalayas in the north to the Vindhya hills in the south. This extended realm was now known as the *āryāvarta*, land of the noble and upright. Certain parts of this territory were inauspicious, especially those inhabited by barbarians (*mlecchadeśa*) where true sacrificial offerings could not be performed and a Brahmin would not accept water.[79] However, regions washed by the Ganges were sacred, especially where the blackbuck (*kṛṣṇamṛga*) roamed free—no ordinary antelope, but the deity of sacrifice itself, which once in a while escaped from heaven, recognizable by its white, black, and yellow marks symbolizing the three great Vedas. Tribes and kingdoms of ancient India, as we shall see in subsequent chapters of this book, staked their claims over the sacred sites of the middle kingdom. The *Matsya Purana*, for example, in the section extolling the virtues of the confluence, imagines sister-rivers Ganges and Yamuna as female guardians waving their blue and white ceremonial yak-tail whisks in praise of Prayag, the king of pilgrimages (*tīrtharāja*).[80]

The Anonymity of Faith

Commoners, without access to the intricacies of Sanskrit syntax and metaphor, may not have the means to fully comprehend the priestly divinations of sacred space and time, but they travel nevertheless through this preordained geography from the depths of their little traditions. This is why pilgrimage to the city of Varanasi strikes the outside observer as different from all other visits to sacred sites in India. The British-Indian civil servant Norman Macleod once said of Varanasi: "I have never seen anything approaching to it as a visible embodiment of religion; nor does anything like it exist on earth."[81] The Orientalist and antiquarian James Prinsep, who earned the distinction of deciphering ancient Indian epigraphy in the Brahmi and Kharoshti scripts, fell in love with the Varanasi waterfront with its ghats and buildings standing by the Ganges's gently curving edge, where after the monsoon rains the water was more than forty-two feet deep and more than a mile across. Musing on the picturesque qualities of the city, he wrote:

> Indeed there are few objects more lively and exhilarating than the scenes from the edge of the opposite sands, on a fine afternoon, under the clear sky of January. The music and bells of a hundred temples strike the ear with magic melody from the distance, amidst the buzz of human voices; and every now and then the flapping of the pigeons' wings is heard as they rise from their crates on the house tops.[82]

Prinsep believed that the bustling crowds of pilgrims at the ghats was one of the key elements of the character of the city:

> Let it be borne in mind that upon the ghats passes the busiest and happiest hours of every Hindoo's day: bathing, dressing, praying, preaching, lounging, gossiping, or sleeping, there will be found . . . on the ghats are concentrated the pastimes of the idler, the duties of the devout, and much of the necessary intercourse of business. In no city of the world is the population invited to a single street or place of recreation by so many attractions.[83]

The ghats of Varanasi are a living panorama of its past, their appeal based on the commingling of myth, antiquity, and history. Many famous ghats from the past, like Adi Keshava and Kotitirtha, are now gone, while Dashashwamedh and Manikarnika have endured. Pilgrims may or may not

James Prinsep, *View Westward from Ghoosla Ghat, Benares* (ca. 1825). From James Prinsep, *Benares Illustrated in a Series of Drawings*, Calcutta: Baptist Mission Press, 1831.

know that the Dashashwamedh is named after the story of Brahma's ten-horse sacrifice by which he gained the power to rule over the ancient king of Kashi, Divodasa, who had driven out Shiva and the rest of the gods from the city, but they would know that the old pipal trees by the river are sacred. They may know the legend of great King Harishchandra, tested mercilessly by the gods, who after one misfortune after another became an outcaste *caṇḍāla* attending the cremation of corpses at the burning ground of Manikarnika, where they say that the fire has never been extinguished, and where—until recently—a Brahmin's pyre could not be started without fire from the hearth of an untouchable. They may not know that most of the ghats and the giant bathing steps to the river were rebuilt during the late Mughal period by the leading Rajput and Maratha houses from western India. They may also notice the sati memorials marking the spots where Hindu widows consigned themselves to the funeral pyres of their dead husbands and abjured the state of widowhood. The composite, eclectic, historic architecture of the Varanasi riverfront is just one dimension of the pilgrim's enchanted world: familiar and beguiling.

The Varanasi riverfront. Photograph by Debal Sen.

Pilgrimages to prescribed sites on the Ganges have always been a mass phenomenon. The fair of Kumbh in Prayag, Allahabad, which is held in the winter months and appeared in *Time* magazine in 2013, recorded eighty million people visiting the site over a period of fifty-five days, making it the largest pilgrimage in the world and perhaps the only human gathering that can be seen in satellite images from outer space.[84] Jawaharlal Nehru in his autobiography speaks of the great mass of peasants who flocked to the banks of the Ganges by the thousands during religious fairs and festivities, representatives of the "naked, hungry mass" that he always had taken for granted before he began his political work as a peasant leader.[85] The sheer magnitude of such numbers makes it difficult to write a granular history of pilgrimage attentive to changes in piety, patronage, the rise and fall of religious orders, and popular forms of piety as instruments of quiescence in the hands of contending regimes. Despite its punishing hierarchies and ritual strictures, popular veneration of the river also give us a glimpse of the Indian subcontinent as an open society, where children, widows, wanderers, ascetics, priests, peasants, traders, soldiers, kings, and landlords could break rank for a moment and mingle with the crowds as part of an indistinguishable mass.

Marx wrote famously not only about religion as the opiate of the oppressed but also about religion as the "general theory of this world"—the universal explanation of the logic and order of things.[86] The altruistic labor that the pilgrim performs for a place in afterlife, however illusory from the point of view of material reason, suggests a different kind of mental and bodily orientation to the inequities and suffering in this world. The pilgrim's journey, to the degree it is idealized, recapitulates a narrative structure borrowed from India's great folkloric and storytelling traditions. It is prefigured by episodes remembered from the epics or worthy deeds of heroic and historical protagonists. Recall the five warrior brothers of the *Mahabharata* who departed from their kingdom for penance in the high Himalayas after the great battle of Kurukshetra, a journey from which they did not return in their mortal bodies. Pilgrims must make peace with their makers, to look for something less in this world and more in the next. In this sense, the course of the river itself—its glacial embryonic form at Gaumukh, its youthful rush before the long descent to the plains at Gangotri, its long and languid meander through the great northern plains, and finally its meeting with the ocean at Sagar Island—provides an overarching allegory of the passage of the *ātman* through the cycle of birth, life, death, rebirth, and deliverance, captured in the functional concept of *gati*, derived from the root verb for motion in Sanskrit—*gaṃ*, which has a range of meanings: going, path, destiny, refuge, escape, and fate. The name of the River Ganges is also derived from this root, suggestive of motion, flow, direction, and force. Ganges as the river of ultimate destiny, in this extended sense, defies the conceptual divide between nature and culture, reminding us once again that life and death are part of the fundamental order of things, and that the river is both the embodiment of Vishnu and the daughter of the Himavant, from a place where the bane of human existence and toil has not sullied the footprints of the gods.

CHAPTER 2

GANGA DESCENDS

At the farthest southeastern reaches of the Indian peninsula, along an extended stretch of sandy shoreline skirting the restless surf of the Bay of Bengal, stand the spectacular rock-cut temples of Mamallapuram. The place is now known as Mahabalipuram, one of the most-frequented tourist spots in India. The earliest of these temples were built by the fabled king Mahendravarman I of the Pallava Dynasty, who ruled during the early seventh century C.E. The legend of his descendant, King Narasimhavarman, who was a formidable wrestler and warrior (*mahāmalla*), lends the place its name in antiquity. Among these fine structures is a giant relief meticulously carved across two boulders more than ninety feet long and forty feet high. This rock face adjacent to a pillared temple is a testament not only to the amazing dexterity of the nameless Pallava stonecutters who breathed life into the stone surface with realistic figures of humans, animals, and celestial beings but also to their attention to the cosmic order of the world, drawn from the deep recesses of Indian mythology.

The relief is known most commonly as the "Descent of the Ganges." Making clever use of a natural cleft between the two adjoining and jutting outcrops of granite as the focal point of the entire panorama, the sculptors have told the story of the tumultuous plunge of the river-goddess from her heavenly abode through the matted locks of Shiva, flanked on either side as she falls by gods, ascetics, and life-size elephants. About halfway down this replica of the river, poised between the two boulders, are figures of two Nagas, mythic creatures of the underworld with human torsos and snake bodies, with their crowns crested by hooded cobras. In Indian Puranic lore, they are creatures associated with Shiva, water, and serpents. During the monsoon rains, water that collects on top of the hilly outcrop cascades down through this channel between the sculpted rocks. With a little bit of imagination, they bring back to life stories about the sacred river that have been retold by Hindus of all manners and persuasions, generation after generation, across the entire Indian subcontinent.

What is most fascinating about this mural carving, apart from its exquisitely detailed figures, is the re-creation of the story of the descent of the Ganges more than a thousand miles away from the floodplains of the actual

"Descent of the Ganges," Pallava rock-cut sculpture, Mahabalipuram, Tamil Nadu, tenth century C.E. Photograph by Gautam Jagannath.

river. King Mahendravarman, the patron, ruled his burgeoning kingdom from the old city of Kanchipuram (also known as Kanchi) located on the monsoon-fed floodplains of the River Palar. It is difficult to guess how many devout subjects from his kingdom made the journey to pay obeisance to the great pilgrimages of the north such as Mathura, Ayodhya, or Varanasi on the banks of the Ganges. Many centuries before Pallava mural arts began to celebrate the advent of the river Ganges on earth, the story told in the Puranas had become commonplace throughout India.

Veneration of the Ganges has remained a perennial feature across the diverse cultural landscape of the Indian subcontinent. Her imagined form as a female deity, the idea of her permanent abode in the celestial heights of the Himalayas, and a transcendent sense of space expressed in the idea of pilgrimage, discussed in the previous chapter, are universally shared. In such an exalted conception, the physical distance from the channel of the Ganges in the north is secondary. As I have argued, most rivers are seen as sacred in the primal image of the Ganges, and in the wider world of the devotee, pilgrimages far and near are intrinsically connected. Long before the ascendance of the Pallavas, seekers of absolution held the city of Kanchi in the heart of the Tamil countryside with the same reverence as some of these holiest spots in northern India.[1] In their artistic offerings and their temple sculpture, rulers of the Pallava Dynasty were merely reasserting a self-evident truth.

This chapter takes up two important and related questions. How is it that among all the natural elements and forces that inspired the awe of ancient inhabitants of India, the Ganges is the one river invested with such mythic salience? And why does it assume such an alluring female form? Most school textbooks on ancient Indian history explain that the desire to see the forces of nature in human form is directly related to the rise of propitiatory rites in a distant Vedic past. Thus the god of rain and the oceans, Varuna, and the sun god, Mitra, were honored and appeased with offerings of food and prayer, much like one's own ancestors, along with many other capricious stars and planets and their sinister forces that one might unknowingly provoke. For many reasons it has been difficult to dispel the idea of such primitive impulses at work in the world of Indian antiquity, and this prejudice surfaces even while scholars discuss the veneration of Hindu deities that date to a much later period. In order to explore the descent of the Ganges, we need to know more about how a complex cultural formation appropriated such primeval animistic and anthropomorphic beliefs, following Durkheim's suggestion that nature worship and animism are related aspects, fundamental to the rise of religion as a collective form of human consciousness and knowledge.[2]

Humans have venerated rivers for their powers of sustenance and devastation from time immemorial. The most famous of these, perhaps, is the Nile, considered a divine gift to the people of Egypt and Sudan. Sometimes she is the deity Hapi in human element, sometimes Osiris, the fertilizer of the land and its vegetation, god of death and resurrection, husband and brother to Isis in Egypt. The Nile is also associated with Nepthe, the spirit of the highlands, and Anubis, the bearer of floods. These bits and pieces of ancient Egyptian mythology that have survived in the epigraphic record are dedications to the power of nature at work—fearsome and life-restoring at once. The Nile defies the arid Sahara to bring the gift of the oxide-rich ferruginous mud all the way from the Abyssinian plateau. It is the very progenitor of the "dark land."[3] It also brings forth the awesome spectacle of the floods witnessed and inscribed in the pyramids of the sixth dynasty or recorded in the hymns from the Theban period. But these are fragments from a civilization long past. The essential myth surrounding the advent of the Ganges is still very much alive in the Indian popular imagination.

We often attach a measure of innocence to ancient myths that appear to have emerged from a time of simple wonderment. In a striking passage, Edith Hamilton, the author most remembered for popularizing Greek and Roman mythology during the 1940s, explained how through these myths we can "retrace the path from civilized man," who resides at a distance from nature to early man, who was surrounded by nature. Myths, she wrote,

> lead us back to a time when the world was young and people had a connection with the earth, with trees and seas and flowers and hills, unlike anything we ourselves can feel. When the stories were being shaped, we are given to understand, little distinction has yet been made between the real and the unreal. The imagination was vividly alive and not checked by the reason, so that anyone in the woods might see through the trees a fleeing nymph, or bending over a clear pool to drink, behold in the depths a naiad's face.[4]

Such a simple prelapsarian explanation can be followed back to the dawn of the study of primitive cultures, pioneered by J. G. Frazer in his wildly popular *The Golden Bough* (1890). Frazer upheld myth as a literal account of the disposition of the primitive mind, much like an inchoate science. A god incarnated in human form was thus not at all a startling discovery for early humans, who sought the help of supernatural powers to make sense of a natural world that they found intimidating.[5] The anthropologist E. B. Tylor was convinced that myths belonged to the primordial state of humanity, at a stage when humans were like children.[6] They permeated the imagination through language itself and brought a semblance of order to the world that our ancestors beheld and the uncertainties of nature they endured. Animism for Tylor was thus simply the expression of an early philosophy of religion, the "childlike science of an early age" that claimed to explain real events in the physical world.[7]

Is it possible, then, that in the ritual worship of the river Ganges elements of a primitive form of veneration have persisted for thousands of years and kept alive the thoughts of the earliest Hindus? Such continuity seems suspect, given the complexity of Indic cultural orders on which the veneration of the Ganges rests. It is tempting to consider, though, despite the condescending primitivism of early anthropology, that forces of nature might have assumed iconic forms in the mind's eye of ancient human cultures, such as those in India. Do elements of a particular conception of natural forces from a remote past still somehow animate the fundamental disposition of its cultural tradition?

Anthropologists who study cognitive evolution among early hominids think that an awareness of natural phenomena might have developed long before tool-making became an established way of life. Steven Mithen has conjectured that separate cognitive faculties arose during different phases of human evolution. The first of these, social intelligence, grew out of relations with members of one's nearest group.[8] A second and later development was what Mithen calls "natural history intelligence," characteristic of hunter-gatherer societies, in which survival depended on reliable knowledge of things such as change of seasons, terrain, and animal migrations. Rituals of obeisance to significant bodies of water belong to this antecedence beyond the realm of history. Mythology may provide clues to a primal relationship between humans and the natural order, and the myths of the descent of the Ganges in the Hindu tradition may echo rudiments of the prehistoric Indo-European society of the great plains of northern India. However, it is nearly futile to try to retrieve the exact juncture at which a vital force of nature such as a river was first brought within the purview of human mythopoeia. We have no access to such a moment in history. Myths, as Ernst Cassirer wrote memorably, belong more or less to an "autonomous order of knowledge," and they represent a stage in the development of human culture far more complicated than an imagined infantile past, and a state of intuitive understanding of the world that is immediate, sensual, and deeper than standard reason or logic.[9]

The Myth and Its Origins

Stories from Indian mythology seem to have endured in rural culture for millennia, buttressed by a myriad of annotations. The noted Indian novelist R. K. Narayan, a fine storyteller, saw the legends of the eighteenth major Puranas retold by the village raconteur as emblematic of the extended moral universe of the Vedas and the Upanishads. For Narayan's storyteller, "stories, scriptures, ethics, philosophy, grammar, astrology, astronomy, semantics, mysticism and moral codes" were interrelated as part of a larger whole—keys to understanding life and existence.[10] This great tradition was alive in the humblest corners of village India, where rural folk who did not have access to classical Sanskrit or the finer esoteric points of scripture could still partake of the lasting moral vision of a higher social order.

The stories, of course, are never quite the same. Not only is each telling slightly different, but many renditions run against the grain of the original epics and their intended moral lessons. In other words, a living mythical past is not sustained by mere replication; far more capricious forces are at work. Poet, scholar, and translator extraordinaire A. K. Ramanujan, commenting on the multiple versions of the *Ramayana* that coexist across the cultures and languages of the Indian subcontinent, saw the prevalence of a particular set of texts in a given culture as a common "pool of signifiers."[11] This was true of oral traditions that sustained a perennial epic imagination with a repository of plots, characters, names, geographies, incidents, relationships, phrases, proverbs, and insults. India remains a culture steeped in stories and storytelling, in which folk and classical traditions have been entwined for a long time. Indian epics like the *Mahabharata* have thus absorbed over millennia an encyclopedic range of folk minutiae—tales, beliefs, and proverbs. As Ramanujan also pointed out, the oral and folk traditions in turn continually redact, refurbish, and redistribute this material across different regional, local, and linguistic registers.[12] In this regard, much like the ancient Greek myths described by Marcel Detienne, the Indian traditions must have been more exegetical than interpretative. Their interlocutors have been engaged ever since in a continuous and lively disquisition, spinning disarming critiques of their own moral standards, symbols, behaviors, and practices.[13]

Perhaps the most frequented source of the story of the descent of the Ganges is found in India's popular and venerable epic, the *Ramayana*. Sage

Valmiki, to whom the oldest surviving version of the epic is attributed, was moved to utter the first lines of his verse near the confluence of the Ganges and its tributary, the Tamasa, when he was distressed by the sight of a huntsman taking aim at two cranes (*krauñcamithuna*) flying together over the sandbanks and bringing down one of the pair, and was moved deeply by its mate's piteous cries.[14] The site of the composition of the *Ramayana* is thus not far from the shores of the Ganges and its floodplains. Today, a section of the River Tamasa, known colloquially as the Tons, runs through the present-day Mau district of Uttar Pradesh, along the fertile plains of the tract between the Ganges and its tributary, the Ghaghara (also known as the Ganga-Ghaghara Doab), about sixty miles east of Varanasi. The domicile of Rama's father, King Dasaratha, in Ayodhya was a city on the banks of the Sarayu, which was most likely the name for a section of the River Ghaghara back then. During his exile, Rama, Lakshmana, and Sita visited the meeting point of the Ganges and the Yamuna at Prayag and crossed the Ganges into the forests of its southern banks. In other words, the Ganges appears in many guises in the epic. It is not just a mythical river that fell from the heavens. It is also part of the natural setting for the story that lends a geographical familiarity and spatial orientation to its larger plot. In Valmiki's *Ramayana*, the story of the Ganges is told by the great sage Vishvamitra to Rama and Lakshmana.

Rama and his brother were sent as young boys to Vishvamitra's forest sanctuary (*āśrama*) to protect the forest-dwelling ascetics from demons who kept disrupting their sacrificial rituals. Toward the end of their stay, the sage decided to accompany them to the principality of Mithila,[15] where a sumptuous sacrifice had been arranged by King Janaka. On the way they crossed the River Shona, and in a day's journey they came to the Ganges. The brothers were delighted to see the holy waters of the Ganges (*puṇyasalilāṃ*) and her banks where wild geese and white cranes flocked.[16] After they had bathed in the river and offered their prayers to ancestors, Rama asked the great sage about the origins of the Ganges. "How does she traverse the three worlds to come at last to the ocean, lord of rivers and streams?"[17]

The sage then narrated the story of the river to these two young princes from Ayodhya. Himalaya, lord of the great mountain ranges of the north, had two beautiful daughters. Their mother was the daughter of Mount Meru, the axis of earth. Ganga was the elder, and the younger one was Uma. During the struggle between the gods and demons, Shiva, the great warrior and leader of the divine army, had taken up the life of an ascetic while his wife, Uma, had taken up a stint of severe penance. The gods were afraid of rushing into battle without Shiva. In their desperation to end the menace of demons, the gods banded together and appealed to Ganga—the river that passes through heaven, earth, and the netherworld—for help. They asked her for an offspring that would be supremely pure and brave, worthy of the looming great battle with the demons. Ganga's father, Himalaya, consented to give her away for this purpose. And that is how, Vishvamitra explained, Ganga came down from heaven to earth. Rama listened attentively to this explanation but was still not satisfied. He wanted to know more about Ganga. Where did she come from? Why does she flow in three separate dimensions?

The sage went on. Shiva, the "dark-throated one,"[18] once desired the goddess Ganga and lay with her for a hundred celestial years, but their union bore no fruit. The gods were concerned about a child of Shiva, the god of destruction, because of his fearsome potency. Shiva assured them that he would refrain from begetting a child, but he could not stop the spilling of his semen. The gods asked the earth to receive his seed, and from this mountains and forests came into being. The god of fire, Agni, helped transform Shiva's sperm into a lustrous white mountain, from which a son, Kartikeya, would be born. However, Shiva's wife, Uma, could not bear this, and she put a curse on the gods so that none of them would be able to have children ever again. Alarmed by this, the gods beseeched Brahma, the forebear of all gods. Brahma told them that Uma's curse could not be undone, and indeed they would not be able to father children; however, Agni, the fire god, should beget a child with Uma's sister, Ganga, the river-goddess, who would grow up to be a great soldier and fight for the gods against demons.

Ganga agreed to this as well, and Agni, captivated by her enchanting beauty, proceeded to consummate the union. Alas, the embryo was too fiery for her to bear. Agni asked Ganga to place it at the foot of the Himalayas. The embryo had the "luster of molten gold," and as it came in contact with the earth it was transformed into various elements.[19] The mountains and the forest were turned to gold. The child was nurtured by the gods, who sent maidens from heaven (Krittikas) to nurse him. The boy was named both Kartikeya (because he had been mothered by the Krittikas) and Skanda, and just as Brahma had foretold, he became a mighty warrior who single-handedly demolished entire clans of demons.

Different versions of the story of the conception of Kartikeya appear later in the *Mahabharata* and the *Brahmavaivarta Purana*, but the Ganges does not feature as prominently in these renditions.[20] The telling of this story in Valmiki's *Ramayana* is thus of great interest to us. Not only does Ganga appear here as the sister of Shiva's consort and daughter of the Himalayas, but there is a subtle interweaving of whims and desires of the gods and the alignment of natural forces. The mountains, forests, water, and elements appear in the guise of deities. The Himalayas are real mountains, but they are also the resting place of Shiva, the reclusive, scandalous ascetic and his fearsome and ghoulish companions. The Ganges is a river like any other. Its banks are skirted with lush forests that shelter birds and animals of every description and provide refuge to ascetics and holy men. Yet the same Ganga is also a daughter of the mountains, an enticing beauty who violates Shiva's meditation and arouses his lust. She is the mother of Kartikeya. She is pure and unsullied, and only she can bear a child sired by Agni, the god of fire.

There was much more to the story that Vishvamitra had started to narrate. As we shall see, Ganga was eventually persuaded to water the plains below the Himalayas—not because of divine caprice after all, but through the acts of a worthy and virtuous man. This man would have to suffer impossible ordeals, yet remain unshaken in his resolve to purge the sins of his ancestors. This was Bhagiratha, who belonged to the royal line of Ayodhya, as did Rama. And because Bhagiratha succeeded in coaxing the celestial river to flow through human habitations and bless the earth with her endless bounty, the Ganges also came to be known as Bhagirathi. In its eastern stretch, the Ganges is still called by that name.

Ganga Comes to Earth

A righteous and brave king of Ayodhya called King Sagara had two beautiful wives, Keshini (the one with lustrous hair) and Sumati (the well intentioned). Sagara was childless. Driven by the quest for a child, he and his wives visited the Himalayas to perform austerities for a hundred human years. Pleased by such dedication, the great sage Bhrigu granted the king a boon. He and his wives would be blessed with children, and he would attain unrivaled fame in the world. One wife would bear sixty thousand sons, and the other would give birth to the one who would carry on the dynasty. The sage left it up to the two wives. Keshini wanted her son to be the scion who would keep the lineage, and Sumati chose to be the mother of a multitude of sons. Keshini gave birth to a son, Asamanja. Sumati delivered a fetal mass from which sixty thousand male infants emerged. This great band of brothers grew up to be strong and handsome men. Kesini's son Asamanja, as he became older, turned into a rogue and a menace to the inhabitants of Ayodhya, seizing children and drowning them in the waters of the Sarayu for sport. His father had to eventually banish him from the city. Asamanja had a son named Amshumant (Angshuman), who became known for his kindness and bravery.

Now Sagara, the aging king, decided to perform the great sacrificial rite of Ashvamedha, in which a royal white stallion was let loose, followed by the king's troops. Wherever the horse ran unchallenged, the king would claim his dominion. King Sagara's horse was guarded by Prince Amshumant astride a great chariot with his bow at the ready. However, Indra, the lord of all the gods, decided to play a trick on Anshumant and his retinue and teach them a lesson. He took the shape of a demon and carried off the sacrificial horse. This was distressing news for the king. A defective horse-sacrifice would surely incur the wrath of the gods and bring disaster on the king and his progeny. Sagara beseeched his sons to pursue and seize the horse by any means possible. "You must dig up the very earth itself until the horse is found," he said.[21]

At this command, Sagara's sons began to tear up the corners of the earth with great fury, molesting and scattering a multitude of netherworld dwellers: gods, demons, ogres, and serpents. In sheer terror they fled to Brahma, praying that he might stop this wanton destruction of the innocent denizens of the underworld (*rasātala*). Brahma assured them that this misfortune had been preordained, and that Sagara's sons would soon pay a terrible price for their misdeeds. Meanwhile, though they had dug up the earth in every direction and destroyed countless creatures, Sagara's sons had not yet found the horse thief. In another version of this story that appears in the *Bhagavata Purana*, the oceans of the world were created by the earth being dug up by Sagara's sons during their frenzied search.[22]

Indra, king of the gods, had actually run away with the sacrificial horse and taken it to the subterranean ashram of the famous sage and philosopher Kapila, where it grazed. Eventually Sagara's sixty thousand sons arrived at this sanctuary and spotted their missing stallion. Thinking that Kapila had stolen the horse and blinded by fury, they tried to attack him. Kapila, livid at this affront, glared at the intruders and, uttering the word "hum," incinerated them instantly into a heap of ashes.[23] The only one left now to look for the horse was the king's grandson, Amshumant. Following the trail of destruction left by Sagara's sons, he quickly reached the netherworld and found the remains of his kinsmen. Overcome by grief, he looked around for water to perform libations for the last rites, but he found no source of water nearby. At this time Suparna, the king of birds who lived in the nearby forest, stepped forward to console Amshumant. He told him not to grieve and to accept the fate that had befallen his clan.[24] He also told him that no earthly water could be offered as libation for deliverance of these poor souls. Their ashes must be washed with the waters of Ganga, the eldest daughter of Himalaya, and only then would the spirits of his slain kinsfolk be set free.

Thus a terrible burden now rested on the progeny of Sagara. How to bring down the Ganges from the faraway heavens to the netherworld? Neither Sagara nor his grandson, Amshumant, could fathom this in their lifetime. Amshumant's son Dilipa took up the cause and attempted to appease the divine forces with a severe regime of austerities on one of the mountain peaks of the Himalayas, which is said to have lasted for 3,200 years. Despite such resolve and penance, Dilipa was not able to receive the right favor from the gods, nor did he succeed in coming up with a remedy for the funerary rites and salvation of his ancestors. All through his life he was preoccupied with one thought: "How can I bring down the Ganges? How can I perform these funeral libations [*jalakriyā*]?"[25] After his death, his son Bhagiratha became the king, and this became the mission of his life as well.

Bhagiratha could not beget any sons. To win the favor of the gods he went to the Gokarna Mountain as a humble ascetic and began a long period of austerity. He kept his arms raised at all times, ate only once, and succeeded in extinguishing all sensual desires. Much pleased with such virtuous behavior, Brahma, the creator of the universe, granted him a special favor. He ordained that the waters of the Ganges would come down from the heavens and wash the ashes of Bhagiratha's ancestors, releasing their spirits for their final resting place. The only condition was that the great Shiva must check the fall of the mighty Ganga, daughter of the mountains, once she plunged to this earth. Hearing this, Bhagiratha, in order to pay homage to the idiosyncratic Shiva, stood on the tip of one big toe for a whole year. Pleased at his painful supplication, Shiva agreed to bear the mighty Ganga on his head. And thus Haimavati, the "snow-clad" daughter of the mountains, assumed the natural form of the river Ganges, fell from the skies with thundering force onto Shiva's head, and rushed into the various coils of his copious and matted locks. This was a mighty labyrinth from which she could not escape, and she splashed about in his curls for a number of years. Shiva was pleased by this and eventually released the river into a pool called Bindu, from where her waters rushed down to earth in a fearsome roar. A host of gods and heavenly creatures came to watch this marvel and pay their obeisance. The *Ramayana* provides a vivid account of the cascade of waters on earth in this passage:

> The cloudless sky was shining with the hosts of hastening gods and the splendor of their ornaments, so that it was illumined by a hundred suns. Filled with masses of twisting snakes, crocodiles and fish, the sky seemed shot with scattered lightning. Now white with a thousand shreds of flying foam and flocks of snowy geese, the sky seemed suddenly filled with autumn clouds. At some points the river flowed swiftly, at others slowly. In some places it moved tortuously or broadened out; at others, it narrowed and sank between its banks only to rise elsewhere. Here and there the water dashed back upon itself, momentarily hurled up into the air, only to fall to earth once more.[26]

Because they fell from Shiva's locks, the waters were absolutely pure (*toyam nirmalam*) and had the power to wash away all sins. All came to bathe in the Ganges, including the accursed who had been banished from heaven, and rejoiced as the waters bore their sins away. At the head of the rushing waters, mighty Bhagiratha rode, mounted on a celestial chariot, followed by gods, seers, and demons. Finally, on its way to the ocean, the waters of the Ganges flooded the crater where Bhagiratha's ancestors had been turned to ashes. Brahma then spoke directly to Bhagiratha, assuring him that Sagara's sons would now ascend to heaven as gods, and the Ganges would be known as his own daughter bearing his very name—Bhagirathi.

Most of these narratives of Ganga's plunge to the earth thus intersperse their mythical accounts with a conscious invocation of natural imagery. The earliest interlocutors of this tale were well aware of the extensive riparian landscape associated with the Ganges: glacial melts, watersheds, meanders, sandbanks, and the long journey of the river to its delta and the ocean.

Variations on a Theme

How old is this story of the descent of the Ganges? Historians have debated the dates of the compositions of the great epics *Ramayana* and *Mahabharata* at least since the advent of antiquarian scholarship during the early period of British rule in India. Today, scholars seem to agree that the compilation of the *Ramayana* as a distinct corpus can be traced back to a stretch of time between the fourth or fifth century B.C.E. and the third century C.E.

Cities and kingdoms mentioned in the *Mahabharata*, located in the upper tracts of the Ganga-Yamuna Doab, seem to have been settled earlier than the lower Ganges plains where the *Ramayana* unfolds. The *Mahabharata* in its present outline probably took shape sometime between the fourth century B.C.E. and the fourth century C.E. Myths of the origins of the Ganges can be found in bits and pieces in both these epics and also in many of the eighteen major Puranas. Such stories, embedded in an extensive oral tradition and recounted in popular memory, clearly acquired a coherent or consistent form centuries before they were rendered into written texts.

The Puranas are said to have been put into writing by the first sages, beginning with Vyasa during a time beyond human reckoning. They contain stories of the creation and re-creation of the universe; cycles of the primordial ancestor of humans, the Manus; the rise and fall of clans and dynasties; genealogies of gods and their deeds that shaped the world; lineages of ancient sages; and accounts of major dynasties.[27] The Puranas existed long before the advent of the Christian era and were written down as texts after the early first millennium C.E.[28] Even afterward they continued as part of the recitative tradition of genealogy, guarded closely by Indian bards and rhapsodists (*sūta*s). At times they were written anew under the auspices of different rulers and tweaked for the pleasure of dynasties that patronized particular versions, especially in the interest of their own favored deities.[29] A vast body of disparate narratives was absorbed into the corpus of the Puranas before the first compilations were put together. They were being restructured and recast as late as the Gupta period and after. The freedom with which the Puranas have been revised and rewritten creates serious limitations to their exact dating, leading to wild variations in the speculative date of individual texts, such as those of the *Bhagavata Purana*.[30]

Nevertheless, they make up the largest part of Sanskrit literature—a bewildering mass of text bustling with genealogy, cosmology, and history—precipitated through centuries of collective recall. They are always in the process of revision, and the currently identifiable versions can be traced not much further beyond the fourth and fifth centuries C.E. Some scholars view them as corrupted forms of older standardized versions.[31] Some evidence suggests that new versions have appeared as late as the eighteenth century. Given such variability and the virtuosity of both oral and written forms, it is no surprise that versions of the story of the descent of the Ganges abound in the different Puranas, as they do in the *Ramayana* and the *Mahabharata*. No doubt these stories are connected. As one may well expect, there is an

Shiva bearing the Descent of the Ganges River, Himachal Pradesh, ca. 1740. Photograph © Museum Associates/Los Angeles County Museum of Art.

organic relationship between these specific episodes and the flow of the major epics. A comparison of versions shows how frequently the same segments of narrative appear. They are a testimony to both the resilience of the mythic form and the shared practices of the storytellers of Indian antiquity.

In the *Bhagavata Purana* attributed to the sage Krishna-Dvaipayana Vyasa, Ganga appears in the form of a river adored by all three major deities: Brahma, Vishnu, and Shiva. She is borne by Brahma in the water pot for his ablutions. She is the consort of Shiva. She seeks sanctuary at Vishnu's feet, from where she eventually emerges, heeding the call of Bhagiratha, who needs her for the last rites of his ancestors. Appearing before Bhagiratha and consenting to his wishes, she asks him how the earth was going to withstand the force of her fall (*ko 'pi dhārayitā vegaṃ patantva mahītale*).[32] Bhagiratha replies that it could only be Rudra (Shiva), out of whose body the entire fabric of creation has been woven (*viśvaṃ śāṭīva tantuṣu*). Bhagiratha, seated on a chariot traveling like the wind, leads the way, and the mighty river follows behind him. He is also the architect of her tortuous path through the mountains of the Himalayas. The popular Bengali rendition of the *Bhagavata Purana* includes a lively retelling of Ganga's tumultuous descent:

> King Bhagiratha sounded his conch
> Ganga rushed forward, but could not find a path
> And thus she lay there for a year
> Grief-stricken, Bhagiratha blamed his fate.[33]

Despite the gigantic hurdle presented by the Himalayas, Ganga's raging waters released from Shiva's own matted locks (*jaṭāgranthi*) could not be held forever. She traversed 5,000 miles at once, arriving near present-day Haridwar, the abode of the seven great sages that appear as the constellation of seven stars (*saptarśimaṇdala*, or Ursa Major). From here, the seven great streams flew in seven different directions.[34] This was just the beginning of Ganga's travails. She was lured to the quarters of the great sage Jahnu at the sound of his propitiatory conch and flooded his ashram. Outraged, Jahnu drank all of Ganga in one mouthful (*gaṇḍūṣa*). In great distress, Bhagiratha sounded his conch once more, and out came Ganga from Jahnu's belly and through his knee, much to the old man's surprise.[35] After she emerged, Ganga paid her obeisance to the mighty sage, addressing him as her father and asking to be called his daughter. Because she came out of Jahnu's knee, she would henceforward be called Jahnavi. Ganga rushed forward toward the ocean. In this path she met her sister river Padma, whose call she heard "as a mother-cow would hear the lowing of a calf."[36] Finally her waters merged into the sea and descended further to the netherworld in search of Kapila's fabled ashram, where Sagara's sons had been turned to cinders. As soon as her waters touched the ancient ashes, the lost souls of Sagara's progeny found their path at last to the abode of their departed ancestors.

In the *Bhagavata Purana*, the story of Ganga's fall to earth is told in a specific episode (*jāhnavīdevyā upākhyānam*) in which Ganga appears as one of the wives of Lord Vishnu along with Lakshmi and Saraswati.[37] Ganga and Saraswati are both incarnated as terrestrial rivers. According to this Purana, it had been preordained in the saga of all creation that Ganga will appear one day from above for the salvation of all humankind (*mānavān sarvvān*) in the form of water on earth at the beseeching of the great Bhagiratha.[38]

> If one merely touches your water
> Sins acquired over a million births dissipate
> Just by gazing upon your pure flow
> One reaches the abode of Vishnu[39]

Lord Krishna says in the *Brahmavaivarta Purana* that Ganges water has the power to absolve the bad fruits of karma acquired by an individual over the course of many lifetimes.[40]

It was always destined that Bhagiratha would be the first to receive and worship Ganga on earth. Humankind had been waiting through 5,000 dark years of the Kaliyuga. Bhagiratha was the dutiful descendant of Sagara, who went through great physical pain and spiritual anguish to rescue the condemned souls of his ancestors by bringing Ganga down to earth. In doing so, he gave humankind a way to wash away the debris of their sinful deeds. The *Bhagavata Purana* also explains why the Ganges is a river so blessed. Not only does she splash around in Shiva's dreadlocks and in Brahma's ablution pot, she also hides for a while in Vishnu's eternal lotus feet.[41] At one point Krishna, Vishnu's most recognizable avatar on earth, embraced the beautiful river on his left flank, alienating his own consort, Radha. The *Skanda Purana* insists that the waters of the Ganges are nothing but the essence of Shiva's potency, created and let loose for all fallen creatures crying out for redemption. The Ganges is therefore Shiva in another form: *śivasvarūpinī*.[42]

Legends of Bhagiratha

It is worth considering here that Bhagiratha—although he had communion with divine forces through his severe *tapas*—was, after all, just a man. Not only did he revive the spirits of his ancestors from their ashen remains and make the divine river flow through a sinful world plunged in darkness, he gave hope to the wretched and the fallen. Regardless of caste, age, sex, or offense, the sacred river does not turn anyone away from absolution by her waters. Touched by her, they do not have to suffer years of penance, make sumptuous offerings to the gods, acquire knowledge of the scriptures, or even secure the mediation of priests. The waters are there for every soul, living or dead. This fact alone makes the story of her descent and the man who made it possible remarkable. But what kind of a character was Bhagiratha, and why was he chosen for this deed in the first place?

The sixty thousand sons of Sagara had to be punished for their destructive ways and their excessive hubris. This was why the king of gods, Indra, secretly decided to lead their sacrificial white horse astray. Furious, they laid the earth to waste looking for the steed. When they finally found the horse in Kapila's ashram, they not only disrupted his mediation and prayers but also insulted and tormented him. The fact that they were smitten and incinerated by the great sage's rage was a fitting punishment. Far from home and deep in the netherworld, they were burned to ashes without a funeral or last rites. Their spirits were denied a final resting place. However, sins in the Puranas are not merely a function of individual culpability and are not even entirely the poison of one's own karma. They are also tied to ancestral deeds. Sagara's sons *inherited* their characteristic flaws.

King Sagara was the child born of King Vahu and Queen Yadavi in the great clan of Ikshvaku. Vahu had another wife who was jealous of Yadavi, and when Yadavi was pregnant, she tried to poison her. By divine grace, she did not die or lose the child. Vahu had many enemies, and they saw this ordeal as an opportune moment to attack him; he had to flee to the forest with his family, including his pregnant wife. Because of the action of the poison, Sagara, when he was a fetus, could not see the light of day for a hundred years. Vahu died after many years in exile, and Yadavi prepared to mount her husband's funerary pyre. At this critical moment another great sage, Ourva, appeared and asked her not to throw herself into the fire. Yadavi was with child, he beseeched, and an act of self-immolation in this case would surely end the life of the unborn. Ourva foresaw that Sagara would grow up to be a good king and a formidable warrior, and he would one day avenge his father's death by defeating his enemies. Thus persuaded, Yadavi decided not to take her own life and went with Ourva to live in his hermitage, where she eventually gave birth to a boy who had survived poisoning. And thus he was named Sagara (*sa gara*), "one born with poison."[43] Sagara was brought up and educated in the ashram. He learned the Vedas and studied the arts of combat. He resolved to recover his father's lost kingdom and vanquished the clans of the Haihayas, Talajanghas, Yavanas, Shakas, Kambojas, and others.

This segment of the story appears in many sources, including the *Bhagavata Purana*, the *Brahma Purana*, the *Padma Purana*, and the *Mahabharata*. What follows has been narrated before in the context of the *Ramayana*, barring some variations. Sagara was unable to father a child and propitiated Shiva with prayer and offerings. Shiva promised him progeny from each of his two queens. One would be the mother of sixty thousand strong sons, all of whom would eventually perish. The other would give birth to the son who would carry his line. In due course both queens were expecting. Sumati gave birth to a beautiful son. The other queen, Keshini, gave birth to a gourd. This gourd was placed in a pot of warm clarified butter, in which it seeded. The sixty thousand sons were begotten from these seeds. The *Brahmavaivarta Purana* states that Keshini gave birth to a ball of flesh, which by the grace of Shiva and the blessing of the Ganges was split into sixty thousand parts, each of which became a son.[44]

As they grew up, Sagara's sons became a band of unruly tyrants. The son from the other queen, Asamanja, was different. Bhagiratha was the great-grandson of Asamanja and the son of Dilipa. All of Sagara's progeny were cursed from the beginning because of the lingering effects of the poison that had been administered to his mother. Even the lineage that was kept alive by Asamanja almost came to an end with Dilipa, who proved to be impotent and could not father any children. As we shall see, Dilipa was not exactly Bhagiratha's natural father. Sagara's sons and grandsons had been tainted by the poison, and all tried in vain to bring Ganga down to earth from the heavens, hoping that she would flood the subterranean realms and the crater that contained the residuum of the sixty thousand sons incinerated during Kapila's wrath. They all failed except for Bhagiratha.

Bhagiratha's patience is considered exemplary in the Indian mythological repertoire, and his self-imposed penance and mortifications as an offering to the gods are remembered as exemplary acts of virtue. However, only because of his pure heart did he persuade the river goddess to break forth on the planet, flood the subterranean realm, and then mingle with the ocean. Bhagiratha did not carry his ancestor's curse because he was not conceived out of an ordinary union between man and woman. Some of the Puranas indicate that he was born out of sexual congress between the two queens of King Dilipa. And because Bhagiratha was conceived without semen, he did not carry the seed or the curse that his father Dilipa had inherited from his ancestors. Bhaigratha was not an offspring who resulted from an ordinary consummation, and this made him spotless and pleasing to the gods.

Versatile Narratives

The ubiquity of the Ganga myth in India demonstrates how popular culture digests and regurgitates narratives without breaking down their basic structure. This is evident in Bengali texts on the Ganges from eastern India, at least from the fourteenth century. They ascribe Bhagiratha's birth to a miracle of double maternity. According to scholars who are interested in aspects of femininity and atypical forms of gender identity in medieval India, the Bengali versions of the *Padma Purana* and the *Ramayana* stand out in their characterization of the genesis of Bhagiratha as a result of the sexual union of two women. A version of this female-centered story appears in some manuscript versions of the *Padma Purana* that have been found in eastern Bengal.[45] It also appears in detail in Krittivasa's widely popular *Ramayana,* written in Bengali, which dates to the late fifteenth century. A similar account appears in Dvija Madhab's compositions extolling the Ganges, which appeared almost two hundred years after the writing of Krittivasa's *Ramayana.*

The episode on creation (*sṛṣṭikhaṇḍam*) that appears in a popular version of the *Padma Purana* collated by Abdul Karim provides a detailed account of Bhagiratha's lineage and place in the Ikshvaku clan.[46] According to this text, Ikshvaku, the progenitor of this line, had a hundred sons, and the family of King Sagara was begotten from Ikshvaku's son Kakustha. The same clan produced legendary kings such as Mandhata, whose name in Sanskrit is synonymous with "antiquity"; King Harishchandra, famous for his sacrifices and suffering; and also King Vahu, father of Sagara, who survived poisoning. Sagara became very accomplished and powerful because he grew up in the ashram of Sage Bhargava, who taught him how to use weapons that spouted fire.[47] Later as a king he commanded the greatest horse-sacrifice with a stallion that emerged from the depths of the ocean. This is the same horse that Indra stole and brought back to the underworld, and the one that Sagara's sons searched for, tearing up the earth. This form of Bhagiratha's genesis is narrated with vivid imagination in Krittivasa's *Ramayana,* one of the most widely circulated and recited texts from medieval Bengal.[48] In Krittivasa's rendition, Bhagiratha's father, Dilipa, spent 1,000 years in severe penance, fasting for long periods and even forgoing water, seeking redemption from the gods for his fallen ancestors. Eventually, Dilipa died without a son to carry on the name of the clan and the dynasty. However, his suffering had stirred the conscience of Shiva, who came down from his abode in Mount Kailasa in search of Dilipa's two queens and appeared in front of them in his original form. He told them, "One of you will conceive from my boon."[49] The queens, surprised to hear this, exclaimed, "But we are widows!" Shiva then asked the two women to make love to each other and assured them that they would beget a child from this unusual union. Acquiescing to this request, the two queens bathed and prepared for this ritual. When one of them had become fertile, they made love. Just as Shiva had promised, a baby was born in ten months' time. However, because the child was begotten without semen, it was nothing but a lump of flesh, with no skeleton. Much distraught at this outcome, the two mothers wept. They could not figure out how to raise this deformed child.

Putra kole kore kāndena duijona
Heno putra bara keno dila trilocana
Asthi nāhi, māṃsapiṇḍa colite nā pāre
Dekhiyā hāsibe loka sakala saṃsāre[50]

The two sat crying with their son in their lap
Why did the third-eyed-one give such a blessing?
No bones, a lump of flesh unable to move
The whole world would laugh at it

Despite his disfigurement, the boy survived and grew up in the care of his two mothers. Because he had been born out of the union of two women, he was called Bhagiratha. One day the famous sage Ashtabakra was passing down the street when he saw the young boy. Ashtabakra was an old man "bent in eight places" and moved with much pain and difficulty. Not knowing the boy's history or condition, the sage thought that Bhagiratha was mocking his movements. Enraged at this, Ashtabakra was about to curse and incinerate the boy with his supreme rage (*brahmaśāpa*), when he realized that in fact he had been born with a curse. His rage turning to empathy, Ashtabakra decided to use his special powers and endow Bhagiratha with a skeleton.[51] This is how Bhagiratha survived.

As Bhagiratha grew up, he decided to ask about his father. People would not tell him the truth. They made fun of him and hurled insults. Shamed and humiliated, he pestered his mothers to tell him the true story of his conception. When he found out about his family and his ancestors, a new resolve arose in his heart. He decided that he would bring Ganga down to the earth and assuage the sins of his ancestors, whose ashen remains lay near the sanctuary of the great seer Kapila. Just as his father had done, Bhagiratha sought out the highest mountains and began to pray to Shiva for help. He disregarded all physical comfort, winter cold, and summer heat. At one point during his austerities, he even submerged himself in water for four months. No one had seen such dedication, such painful penance. Eventually, his suffering began to move the great gods Brahma and Vishnu. As the text indicates, the one holding the *cakra* (Vishnu) dissolved (*drabarūpa hailena nije cakrapāṇi*).[52] This idea of "dissolving" out of love for the humble human supplicant is noteworthy, for it marks the text with an emotion that is typical of medieval devotion, often described as *bhakti*, which had begun to flourish in Bengal around this time.

Thus the heavenly Ganga, the essence of Vishnu's dissolving kindness, plunged earthward from *brahmaloka*, the seat of the gods. First she fell on the great mountain Sumeru, the highest point on earth, and resided there for twelve years, locked in a dark, vast cavern. There the waters of the river thrashed about, with no place to go, wrapped in blinding darkness. Finally Ganga pleaded for release, and asked Bhagiratha, the man who had precipitated her fall, "Which way out?" Who could cut through ice and rock and let her waters out of this lofty prison? Ganga asked Bhagiratha to ask for Indra, the king of gods, to lend his mascot, the great white elephant Airavata, who would be able to tear up the side of the mountain with his gigantic tusks. The mighty Airavata, lord of elephants, came trumpeting to the rescue. However, bedazzled by the divine beauty of the goddess Ganga, the rutting beast desired her. Here, Krittivasa's text takes on a distinctively ribald tone. Ganga told the elephant that she would consent to lie with him if he could withstand more than two waves. Blinded by lust, Indra's elephant tried to ride the mighty waves of the Ganges and nearly drowned. Rescued from the water and duly humbled, Airavata cut the great mountain Sumeru into four giant sections with his enormous tusks, and the Ganges fell into four distinct streams: Vasu, Bhadra, Shveta, and Alakananda. Vasu proceeded to flow toward the eastern sea, while the other three rivers fell straight down to earth, then cascaded in three different directions. Having almost drowned, Airavata, exhausted and having fulfilled his task, beat a hasty retreat.

Krittivasa's version of the Bengali *Ramayana* is now a standard text. Looking at this detailed account of Bhagiratha's genesis, penance, and the coming down of the Ganges to the earth, one can form an idea of how widespread this version of the story had become in eastern India. It sets the stage for other texts in Bengali from this period onward, and the story became even more embellished with details and characters, more patchwork in the fabric of popular imagination surrounding devotion to the Ganges and its waters in various forms. Another, more concise text from Bengal, the *Gangamangala*, composed during the late sixteenth century and attributed to a poet named Dvija Madhab from the region of Saptagram and prevalent in eastern Bengal, tells the story of Bhagiratha's birth and his feat of bringing the Ganges to earth in witty and colorful language.[53] We do not know enough about this poet to say that this is the same person who wrote the well-known *Mangala Candira Gita* around the same time. The text is quite closely related to the main story of the Puranas but is distinctive—its style of storytelling is much like that of other Bengali ballads of the Mangalkabya genre, showing how adaptable and ductile the story of the Ganges can be. Dvija Madhab narrates the story of the birth of Bhagiratha as part of the miracle of the descent of the Ganges to earth, which starts with the inability of King Dilipa to produce an heir. His heart was not into the affairs of the kingdom. Leaving behind his two queens, he went away to meditate in solitude. The two queens in search of a son started worshipping the sun. On the strength of their prayers, the Sun God appeared and granted them their wish. The queens received a charmed capsule, which would increase their sexual desire (*madanamodaka baḍi*).[54] After ingesting the pill, the queens became infatuated with each other, and afflicted by desire they made love. As a result of this union a boy was born, and because of his birth from female sexual organs, he was named Bhagiratha.

The work of myth is never done in a culture steeped in storytelling, and rural Bengal makes a convincing case. Such conscious recasting of the Ganga-Bhagirathi myth is evident in a remarkable text from 1855, Duragaprasad Mukhopadhyaya's *Gangabhakti Tarangini*, which shows that the Ganges myth was live and evolving during the height of British rule in India. Durgaprasad hailed from the city of Sanskrit learning and Vaishnava devotion, Nabadwip on the Ganges. In the preface to his work he says that he was inspired by a dream his second wife, Sati, had about the Ganges.[55] In his long and lively account in rhyming verse, Durgaprasad adds new cities and markets to the course charted by the Ganges to earth, including the pilgrimage of Kalighat in Calcutta, seat of the veneration of the goddess Kali, and he also includes the abode of Jagannatha in Puri, Orissa, as a domain somehow touched by the grace of the river. Here Ganga is depicted as the daughter of Daksha and Prajapati who was conceived on earth by the celestial nymph and danseuse Menaka after her banishment from heaven and long penance. Durgaprasad describes the pregnancy in great detail, because in Menaka's uterus Ganga assumes the female human form for the deliverance of sinners. The entire world rejoices with the impeding news of Ganga's birth, and all mountains known to creation—Mainaka, Kailasa,

Ganga in a Bengali Hindu almanac, Calcutta. *Sanatana Hindu Panjika*, 1892–1893. Block by N. L. Dutta, in electroplate. Photograph by Rajat Sur.

Vindhya, Govardhana, Gandhamadana, and Sumeru—come to pay their obeisance. Ganga's arrival is also marked by the appearance of signs and miracles: the blind are restored to sight overnight and cripples get up and walk. Ganga's childhood is also a portent of her greatness, and the tumultuous sporting of her waves is described as the manifestation of her eternal spontaneity (*līlā*).[56] In the episode dedicated to Ganga's matrimony with Shiva, Durgaprasad gives a vivid account of the feminine rituals of rural Bengal (*strī-ācāra*).

Led by Bhagiratha, Ganga tumbles out of heaven from Brahma's water pot in three streams: Mandakini in heaven, Bhogavati in the netherworld, and Alakananda on earth. With the help of Indra's elephant, she bursts out of the Himalayas and follows Bhagiratha through the northern Indian plains. Ganga carefully skirts the holy city of Kashi and sallies forth toward the ocean. Here Durgaprasad recounts a familiar geography of the major cities of the present-day Ganges valley: Munger, Bhagalpur, Kahalgaon, Rajmahal, and entering the great Bengal delta, Nabadwip, Ambika, Shantipur, and all the way to Kalighat in Kolkata. In the course of his description, Durgaprasad gives vivid details of fairs and markets and the bustling trade of tobacco and betel-nut.[57] When she finally arrives near the island of Sagara, where the cinders of King Sagara's accursed progeny are kept, a great shining chariot emerges from her waters, which bears Bhagiratha like a newly anointed king; he is finally able to make an offering (*tarpaṇa*) on behalf of his forsaken ancestors and deliver them to heaven (*vaikunṭha*), with the help of two hundred thousand golden pots of Ganges water.[58]

An Enduring Icon

Bhagiratha's steadfastness and resolve to bring down the river and propitiate the spirit of his ancestors has been celebrated across many walks of Indian literature. Considered an exemplar of poetic composition, the celebrated *Gangavataranam* of Nilakantha Dikshitar, scholar, poet, and philosopher of seventeenth-century Madurai in Tamil country, on the descent of the Ganges is a composition of eight cantos and is still studied today along with the greatest craftsmen of Sanskrit verse such as Kalidasa and Bhavabhuti. Nilakantha's poem is not just an ode to the river but also a rousing tribute to King Bhagiratha, who donned the loincloth of a holy mendicant (*muniveśa*) and embraced the long and difficult penance that finally disturbed the equanimity of Lord Brahma and caused the mighty river to inundate earth with her resplendent purity.[59] In Nilakantha's story, the celestial river, pleased by Bhagiratha's entreaties, agrees to his request but warns him that her plunge to earth might drown and destroy Brahma's great creation (*jagadaṇḍamaṇḍala nimajjayantīva*).[60] In this version of the myth, Ganga enters the mortal realms behind a triumphant Bhagiratha, who is conducted through the crowded streets of the cities in kingly glory. Nilakantha's description evokes images of royalty that must have been typical of his time. This is how he describes Bhagiratha's swift chariot, with the river making its course by following his path:

> *atha sannihitaṃ bhagīrathaṃ saritaṃ tāmapi tadrathānugāṃ*
> *avalokitumicchatāṃ nṛṇāmatulastatra babhūva saṃbhramah*[61]
>
> And right behind Bhagiratha, the river trailed his chariot
> Countless people who wanted to see this were filled with awe

Roughly contemporary to Nilakantha, another acclaimed figure, Jagannatha Panditaraja, a great scholar of linguistics, poetics, aesthetics, and philosophy, wrote a great ode to the Ganges, the *Gangalahari* [Flow of the Ganges]. Jagannatha arrived at the Delhi court as a young man from Telengana during the reign of Emperor Jahangir and became a court poet. Emperor Shah Jahan loved hearing his compositions and granted him the coveted title of Panditaraja, "king of scholars." Jagannatha was exiled because of his connection with the Mughals, and also because of his marriage to a Muslim woman of Shah Jahan's court. The cantos of the *Gangalahari* are a testament to Jagannatha's poetic virtuosity, but they also include some of the most moving devotional eulogies ever offered to the Ganges. They also suggest a transformation of the figure of the Ganga from a young maiden, a symbol of fecundity and bounty, to a maternal figure of forgiveness and compassion. In Jagannatha's imagery, Mother (*jananī*) Ganga has forsaken the lotus feet of Vishnu and the refuge of Shiva's heavenly locks to deliver the fallen (*patitajananistāraṇa*).[62] She is as fair as the autumn moon, seated on a white Makara and crowned by the white serpent and lunar crescent; two of her hands hold the earthen pot and the lotus (*kumbhāmbuja*), and the other two are folded in the gestures of blessing and *abhaya* ("fear not").[63]

These poetic images vary considerably in time and space, which is also a function of localized folk traditions. Examined more closely, however, at least two archetypes seem to emerge that support the overarching icon. One is an elusive and enigmatic figure, resembling the fickle deities of fortune in the Indian tradition. The other is an embodiment of the natural plenty, all-embracing and all-forgiving. The duality is evident in the classic corpus of the *Mahabharata*, whose natural setting is along the westernmost part of the Ganges plains. The Ganges makes a number of appearances in the *Mahabharata* in the guise of both a goddess and a beautiful woman, and in her beguiling form, she becomes an instrument of fate. Take the example of the story of King Mahabhisa, who in the company of the gods in heaven chanced upon Ganga naked, and she returned his glance.[64] Watching this unseemly exchange, Brahma became annoyed and banished Mahabhisa to earth. He was reborn as Santanu, and King Pratipa became his father. At this time Ganga decided to help the eight Vasus cursed by the sage Vasishtha, who had altered their sex and turned them into men. To regain their status as nymphs, the Vasus asked to be reborn as Ganga's own children, because as celestial beings they could not be born in the womb of an ordinary human. They also asked their future mother to kill them at birth by drowning, so that their spirits could be released. Ganga agreed to this, but also requested that one of her eight children be allowed to live.

One day, when the devout King Partipa of the great Kuru clan was enrapt in meditation, at the source of the Ganges (*gaṅgādvāra*), the river assumed the form of a beautiful maiden (*strīrūpadhāriṇī*), rose from the water seductively, and sat on King Partipa's right thigh to entice him, breaking his concentration.[65] Pratipa asked her what she wanted, and Ganga said that she desired him. Pratipa, however, had taken a vow of abstinence and refused to be moved by lust. He pointed out that Ganga was sitting on his right thigh, which was meant for children and daughters-in-law. The wise Pratipa asked her instead to take his son as husband when he came of age. Ganga consented to this, as long as her visit was kept secret and no one questioned her actions. Before he died, Pratipa told his son that one day a beautiful woman was going to seek him out, and he must accept her as his wife. And in due course, roaming the banks of the Ganges during a hunting expedition, Santanu came upon the same exquisite (*paramāṃ*) woman who had appeared before his father.[66] Smitten by her beauteous form, Santanu asked her to marry him, thus fulfilling his father's last wishes. Ganga became his wife and was soon with child, and when it was born she promptly drowned it in the river, exactly as planned. In this way Ganga disposed of the first seven sons, but the eighth time Santanu intervened. The boy who survived grew up to be the great warrior Bhisma.

These descriptions of the river in human form, an amphibian, fickle, voluptuous beauty, who wants to bear the sons of the king only to take their lives by tossing them back into the waters, paint her as a composite figure that straddles the divide between water and earth. The form in which Pratipa and his son Santanu see her is worth remarking: she has a resplendent body and wears a gossamer fabric, shining like lotus-hearted Vishnu (*padmodarasamaprabhā*), the goddess Shri incarnate. This invocation of the goddess Shri is significant. Shri and Bhumi are both goddesses of fortune; they have different origins, but they become largely synonymous in the Upanishads.[67] They are both consorts of Vishnu and precursors of Lakshmi, and in the *Mahabharata,* their qualities are re-embodied in the figure of Draupadi, the principal wife of the five Pandava brothers, whose dishonor sows the seeds of war and destruction.[68] Shri as the goddess of prosperity rises from the lotus, shiny and moist, an embodiment of the living verdure, redolent of earth and manure, slime and mud. The litany of references to water and earth bodies suggests that Ganga as a deity belongs to a hybrid pantheon of fertility figures tied to the history of the dispersal and assimilation of animistic cults and guardian deities. As we shall see later in this book, they found renewed expression in everyday culture as Jainism and Buddhism redefined the languages and affects of popular piety, contributing to the historical depth and iconic complexity of Ganga as a feminine form, part human, part water.

DIGGING OUT OF PREHISTORY

Prehistoric mounds are still waiting to be excavated along the banks of the Kali Nadi, a river that runs through the Etah district in Uttar Pradesh in northern India. The Kali Nadi originates in the western part of the state in the northern uplands and empties into the main channel of the Ganges in the Kanauj district. It is a heavily polluted stream that serves major industrial cities such as Muzaffarnagar, Meerut, Ghaziabad, Bulandshahr, Aligarh, and Farrukhabad, picking up the refuse of the distilleries of sugar processing plants, fertilizer factories, and paper mills spouting smoke, along with the raw sewage that spews from these increasingly congested urban clusters. The archaeological site of Atranjikhera is located on the right bank of this ailing and sputtering river, roughly eight miles north of the old Grand Trunk Road, the first national highway of northern India.

The importance of this site was first noted in the nineteenth century by the indefatigable tracker of Indian antiquities Alexander Cunningham, of the Royal Engineering Corps. A pivotal figure behind the creation and early exploits of the Archaeological Survey of British India, Cunningham studied the ruins of Atranjikhera during his extensive tour of 1862. He believed that just as Pliny the Elder had followed in the footsteps of Alexander the Great, he would follow the trail of the great Chinese monk and explorer Xuanzang, who as a Buddhist pilgrim traveled widely through Harshavardhana's kingdom during the early part of the seventh century B.C.E.[1] At the time of Cunningham's tour this was an unremarkable place. Herds of cattle grazed near the ancient mounds on the flats of the Kali River, near the old and fabled kingdom of Kanauj, settled by pastoral Ahirs and Bhars.[2] This was also the spot where the old village of Sikandarpur Atraji once stood, an important hamlet recorded in the revenue rolls of the Mughal emperor Akbar that had all but disappeared. Although the coordinates did not exactly match Xuanzang's descriptions, Cunningham was convinced that Atranjikhera was the site of the extensive city and settlement of Pi-lo-shana acclaimed in the account of the Chinese explorer, who measured it to be about two miles across, situated on the road to the Buddhist pilgrimage of Sankisa.[3] This is where Xuanzang thought the great Mauryan emperor Asoka laid down one of his largest stupas, more than 100 feet tall with an attached monastery and adorned by a lofty tower. In the light of more recent archaeological findings, the settlements of Atranjikhera turn out to be of even greater significance than what Xuanzang noted.

Subsequent excavations at this site, which began in earnest after 1962, exposed layers of early historic settlements that show how Atranjikhera was occupied almost continuously from around the early second millennium B.C.E. all the way to the third century C.E. The initial excavation had revealed at least three successive periods of occupation based on the occurrence of pottery fragments: ocher-colored, black and red, and painted gray.[4] These overlapping deposits, dating from the New Stone Age to the rise of the Buddhist empires of the trans-Himalayan steppes, make this find particularly noteworthy. Recent archaeology of the upper and middle Ganges plains has unearthed comparable sites at Chirand, Jakhera, and Jhusi, and now specialists can piece together the evidence from this long line of settlements, which points to the gradual rise of a complex order of civilization in the alluvial basin of the Ganges. These discoveries, drawn from the annals of Indian archaeology, are much like those from any other site of comparable antiquity in the world. Some of the story still remains shrouded in doubt. Material artifacts and the residue of seeds, plants, and bones tell us something about life in early human habitats, but a fuller history requires long, patient, systematic study. Archaeological findings such as hoards of copper implements or bits of pottery, however, lend a material heft to certain claims over a nation's past and contribute to the ebb and flow of speculation lavished on such vestiges. This chapter takes up the archaeological evidence and scholarly debates over the early historic settlement and the domestication of plants and animals in the Ganges valley, from tool making, hunting, gathering, and foraging to the rise of agriculture and the first cities, kingdoms, and empires.

The excavations of Atranjikhera represent milestones that are both promising and illusive. The ceramic records of this site have attracted the attention of enthusiasts of Indian prehistory since the findings of the noted Indian archaeologist R. C. Gaur were published. Gaur's excavations revealed a remarkable continuity in the succession of settlements at the site clearly marked by layers of distinctive pottery. This sequence begins with ocher-colored ware and black and red ware and leads into a long period where the settlers learned how to fire a more durable and sophisticated series of ceramics known as painted gray ware (PGW). Atranjikhera has also disgorged an even more accomplished set of ceramic artifacts known as northern black polished ware (NBPW), associated with later historic kingdoms leading up to the period of the Maurya Empire. On top of this layer lie forms of red ware associated with the trans-Himalayan steppe empire of the warlike Kushans, followed by the common glazed ware of later medieval kingdoms. Some of the earliest evidence of iron is also found here, and perhaps the earliest recorded remains of an entire chariot dating back to around 200 years after the death of Gautama Buddha. Atranjikhera, in other words, seems to hold the enigmatic keys to layers of the archaic past deposited in the valley of the Ganges.

Many scholars doubted the significance of shards of pottery in the study of complex settlement patterns. However, Gaur and his followers appeared convinced that they could verify basic facts about the spread of Indo-Aryan culture into the valley of the Ganges. The conjecture was that the original Aryans were outsiders who had migrated from the valley of the Indus and its tributaries to the valley of the Saraswati, associated with the present-day Ghaghara, which is an important tributary of the Ganges. "On the basis of the present evidence," Gaur wrote, "it appears that the Black and Red Ware people did not belong to the Sapta Sindhava region and their cultural complex does not reflect Vedic culture."[5] He thought that the people who threw the black and red pottery likely belonged to one of the "non-Aryan" tribes much farther south, possibly as far as the valleys of the Narmada and the Mahanadi Rivers. During the later phase of the Vedic civilization, he argued, the Aryan pastoralists spread into the Ganges valley with their iron implements, around 1200 B.C.E.[6] Aryans who composed the *Rig Veda* were the rightful purveyors of the burnished gray ceramic. In fact, "PGW culture" was a phase of Aryan culture, and along with Atranjikhera, excavation of sites such as Hastinapur, Noh, and Ahichchhatra confirmed this picture.[7] This pottery, which soon became known widely among Indian archaeologists as evidence of PGW culture, seemed capable of unlocking the secrets of the Aryan migrations and settlement in India. Carbon dating of PGW samples could show how Aryans occupied the upper and middle Ganges valley when the extensive use of iron artifacts began to change the course of Indian civilization. They also seemed to vindicate the excitement generated by the excavations of Hastinapur, mentioned in the *Mahabharata* as the ancestral site of the Kuru clan.

These claims were not substantiated by archaeologists in subsequent decades. The old theory of Aryan migration fell into disrepute, along with the once-dominant triumphal image of Indo-Aryans as nomadic, chariot-racing, iron-wielding, conquering tribes.[8] Some scholars have suggested further caution in ascribing the full attributes of civilization to these settlements based on the evidence of pottery alone.[9] And yet the advent of the Iron Age in India is still popularly associated with these gray painted shards, located in the imagined birthplace of the *Mahabharata* and a host of other sites mentioned in the Indian epics strewn across the great valley, including Rupar, Kurukhshetra, Panipat, Purana Qila, and Mathura. Some of these assertions about the smelting and use of iron are discussed later in this chapter. Political debate and controversy over claims and counterclaims made over the antiquity of a specifically Hindu civilization and the historical or archaeological veracity of its epics such as the *Ramayana* and the *Mahabharata* have made this a prickly and sometimes daunting field of study.

We now have a generally consistent idea of what Neolithic pottery looked like in parts of the upper Ganges valley. Decades of archaeological digging in the mounds of Chirand, located in the Saran district of Bihar near the meeting point of the Ganges and Ghaghara Rivers, a site with overlapping Neolithic, Bronze Age, and Iron Age settlements, have yielded a trove of artifacts. Red, gray, black, and black-and-red ware were all mostly handmade from finely smoothed clay, occasionally with the addition of mica. Some appear to have been made by crude prototypes of the potter's wheel. They may also be dabbed with appliqué, adorned with incised and punctured designs, painted with a coat of red ocher, and feature rhythmic surface scratches of linear or more complicated geometric patterns.[10] Similar caches of pots, vases, bowls, and spouted vessels have been located farther west in Senuar, nestled in the lush agricultural plains of Kudra at the foot of the Kaimur Hills near Sasaram, Bihar, where the Ganges makes a sharp, abrupt bend.[11] Many of these vessels are red, polished red, or burnished gray; some are rusticated or sport chord impressions. There seems to be a marked difference between these shards and PGW. The general assumption seems to be that these were made by people who had achieved a comparatively higher standard of living.

Archaeologists have long debated the relationship between pottery and sedentary ways of life. In the Near East, settled agriculture evolved from long-term familiarity with steady staples such as wild grasses, wheat, and barley, which seem to have been pulverized and consumed long before the manufacture of pottery became known or widespread.[12] Evidence from sites in the Middle Nile basin dated to the mid-ninth millennium B.C.E. studied by the eminent archaeologist Randi Haaland has revealed the gradual evolution of a diet of fish and shellfish, and millet gathered through foraging. The search for subsistence led to the emergence of pottery for cooking and storing grains. As Haaland points out, pottery allowed hunters and foragers to "occupy a broader ecological niche."[13] Archaeologists have speculated that the widespread adoption of pottery led to a division of labor by gender that affected patterns of child rearing, searching for edibles, and cooking. Female gatherers possibly became the systematic grinders of seed.[14] In this regard, a good index of sedentary ways of life would have been the intensification of hearth-centered female activities related to material ritual aspects of cooking and distributing food.[15]

In the region between the Nile and the Sahara, pottery seems to have radically extended the use of food grains such as sorghum, millet, and teff, and according to Haaland it may well have changed the demographic balance. Long migrations are hard on pregnancy and child rearing.[16] Pottery allows long hours of cooking, and boiled food permits infants to be weaned much earlier, shortening the period of postpartum amenorrhea and the space between successive childbirths, reducing the period of nursing and influencing the fertility index of women. In eastern Africa, such changes would have resulted in more intense use of plants such as sorghum.[17] Similar changes would have taken place in the alluvium-rich valley of the Ganges, with its rich diversity of climates and ecological niches.

Pottery and Prehistory

Who were the people who perfected the kiln techniques of PGW? We are still somewhat in the dark. PGW marked a significant milestone in the ceramic arts that used fine-grained, high-quality clay. Its color ranges from a silvery to a deep steely gray, and its uniform texture indicates a degree of sophistication in firing techniques. The color of ceramics is usually determined by whether the fuel is slow-burning or produces a quick flare, and also by how rapidly the air flows through the kiln and the conditions of reduction inside the kiln. Temperatures between 700°C and 900°C, along with the availability of oxygen at a particular stage in the firing process, determines whether a pot will turn out red or gray.[18] During a certain stage of the settlement of the upper Ganges valley, PGW, especially bowls and dishes, became very widespread. Most PGW is remarkable for its uniform consistency and its delicate texture and lightness, achieved most likely by the use of swift-turning wheels.[19] Its distinctive color is a result of the presence of ferrous oxide produced by the reduction of the iron by gases generated in the kiln.[20] It is also remarkably free of impurities, produced in uniform kilns, and easily

Painted gray ware from Sonkh, Uttar Pradesh, ca. 1000–600 B.C.E. Government Museum, Mathura. Photograph by Biswarup Ganguly.

recognized by its distinctive surface and decorations in geometric designs rendered in black or dark brown paint.[21]

Remains of such earthenware, first identified during the archaeological excavations of Ahichchhatra in 1946,[22] were confirmed by new finds in the city of Hastinapur, one of the more popularly known sites associated with the *Mahabharata*. The discovery sparked new interest in the migration of the later Indo-Aryan pastoralists into the Ganges valley. The renowned Indian archaeologist B. B. Lal, who had worked on Hastinapur and the archaeology of the *Mahabharata* from 1952 to 1955, found PGW sites not only in the interfluvial Ganga-Yamuna Doab but across the breadth of the Indo-Gangetic plains. Curiously enough, although traces of iron were not found at the Hastinapur site, Lal seemed confident that the place had been claimed by people who belonged to a PGW culture.[23] This pottery has subsequently been traced from the Himalayan foothills to the plateau of Malwa in central India, and from the gravelly bed of the Ghaggar-Hakra River near Bahawalpur in present-day Pakistan and northern Rajasthan all the way to the ancient city of Kaushambi, near present-day Allahabad. The main points of concentration are along the basin of the Sutlej, a tributary of the Indus, and the upper Ganges plains, yielding dates that seem to suggest a long and unbroken stint from 1100 to 500 B.C.E. Potters were possibly turning out the same kind of PGW as late as during the lifetime of the Buddha. Ultimately this would be phased out by the arguably finer style of NBPW pottery, which shows up in sites along the course of the Ganges during the rise of the city-states.

Archaeological pursuits in most parts of the world are animated by nationalist sentiments, and they often become a scramble for tangible proof of cultural or genetic heritage. PGW has long been associated with the imagined landscape of the Indian epics and an abiding fascination with horse-riding, iron-wielding Aryans. In a definitive set of papers published on the archaeology of the PGW layers, A. Ghosh, a leading Indian archaeologist, remarked of this pottery that "Aryans carried it with them wherever they went."[24] Decades later, the search continues for the early migrants of the Ganges valley from Iran and Central Asia, whose ancestors once roamed across the great steppes far beyond the trans-Himalayan mountain passes. Given the nature of the available archaeological evidence, such conjectures have placed an unfair burden of proof on pottery remains.

During the mid-1970s, teams from the Indian Archaeological Survey found a multitude of iron objects in the PGW layers in a place called Jakhera, also in the Etah district. The teams found axes, spearheads, and fragments of sickles, arrowheads, daggers, chisels, and nails.[25] These finds, along with similar sites in Kurukhshetra and the valley of the Sutlej, one of the major tributaries of the Indus, where PGW and iron remains of horses were also found, lent credence to the idea that a later and vibrant urban stage of Vedic culture took shape around 800 B.C.E. and was followed by a more mature "later Vedic" culture that relied on iron and horses.

Claims of Aryan migration championed by enthusiasts looking for the antecedence of a distinctive Vedic Hindu antiquity did not go unquestioned. Many saw this use of selective archaeological evidence as not only a deliberate twisting of facts but also a tendentious attack on the historic diversity inherent in the foundations of Indian civilization. One of the most prominent historians of medieval India, Irfan Habib, in a feisty rejoinder to the debate on the significance of the archaeological evidence drawn from PGW, pointed out that some investigators, in their enthusiasm to find corresponding layers of gray ware and iron implements at sites mentioned in the *Mahabharata*, had strayed far from scholarly objectivity and discretion. Habib complained that they had not been fully forthcoming about their evidence or had unjustifiably pushed back the chronology of the PGW to suit their conjectures. The "Aryan appropriation of the Indus culture," he argued, had "increasingly become a preoccupation with many Indian archaeologists."[26]

The search for a graven, material basis for the Indian epics may seem even more misguided if we accept the idea that the stories evolved and changed significantly over time. It becomes almost impossible to tell how and when the *Ramayana* and the *Mahabharata* assimilated stories and precepts from the Vedic age.[27] In fact, as Hemchandra Raychaudhuri, one of the founding figures of the study of ancient Indian history, once noted, the main body of the northern Indian epics, the Puranas, whose earliest compilations cannot be placed further back than the third or fourth century C.E., are notoriously unreliable as sources for the verification of actual dates and must be corroborated by Brahminical literature from an earlier period, such as the *Atharva Veda*, the Brahmanas, and the main body of the Upanishads.[28] Also, this emphasis on the Vedas and the Vedic age has led archaeologists to neglect important and relevant Buddhist sources such as the *Anguttara Nikaya*, a historic section of the *Sutta Pitaka*, one of the great "baskets" of Theravada Buddhist liturgy.[29]

Some of this controversy has also been fueled by selective use of radiometric data based on the organic decay of the carbon-14 isotope. Suffice it to say that radiocarbon dating has not yet been put to optimal or systematic use in the South Asian context. Some of the radiocarbon and thermoluminescent analyses in the calibration of PGW have been deemed either unconvincing or suspect.[30] Excavations of remains of urban settlements of the Late Harappan period ending around the beginning of the first millennium B.C.E. remain in a very early stage. To complicate matters, a form of PGW has been found in a place called Cholistan located on the dry Ghaggar-Hakra fluvial belt in present-day Pakistan, indicating habitations belonging to the mature phase of Harappan culture.[31] We do not have a clear idea of exactly who these settlers were, except that they were beginning to abandon these sites around the first millennium B.C.E.[32] No analysis of PGW has been able to show whether the settlements that left behind PGW layers in the upper Ganges valley overlapped with the last phases of the great urban civilization that persisted in Harappa. More-recent chronometric analyses have generally pointed toward much later dates for this phase of ceramics, corresponding more closely to the age of the new city-states of the Ganges valley associated with NBPW mentioned earlier.[33]

The remarkable sophistication evident in the production of PGW seems to indicate a more telling set of circumstances about social differentiation and the concentration of wealth that must have been taking place with a new basis for the material and ritual divisions of labor. Specialized craft production by artisans and manufacturers was most likely tied to trade and consumption. As Brian Hayden argued suggestively, most societies make a rough distinction between practical and prestige goods, and the patronage of highly skilled and expensive pottery would have been occasioned by the demand for prestige technologies.[34] If Hayden is right, in early Vedic India as well, the rise in the quality of the potter's art would indicate a host of related factors: achievement of stable food surpluses; advances in the mining and working of metals; the sophistication of art; and the knowledge, classification, and domestication of plants and animals.

Vestiges of a Lost River

Such confounding archaeological puzzles are further vexed by more recent, and not yet fully substantiated, claims among some Indian scholars that the late Indus Valley and early Vedic civilizations were one and the same, and that this civilization emerged along the beds of the now-desiccated ancient channels of the Ghaggar-Hakra basin, once known as the

Saraswati Valley.[35] In this view, the Saraswati River mentioned in the Nadistuti Hymn of the *Rig Veda* was a major perennial river connected to both Sutlej and the Yamuna that ran parallel to the Indus River all the way to Gujarat.[36] The so-called lost-Saraswati controversy has reignited the debate about the end of the Harappan civilization and the beginning of the Vedic civilization of the Ganges. Were aspects of the mature Vedic cultures of the first millennium B.C.E. derived from a late version of the Indus Valley culture? Historians of the decline of the Harappan phase of the Indus Valley civilization such as Shereen Ratnagar have described the two as essentially disconnected, separated in both time and substance.[37]

Emboldened by new satellite imagery of long-abandoned channels and dry riverbeds in the vicinity of the Hakra River and in some sections of the Thar Desert, a few Indian archaeologists now argue robustly about the primacy of a parallel river system between the Indus and the Ganges in far antiquity.[38] Speculations about the exact location of this river system abound today, especially on the Internet. Amateur enthusiasts are keen to connect the Ghaggar, which flows alongside the Sutlej (one of the main tributaries of the Indus) to the present-day Hakra River, which empties into the Rann of Kutch and seems to have been connected to settlement sites in Cholistan on the present-day India-Pakistan border. The location of settlements along the path of this lost river has led to the assertion that there was a distinct "Saraswati Valley" civilization; some say that this should be seen as part of the Indus Valley culture, and therefore the civilization should be renamed as the "Indus-Saraswati Civilization." While there is archaeological evidence of sites in the regions between the Indus and the upper Ganges valley, there is no basis to suggest that they were urban clusters similar to Harappa and Mohenjo Daro. The sites found in Cholistan bearing shards of identifiable Hakra pottery seem to have been ancillary campsites, suggesting temporary shelters for nomadic pastoralists.[39] These may have been outlying areas of the Harappan territory watered by the Saraswati river and its channels, an entire swath of land that has now been claimed by the Thar Desert, and they might have yielded agricultural produce for the urban dwellers of the Indus Valley.[40] The increase in the number of sites from this later, mature phase of Harappan city-states and their subsequent desertion, suggesting that people were gradually migrating toward the Ganges and the Yamuna, does not justify the elevation of the Ghaggar-Hakra sites as a parallel nucleus of early South Asian civilization, let alone the wild but popular contention that the original Aryans were not outsiders after all but inhabitants of northern

India.[41] All such claims have been challenged, refuted, and revised by experts in the field.[42] If we are indeed closer to mapping the actual course of the Saraswati River, the scant nature of the archaeological evidence presented so far has not yielded any major revision of our understanding of the end of the urban civilization of the Indus and its connection to the rise of sustained settlements in the upper Ganges plains.

We also know very little about Saraswati as a female icon in the Vedic pantheon. She is generally seen as a goddess or literary and artistic muse and not always represented as a geographical feature. Some historians speculate that the mythical rivers of early Sanskrit hymns in praise of natural elements might well have referred to the rivers of Sind or Afghanistan. It seems futile to labor over archaeological proof in support of mythology. Myths are allusive and open-ended by nature, and these attempts have led to a frustrating tangle of political debates about the origins and identity of Indian civilization.[43] While the novel theory of Aryan outmigration from the Indian subcontinent has kept the pottery debate alive, we still know very little about the makers of PGW who spread across the area known in Sanskrit as *madhyadeśa* ("middle country"), referring to the land between the basin of the Ghaghara and the Ganga-Yamuna Doab. We do know, however, that at some point, the ocher pottery from the Late Harappan era of the Bronze Age was replaced by early black-slipped and gray pottery made by the same kinds of people in Atranjikhera. They lived in wattle-and-daub houses lined with bricks under thatched roofs held up by posts, had access to a relatively sedentary agrarian economy, knew the use of iron and horses, and specialized in a number of distinct crafts. New settlements may have arisen in the wider trans-Ganges region as the old cities in the core of the old Indus Valley declined, with new forms of social and economic organization.[44]

The political wrangle over the origins of Indian civilization notwithstanding, a host of intriguing questions still bedevil the archaeological bridge between the copper and bronze cultures of the Late Harappan civilization of the Indus Valley and the rise of what is still referred to as the Indo-Aryan culture of the upper Ganges plains.[45] New and rapid advances have been made in the dating of sediments and pottery through the technique of thermoluminescence, in which crystalline material previously exposed to sunlight or fire is exposed again to intense heat or light, enabling scientists to measure the accumulated radiation. Especially in cases where radiocarbon dating is not feasible, thermoluminescence helps approximate the date of ancient ceramics. Hoards of copper coins and the rude, early samples of the ubiquitous ocher pottery dug up at Atranjikhera and the neighboring sites of Bahadarabad and Hastinapur have yielded a chronological range of 2650 to 1180 B.C.E., with a broad margins of error.[46] This would suggest that the settlers of the middle Ganges floodplains were coeval with the settlements of the Late Harappan period, although it remains unclear whether any significant exchanges took place between purveyors of Late Harappan pottery and those who adopted PGW.[47] We do not have a comprehensive picture of Late Harappan architecture, crafts, and pottery, but new finds have brought to light certain important changes in the manufacture and distribution of artifacts such as black and white agates, glass beads, finely glazed faience jewelry, and tablets, as well as new kind of kilns, suggesting a new phase in the Bronze Age culture of northwestern India. There is enough reason to believe that these communities were quite different in character from those in Mohenjo Daro and early Harappa.[48] Other than the one exception of Atranjikhera, the archaeological evidence gathered from the sequence of deposits of the painted ashen or PGW mentioned earlier, coupled with carbon-14 radiocarbon analysis, shows that these subsequent layers of occupation are likely to have happened much later, between 800 and 400 B.C.E., that is, from the margins of prehistory to history, well into the period of the great city-states and tribal confederacies.[49]

First Settlements in the Valley

Although not always easy to visualize, geological changes have always affected the broader sweep of human history. The story of the Ganges basin is no exception. More than one and a half million years ago there was an unprecedented rise in the number and intensity of glaciers in the Himalayas, which coincided with a massive tectonic movement—a major uplift of the crust and the upper mantle of the earth that drove the snowline much farther south than it is today, across the lesser Himalayan ranges and the Siwaliks, the chain of mountains that rises from the plains. The Siwaliks run for at least 1,000 miles across the northern part of the Indian subcontinent, from the edges of Kashmir to southern Nepal, rising to heights of 4,000 to 6,000 feet. The Siwaliks, named after the mountain-dwelling Shiva's matted locks, belong to the last phases of the geological formation of the Himalayas that began during the period of the mid-Miocene and continued until the beginning of the Pleistocene. They are deposits of relatively fragile sandstone and rock conglomerate, mostly detritus of the Himalayan range that towers over them, and are bounded in the south by the principal tectonic plate and skirted by a broad stretch of coarse alluvium known in India as the *bhabar* land. The first hominine fossils of South Asia have been all found in the northwestern reaches of this sprawling formation.

Geologists believe that the Siwalik waste deposits were created by the subaerial, or surface, erosion of the Himalayas, mostly the ancient alluvial deposits swept down by the myriads of rivulets and streams. Rivers break down rocks physically and chemically and carry these fragments to more level basins, where sediments are eventually disgorged. Rivers such as the Ganges and its tributaries have eroded the Himalayan massif for a long time, systematically bearing away remnants of carbonate, sedimentary, metamorphic, and igneous rocks from the Lesser Himalayas, as well as the sedimentary rocks of the Sub-Himalayan Range, transporting this copious gravel across the northern Indian basin. Among these are found large and small clastic sediments such as quartz crystals (silicon dioxide) embedded in sandstone, as well as feldspar and sections of limestone. There is also black giant shale with plant fossils, and gray and brown sandstone with relics of the marine life that once abounded in the Sea of Tethys, the bed of which now forms the lofty crest of the Himalayas.

The present drainage basin of the Ganges is a southward extension of the foreland basins that emerged after the Siwaliks were formed, shaped by a river system that has pulverized rock and boulders and deposited their sediments into vast alluvial floodplains. Gaping, precipitous gorges, such as the one cut by the Ganges, are a testament to the enormous power of water in ceaseless and rapid motion. Tributaries such as the Gomti, Ghaggar, Gandaki, and Koshi that flow into the main channel of the Ganges, transporting copious alluvia, ultimately created these enormous fan-shaped sedimentary formations that geologists call fluvial megafans. Studies of the Ganges megafan have revealed a rich history of successive landforms, records of past monsoons, and a staggering diversity of flora and fauna that give us intriguing clues to the emergence of a secure ecosystem based on the rhythm of the seasonal monsoons that would one day support human settlers.

The remains of earliest hominines found in South Asia suggest that they might have spread through the plains of Sind and the Thar Desert. These were likely *Homo erectus* or *Homo heidelbergensis,* whose records exist in the region of the Narmada River flowing from central to western India. Surprisingly, the great alluvial plains of the middle Ganges valley do not contain early remains of tool-making, or Acheulean, humans.[50] In order to understand this, we must consider the broader geological context that made these tracts unsuitable for early hominid settlers. Significant natural hazards may have delayed the arrival of people in the Ganges basin. Sometime between the Early and Middle Pleistocene period, about seven to eight million years ago, this area went through a major tectonic stir that dramatically increased the amount of sediment in the lower reaches of the Ganges, especially in the Bengal Basin. It created conditions that affected Himalayan habitats from Kashmir to Tibet, weakening the sway of the southwest monsoon and creating dry and cold spells that depleted the biomass of the sub-Himalayan terrain and disturbed the ecosystem of the older savanna belt. Possible niches for human habitation in the Siwaliks and the Himalayan foothills may not have survived these altered conditions, which may begin to explain the noticeable lacuna in the sequence of hominine occupation during this phase. The geological evidence suggests the emergence of powerful bed-load streams and remnants of sedimentary rocks with individual clasts or accumulated fragments of older rocks cemented together in the fine-grained matrix. Fossil remains suggest a significant southward migration of fauna, especially large grazing mammals, into the more stable basin of the Indian peninsular craton.

A number of speculations exist as to why the early human settlers of the Indian subcontinent—often identified as bands of Lower Paleolithic settlers of the Soan Valley or the Pabbi Hills of the Upper Siwalik range (both located in present-day northern Pakistan), who were rough contemporaries of the Acheulean settlers of Europe and West Asia—were unable to colonize the Indo-Gangetic floodplains. The evidence is scant, and key pieces are missing in what appears to be a complex puzzle. The evidence for consistent human occupation of the peninsula dates back to about the Middle Pleistocene phase, and we cannot prove that they were present in the northern part of the Indian subcontinent any earlier, although scholars of early human migration believe that northern and western India lies in the favored migratory corridor between southwest and southeast Asia.[51]

Anthropologists strongly disagree about the migration of descendants of the early *Ramapithecus* who appeared in the Siwaliks during this period. Certain subgroups of *Dryopithecus* have been christened with Indian-specific names such as *Dryopithecus punjabicus*. Evidence from stone tools has been inconclusive. Significant variations exist in stone tool assemblages associated with early hominine dispersal out of the African mainland, and it has been hard to detect the specific adaptive and functional differences among them, or indeed particular cultural or biological characteristics associated with Early Paleolithic stone industries. Flake and flake fragments found at sites in India and Pakistan are highly variable, and many of them have been subsequently retouched, posing further problems for accurate dating. Thus, many questions remain unanswered about why fossil remains of these tool-making hominids have not been found in the lower latitudes of the Indian peninsula during this period, and early human settler groups do not appear before 700,000 or 800,000 years ago.[52] The ecological and climatic factors behind this are also not well understood. Some scholars suggest that these early humans may have dispersed along the Siwalik corridor all the way to Southeast or Central Asia.[53] Some suggest that a few might have also migrated southward to the old basins of the Indian peninsula, along with other contemporary mammals—small rodents like gerbils, grazers such as antelopes, and carnivores—as their fellow travelers. Curiously, however, they seemed to have avoided the great stretch of the country along the course of the Ganges.[54]

Studied across a broad swath of geological time, the climate and ecology of the sub-Himalayan terrain appear to have changed dramatically a number of times between the Miocene and Pleistocene, when the descendants of the early hominids, most commonly *Ramapithecus*, made their home in the Siwalik mountain ranges. The earliest hominid fossils that have been found in these foothills of northern India and Pakistan, dating back between seven million and five million years ago, suggest that they had access to perennial freshwater springs and water channels, and abundant rocks to fashion primitive stone tools. The evidence from the Miocene period shows that much of northern India was under an extensive forest canopy of angiosperms. When the first hominids appeared in the valleys of the Siwalik mountains, the climate seems to have been humid and warm, supporting plentiful fauna including primitive elephants, giraffes, sloth bears, tortoises, and bovid herbivores. This ecology kept shifting during the course of the Pleistocene, yielding specific niches: riverbank, forest, savanna, steppe, and veldt. Imagine a typical landscape of dense evergreen forests, interspersed with open woodlands and grassy plains and traversed in every direction by meandering streams.[55]

In a recent survey of the available evidence, Robin Dennell has suggested that the wide, flat floodplains of the Ganges or the Yamuna that supported open grassland must have had little natural cover for these prehistoric people who did not have access to significant projectile weapons.[56] They would have shared these open spaces with large carnivores of the Late Pleistocene, especially the "short-faced" giant hyena (*Pachycrocuta brevirostris*) that seems to have preyed on them, as is evident from fragments of anthropoid jawbones and skulls.[57] They competed with these carnivores as top scavengers and would also have had to contend with other deadly predators, such as saber-toothed tigers and panthers. Dennell also suggests, quite plausibly, that this open space was too large a habitat for groups habituated to smaller forested or hilly environments, and they were unable to make headway in this new and much enlarged foraging radius, which might not have provided in any one spot all the basic resources required for survival such as stone, water, carcasses, and plants.[58] Stone tools of this period were typically made from small pebbles of quartzite, limestone, schist, cornelian, porphyry, flint, or chert. The great valley may have had a paucity of stone suitable for flaking, and the new migrants might not have been able to harvest the stones from the beds of deep, torrential rivers. Given the state of research in the field, it may be safe to conclude from the finds of Riwat or the Pabbi Hills that the denizens of the Siwaliks may not have occupied the Indo-Gangetic basin in the Early Pleistocene in any great numbers, although a few small settlements may have been attracted to areas where stones could be found near riverbeds.[59] In other words, the lush, watered course of the Ganges across the northern Indian forests was not an obvious corridor for the extension of settlements. In this sense, the fossil record from the banks of the Ganges is very much like that of the Nile, which did not support extensive settlement of hominid groups along its course during this same period.

The links between these tool-making sites in India and contemporary Acheulean sites of Europe and the Middle East have not been clearly established. Early Stone Age humans in India presumably did not know how to make hand axes and cleavers, a characteristic that led Hallam L. Movius in 1948 to draw a historic divide between Africa, Europe, and the Middle East on the one hand and East Asia and Southeast Asia on the other, arguing that early prehistoric tool technologies forged triumphantly ahead in the former regions. This "Movius Line" also separates northern India from southern India, where Acheulean hand axes have been found at a later period.[60] Whether the available fossil evidence and the missing component of the handheld ax warrant this idea of uneven evolutionary progress is debatable. Hominids from Africa who made it to the northern Indian foothills may have branched off at a much earlier stage in the development of tool making.

Hunters, Gatherers, and Foragers

The Ganges plains were settled long before the first millennium B.C.E. by Late Stone Age hunter-gatherers. These settlements fall within the range of extensive Mesolithic sites that sprang up around the same time across the Indian subcontinent, from the dry plains and rocky terrain of Gujarat and Rajasthan to the hills and forests of Central India, where archaeologists have located significant caves and shelters.[61] The groups that eventually settled along the course of the Ganges, mostly by the shores of ancient lakes, adapted to a different environment, marked by high rainfall and dense forests. In much of the scholarly discussion, the word *Mesolithic* turns out to be a bit misleading. There is no hard and fast line separating the users of large and small stone implements, and outside a small circle of specialists, the Mesolithic versus Neolithic divide may seem an arbitrary and academic distinction at best.

The course of the Ganges shifted more than 34 miles to the south during the Late Pleistocene. Along the shores of oxbow lakes near these prehistoric

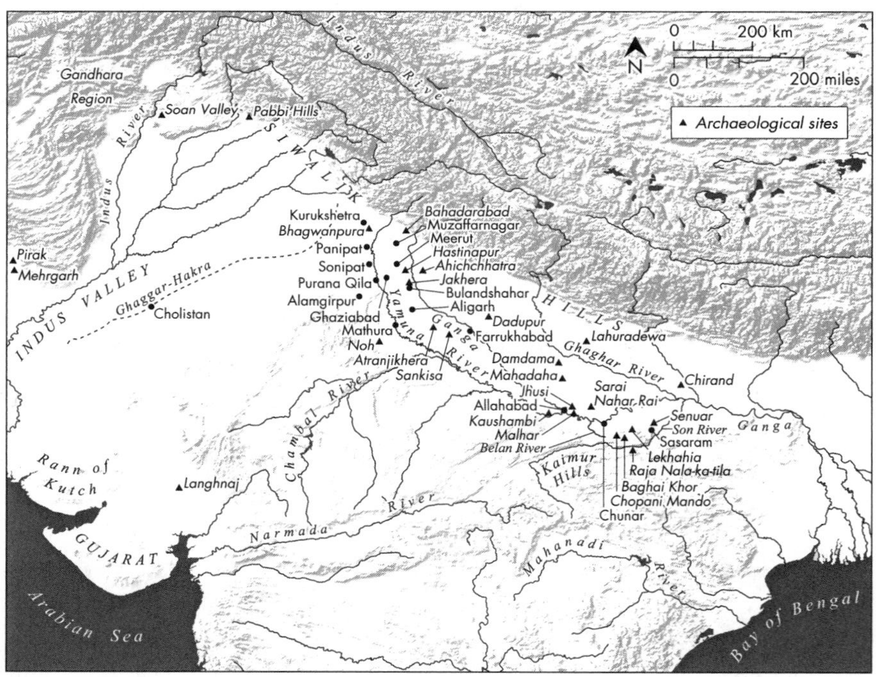

Major archaeological sites of the Ganges basin.
Map by M. Roy Cartography.

channels, which now lie along the northern border of the Mirzapur district in Uttar Pradesh, cemeteries with Mesolithic human remains were first found in the late 1960s. In nearby Sarai Nahar Rai, in the Pratapgarh district, some of the most detailed evidence of human habitation was discovered later—not only dwellings, communal hearths, and graves but also expertly retouched blades and scrapers made out of fine-grained stones such as chalcedony and carnelian.[62] Radiocarbon dating places these artifacts in the range of 8000 to 9000 B.C.E. Around two hundred of these Mesolithic sites have so far been unearthed, dispersed across the valley of the Ganges. The archaeological record shows that many of these were stable and long-term habitations, marked by a plethora of tools and weapons made out of antlers, bones, and blades, especially an elaborate range of scrapers and burins.[63] Their makers knew the use of hammers, grinding stones, and querns, implements pointing to a steady diet of grains and seeds. The fact that these were primarily hunting, gathering, and foraging communities is also proven by studying their bones and dental remains. These people had the basic anatomy shared by all modern *Homo sapiens,* although they had larger craniums and more substantial molars. They were possibly taller than present-day Indians. They seem to have been robust and healthy, living well into their fourth and fifth decades.[64] Their knowledge of working with fire, stones, wood, and bones was peculiarly adapted to the ecology of the Ganges plains, and it would be hard to speculate how such technology spread through this region. In many ways, the archaeological record left by these hunters and foragers tends to support the theory that the evolution of early human society and technology followed a mosaic pattern rather than being the result of a simple linear progression.[65] Advantageous facets of culture and innovations in methods of tool making and hunting must have found their way through contacts between small bands much more rapidly than from the specific genetic characteristics of the group from which they might have emerged.[66]

We can form some preliminary ideas of what these lake-dwelling peoples might have looked like. They wore skirts woven of grass or leaves, and ornaments, rings, and necklaces made out of animal bones and antlers. Rock paintings from some of these shelters indicate the general patterns of their subsistence, with scenes of fishing, trapping, and foraging for plants.[67] Settled lifestyles must have led to more elaborate practices of mortuary rituals. Skeletal remains found at both Sarai Nahar Rai and Mahadaha reveal that the dead were placed in shallow graves with their heads facing west and their right forearms placed at an angle across their abdomen. Offerings of aquatic shells have been found in these crypts.

The climate of the Early Holocene became increasingly humid, and evidence from the northern Indian plain suggests an abundance of plants and animals. Animal remains suggest that the settlers of the valley hunted antelopes, wild pigs, elephants, rhinos, Indian bison, and wild buffalo, along with a range of fishes and birds. This lush environment would have supported a number of hunting and foraging pathways and demanded new skills of planting and harvesting. The ecological implications of these patterns of subsistence are still not fully understood.[68] It is estimated that there must have been a significant rise in the population, as nomadic hunters and gatherers turned to more sedentary patterns of livelihood. Newly revised radiocarbon and thermoluminescence dating place these within the first millennium B.C.E., much later than what earlier scholars had estimated, that is, just around the time of the rise of the first cities across the Ganges plains. These Mesolithic settlements, where people practiced much older patterns of hunting and foraging while new groups of sedentary agricultural specialists were busy tilling the land and erecting early fortified habitations, might have been relatively recent.[69] The aggregate chronological evidence gathered from excavated sites such as Baghai Khor, Damdama, Lekhahia, and Mahadaha suggests that early foragers and later agriculturalists existed during the same period, with some sites, such as Sarai Nahar Rai, still occupied as late as the first millennium B.C.E. However, while some scholars imagine that there might have been significant interaction between hunter-gatherers and later agriculturalists, including some forms of trade and exchange, the evidence does not support this conjecture with any kind of certainty.[70] During this period, these communities might have been relatively homogenous and isolated. Later, in the period known as the second urbanization of South Asia, when fortified cities emerged all across the greater Ganges plains—judging strictly from patterns of dispersal of pottery—the interaction and exchange between hunter-gatherers and sedentary groups became much more routine.[71]

Agriculture and the Rise of Sedentary Habits

In the deepest part of the Ganges plains, alluvium deposits reach three miles or more from the surface of the earth, a result of silt-laden currents flowing for nearly half a millennium. Palynological evidence—studies of the remains of plant pollen—suggests that the valley was once an open expanse of thickets and forests. Paleolithic settlements were attracted to this riverside landscape of dense flora, with highly fertile soil, natural levees, myriad bodies of water, and a high water table. Meanders and the widespread reticulation of channels of the Ganges offered a plethora of microclimates supporting a great diversity of flora and fauna. From the Middle Paleolithic sites of Kalpi in the valley of the Yamuna River to the denser settlements that sprang up during the Early Holocene, analysis of surface sediments and cultural pollen shows that a slow turn toward planting and harvesting might have put humans on the eventual path of agriculture almost 15,000 years ago. It is no accident that wild rice began to appear at the side of human encampments during the very early phase of Neolithic culture.

Although the rise of sedentary farming was a fundamental change in the orientation of humans to the natural world, the factors behind this transition are not well understood. Once seen indispensably as part of the Neolithic "revolution"—fastened to that iconic image of the post-Paleolithic polished stone ax in Europe—cultivation of the soil as a way of life is now seen more as an extension of pastoral culture, which always included small amounts of farming.[72] In terms of stone tools as an index of progress, this was not simply a passage from chipping to grinding but the gradual extension of control over food and nutritional intake, command of natural resources, and steady progress in the domestication of plants and animals. The history of the rise and spread of the first agrarian civilization in the Ganges valley in this instance is undoubtedly part of a much wider history of the convergent aspects of modern human civilization that took shape about 10,000 years ago, which also led to a fundamental change in the biodiversity of our planet.

Gordon Childe, one of the pioneers of Old World archaeology, summed up this momentous change in human prehistory in his popular book *Man Makes Himself* as the "first revolution that transformed human economy and gave man control over his own food supply."[73] Childe wrote emphatically about the domestication of basic food crops such as wheat and barley that helped sustain the ancient civilizations of the Mediterranean, Mesopotamia, and India. He argued that this Neolithic revolution must have happened where wild precursors of latter-day plants and animals tended exclusively by human beings can be found. Childe was partly responsible for the basic rule of thumb summed up in the phrase *ex Oriente lux*: early farming rose in the Near East and the Fertile Crescent and thence came to Europe before making its way eastward to the Indian subcontinent. Farming changed the patterns of migration and habitation, creating more dense clusters of settlement, and thus went hand in hand with an urban revolution derived from the intimate bonding of humans, food crops, and animals: barley, sheep, and goats. These settlements were like oases in the wilderness, where agriculture first found its footing and where dibbling sticks, primitive hoes, and scythes became the tools of life. Childe imagined that once herds began to graze on planted fields next to human habitations, the annual cycle of planting, harvesting, and letting the land remain fallow had begun.[74] This is how Sumer rose in Mesopotamia from the fecund alluvium of the Tigris and Euphrates around 3500 B.C.E.; how Egypt began along the green band of the Nile along the Sahara in 3000 B.C.E.; and presumably how the lesser-known experiments of Mohenjo Daro and Harappa began by the banks of the Indus flowing through the Indian subcontinent during the same era.

This view of the rise of farming in the Fertile Crescent and its spread to Europe at the end of the Great Ice Age toward the end of the Pleistocene, and the inexorable march of Neolithic farmers toward civilization, is not entirely accepted by present-day scholars. We now understand farming and the cultivation of the soil as part of the larger imprint of human society on plants, animals, and the ecology as a whole. In this regard, agriculture was not simply a singular transformative event—or even a series of accidents and discoveries—but rather a culminating point in the long and variegated history of the dispersal of humans across landscapes with an uneven distribution of natural resources. One of the primary founders of this kind of thinking was the geographer Carl Sauer, who argued persuasively that every cultivated plant has a dual past—a natural history and a cultural history—and that this duality is integral to the complex story of the diversity of foraging and planting on the planet.[75] In a set of lectures published in 1952 as *Agricultural Origins and Dispersals*, Sauer popularized the distinction between roots and seeds in the history of the evolution of human farming, arguing that root crop agriculture, suited to tropical conditions, was prevalent in the lower parts of South America, while seed farming became the norm in Mesoamerica. Sauer suggested a simple example to illustrate this inference. In places where planting and harvesting were the dominant modes of subsistence, root crops were always seen as supplements to the diet, which was why Mesoamericans foraged for wild roots rather than planting and tending them. This insight remains relevant today in considering the basic question of why certain wild plants evolved into planted crops for prehistoric humans.[76] Sauer argued that the geographical dispersal of major crops across diverse ecological horizons was not just natural happenstance or the result of independent human discovery but the outcome of the gradual diffusion and adaptation of ideas of long-term survival.[77]

It seems plausible to consider the rise of agriculture as a result of permutations in the mix of foraging and farming, and it is difficult to believe that the systematic cultivation of food crops in the Ganges valley was entirely a result of either alien colonization or cumulative acculturation. The old Victorian classification of the Mesolithic period in human history as distinct from the Neolithic has been somewhat discredited, and scholars are no longer committed to the fixed idea of a fundamental divide between nomadic and sedentary ways of life. In fact, new archaeological evidence from the Mediterranean shows that hunters and gatherers had been following herds of herbivores for a very long time, closely enough to take the first step toward husbandry, especially evident in the mix of foraging and farming communities not only in the islands of the Greek archipelago such as Cyprus, Crete, and Peloponnesus but also in many other parts of the Mediterranean world.[78]

Some of the earliest dates for sustained farming come from the Levant and the region of Iraqi Kurdistan south of the Zagros Mountains. Archaeologists have noticed a comparable degree of sophistication in sites along the eastern Iranian plateau and Baluchistan, such as Mehrgarh, which were the precursors to the widespread farming settlements that supported the complex urban centers of the Indus Valley civilization. However, despite such advances in the far northwest of the subcontinent, the early pastoral and farming communities of peninsular India seem to have come together much later, after 2500 B.C.E. Domestication of animals and farming thus spread quite unevenly across the Indian subcontinent.[79] The early farming villages of eastern India, Mesolithic communities of central and southern India, and advanced copper- and bronze-wielding town dwellers of Harappa may have coexisted for a long time without significant technological cross-diffusion.

Recent scholarship on the rise of agriculture in India suggests that the harnessing of animals might have evolved independently in the subcontinent.[80] Some key mammals of the Late Pleistocene, such as the wild aurochs (*Bos primigenius*), ancestor of the present-day cow, and the smaller Indian bovid (*Bos namadicus*), progenitor of the present-day Indian bison (*Bos guarus*), might have been tamed and adopted within India, especially as Indian cattle severed their ties from the rest of the world's herds as early as

200,000 years ago. A similar case has been made for the indigenous adoption of rice as a staple. An early version of the wild perennial rice (*Oryza rufipogon*) was abundant in India, especially in the valley of the lower Ganges. The main Indian variety of rice, *Oryza indica*, was domesticated separately from the *Oryza japonica* prevalent in East Asia and Southeast Asia and shares a different phylogeny. The planting of rice started most likely as an extension of natural foraging. The semiarid plains of the Ganges became much wetter toward the end of the mid-Holocene period, and new forms of wild rice began to appear around lakes and ponds along the course of the river and its tributaries. The terrain and the climate now supported the monsoonal rather than marshland species of rice.

Much like the debates surrounding the use of advanced stone tools, differing viewpoints on the rise of settled agriculture tend to either debunk or uphold the thesis of exogenous influence, especially the migration and settlement of the so-called Indo-Aryans. A further complicating problem is posed by the fact that many Mesolithic sites that have been associated for a long time with hunter-gatherers are now seen, in the light of fresh archaeological evidence, as oriented also toward herding and seasonal migration. Stone tools give us some but not all of the clues to the transition from foraging to cultivation. Most archaeologists and scholars of early hominine cultures question the idea of a singular and definitive Neolithic revolution in the post-Pleistocene epoch that was once supposed to have transformed the reindeer herders of Europe into harvesters of wheat and barley. In other words, there is no reason to assume that agriculture as a way of life offered advantages to early humans over other forms of subsistence. In fact, we might even question the very evolutionary premise that all human societies should have gravitated inevitably toward settled farming.

Lewis Roberts Binford, one of the best-known American archaeologists of his generation, looking for factors that led to the rise of sedentary lifestyles among early human cultures, laid the ground for a new debate over the origins of farming. Binford argued against the widely held consensus that people adopted agriculture as a way of life only when they were forced out of environments of plenitude, or what he called little "Gardens of Eden" where food was abundant.[81] Human groups did not simply become sedentary because of the relative ease with which they were able to acquire food. The quest for subsistence, in other words, was never singular, and forms of hunting, foraging, and nomadic habits were complex responses to different environmental and climatic challenges. Binford posed an intriguing but plausible hypothesis to archaeologists wrestling with the question of the evolution of agriculture: human groups do not necessarily remain fixed in one place until they are forced out by external circumstances. The prehistoric record seems to suggest that what may seem natural at first glance—that is, a predisposition toward settlement—may not be natural after all. The movement and resettlement of hunters and foragers is based on a complex interplay of environmental factors and population density. While many of Binford's original observations have been questioned, his basic premise—that digging and planting were the most crucial human adaptations that changed the fate of the planet—remains relevant, especially to our discussion of how agriculture arrived in the upper and middle Ganges plains.

The change from an itinerant hunting and foraging lifestyle to sedentary cultivation, following the seasonal cycle of sowing and reaping, marks a pivotal change that altered the equation of biology and habit. It is still not clear what factors triggered this quantum change, but it led to a new society of primary producers who eventually cleared significant swaths of scrub and forests in the Ganges valley and made possible the complex patterns of interdependence that ultimately led to war, trade, and cities, laying the foundation of early civilization in northern India. The rise of agriculture was also made possible by the systematic use of iron implements. There were surely different pathways to such settlements, but a steady and reliable supply of food must have bolstered the average nutritional intake, reducing the normal period of lactation among females, extending the possibility of child bearing, and effecting a significant increase in the size of population clusters. Once the tools and techniques of agriculture had been instituted, the settlers could access a vast stretch of old and new alluvium and a host of microclimates, including that of the Ganga-Yamuna Doab, which may well be described as the second fertile crescent of the subcontinent after the segment of the Indus that helped sustain Mohenjo Daro and Harappa.

Early Holocene foragers begin to appear in the archaeological record of western Asia and the Indus Valley with digging sticks and mortars and pestles. These people laid the foundations for a pattern of tillage that survived into the period of documented history. Agricultural villages became widespread in eastern Iran, Turkmenistan, and Baluchistan, and similar hamlets sprang up in the valley of the Indus, foreshadowing the agricultural revolution of the subcontinent. The inhabitants of the cities of Harappa, for instance, had developed the first prototypes of cotton. We do not know much about the Early Paleolithic period, but evidence from the banks of the southern tributaries of the Ganges, such as the Belan and the Son, suggest that blades were used widely for plant foraging and processing.[82] Some inhabitants of the Ganges valley were harvesting wild rice by the Early Holocene. Why did foragers in the upper reaches of the Ganges valley adopt new routines of planting and harvesting, especially after the second millennium B.C.E.? Were such habits introduced from farther west?

We know that during the Late Mesolithic period, the plow appeared in the Harappa region of the Indus Valley. Early signs of this transition to a more sedentary pattern of living are becoming clear from the archaeological evidence gathered from the Neolithic sites of Mehrgarh, near the Bolan Pass, west of the Indus River valley in Pakistan, and Langhnaj, Gujarat, in western India. We still do not know whether new techniques of agriculture spread simultaneously east and south or to what extent they might have displaced hunting and gathering populations. The adoption of sedentary lifestyles associated with agriculture changed the patterns of nutrition and aging, which can be detected from human skeletal remains. New research on Mesolithic sites unearthed in the Ganges valley suggests that there might have been substantial population growth following the wetter climate of the Early Holocene, which saw a rising abundance of flora and fauna. Vestiges have been found of spotted deer, blackbuck, elephants, hippopotamuses, gaur (the Indian bison), the Indian sloth bear, and large carnivores such as tigers.[83]

The exact relationship between the proliferation of these sites where both large and small stone tools were being developed and the growing diversity of foraging pathways is not yet fully understood by paleontologists and archaeologists.[84] Surely, different bands of hunters and foragers were staking their claims in the valley and marking out substantial habitations for long-term settlement, especially along extinct oxbow lakes and the beds of smaller streams. Such people have left behind the remnants of elaborate cemeteries, along with communal clay hearths, huts, and burial grounds. Particularly promising are the extensive skeletal remains that have been found in shallow rectangular graves in Sarai Nahar Rai and Mahadaha, both in the Pratapgarh district in Uttar Pradesh. These burials suggest not only a specialized and consistent pattern of mortuary rites but also the emergence of livelihoods adapted to local plants and animals and changes of the season.[85]

Analysis of the bone fragments of hog deer (*Axis porcinus*) of the grass and scrub jungles and marsh-dwelling swamp deer (*Cervus duvuaceli*) found near these settlement sites suggest that many of them were occupied during both summer and winter months.[86] These settlers also dug pits for storing grain. The butchering and rendering of animal carcasses must have been a highly specialized craft, and a plethora of charred animal bones, bone arrowheads, and bone tools and ornaments suggest easy access to fauna, especially to aquatic birds, turtles, and fish. Research on the bioarchaeology of human remains from these Mesolithic dwellings associated with this lake culture has yielded a number of suggestive patterns.[87] Cumulative data on dental and skeletal pathology also show the evolution of subsistence patterns. The shape and attrition of teeth, especially a process of decay known as enamel hypoplasia, suggest a periodic ebb and flow of resources and the recurring incidence of famines and feasts. The seminomadic hunter-gatherers of the Ganges plains were adapting to a radically new, resource-rich environment very different from that of the urbanized people of the early Mehrgarh or Late Harappan cultures.[88]

Excavation of these burial sites, not only at Sarai Nahar Rai and Mahadaha but also Damdama, Baghai Khor, and Lekhahia over many decades, has yielded a considerable array of human skeletal material, outlining a clear temporal overlap between agriculturalists and foragers. We know that the Mesolithic site of Sarai Nahar Rai was occupied well into the first millennium B.C.E. In other words, the lake- and forest-dwelling settlers of the Ganges valley were there during the beginning of the period known as the second urbanization of South Asia, when new fortified cities began to emerge along the river. How did this uneven development take place? What were the imperatives that led some groups to defend the boundaries of their settlements? The archaeological evidence suggests that hunting and gathering continued unabated alongside dense pockets of agriculture.[89] Given the expanse of the temporal and spatial horizons that we are considering, it is quite possible that no one pattern of livelihood would have dominated this landscape, and our categories of classification—hunters, foragers, pastoralists, or agriculturists—may prove to be deficient and imprecise. Much like sites along the Vindhya Hills in central India, the Ganges sites show evidence of diverging subsistence patterns distinguishing hunter-gatherers from agriculturalists, which is also clear from the use and distribution of ceramics.

A Prehistoric Ecological Setting

The geology of the Ganges basin provides the axial timeline for the succession of alluvial activity that once made up the middle and lower reaches of the basin. Near the channel of the River Son, which empties into the Ganges channel near the village of Hardi Chapra, the plain rests on a deep shelf of primordial Vindhya bedrock overlaid with several strata of alluvial accumulations. These layers preserve a record of distinctive periods of geological activity. This immense alluvial expanse, with little rock in its vicinity, is mostly composed of Late Pleistocene and Early Holocene alluvium (Indian *bhangar*) and new silt (Indian *khaddar*) expanding southward with the main and subsidiary channels of the Ganges, along with its array of tributaries.[90] These deposits also contain a rich sequence of archaeological records of successive human settlements from the Early Acheulean to the Neolithic. Toward the peninsular craton girdling the valley of the Son lies what geologists call the Toba ash bed, formed by one of the largest volcanic eruptions on earth that took place 74,000 years ago; its tephra enveloped much of the eastern parts of the Southern Hemisphere. New studies show a widespread distribution of volcanic tuff, but we do not yet have a clear picture of how this might have affected the flora, fauna, and hominine cultures of the region.[91]

This volcanic ash bed lies amid extensive alluvial formations from the Quaternary period bustling with faunal and early human remains. The lowest layers carry Acheulian tools and flakes from the Lower Paleolithic period resting on the Lower Proterozoic basement, which is mostly a bed of gravels belonging to remains of alluvial fan and debris. The next stratus is alluvial sand, clay, and parts of the old ash bed acting as a channel fill. The next two layers are also alluvial, and they harbor artifacts from the Middle to Upper Paleolithic. The uppermost deposit is a result of aggradation, which is the elevation of the land due to the accumulation of sediment greater than what the channels of the river are able to bear away. This activity results in a kind of geological terrace dating to the period between the Upper Pleistocene and Holocene, containing Late Paleolithic, Mesolithic, and Neolithic artifacts. At several sites along the Ganges, open sections of the riverbank show four distinct layers of sedimentation, suggesting that the climate was changing from wet to dry during the Late Pleistocene. The river system itself was meandering south, leaving behind a maze of oxbow lakes. The sand and silt from these beds of alluvium indicate the beginning of the milder climates of the Holocene, when marshes and fluvial beds inundated vast grasslands. These conditions might have lured the new hominid migrations from the Vindhyas in the south, adding to the peopling of the Ganges valley during the Late Upper Paleolithic and Early Mesolithic periods. The earliest evidence of human artifacts along the river comes from the Middle Paleolithic site Kalpi, located in the valley of the Yamuna River south of the Indo-Gangetic plains and dating to about 45,000 years ago. Here, small choppers fashioned out of pebbles, cobbles, and bone tools indicate the beginning of settlements from the Early Holocene period onward. We also find a number of vertebrate fossils, bones of horses, elephants, hippos, crocodiles, and turtles worked on by human tools, along with some human femurs with cut marks suggesting that the settlers preyed upon each other as well. In this manner, the prehistoric alluvium that accreted over half a millennium, reaching to a depth of more than three miles, carries the clues to the rise of complex societies in the valley.

What was the landscape of this extended basin like when humans had begun to develop foraging, hunting, fishing, and planting as interdependent forms of livelihood? The evidence from pollen suggests the presence of extended grasslands dotted with thickets and forests, interspersed by the rise of natural levees and alluvium. There would have been an abundance of water bodies and a high water table, with a highly reticulated expanse of old and new channels, many of the old meanders forming large and independent oxbow lakes. This abundance of flora and fauna must have attracted the first great settlements. The study of surface sediments, especially cultural pollen, transports us back to a period about 15,000 years ago and shows the use of naturally occurring grains and seeds that filled the baskets and querns of the hunters and gatherers. In fact, wild rice seems to appear at the very early phase of Neolithic cultures of the Ganges valley. African crops likely spread to the Indian subcontinent during the period of the Indus civilizations through oceanic trade routes during the late third millennium B.C.E.[92] During this time, large millets from the African savannah and the Horn of Africa arrived, to be cultivated during the monsoon, and summer crops supporting barley and wheat. A wide range of cultivars thus became available to the early settlers of the Gangetic savanna.

We are still searching for clues as to what triggered the domestication of livestock and adoption of patterns of sociability that would eventually lead to settled agrarian society. The evidence for the evolution of seasonal habits that are required for planting, harvesting, and the care of seeds that would provide a steady, perennial source of nutrients remains unclear. What kind of ecological conditions were ripe for this turn to the cycle of seeds and grains? Clearly, the transition from foraging, fishing, trapping, hunting, and scavenging to pastures and sedentary, agrarian habits happened over a very long period of time. However, it is reasonable to assume that both herding and planting involving larger communities with many mouths to feed would ultimately lead to the clearing of scrub and forest and the expansion of territory protected against pests and predators. We have to imagine a long, uneven, and complex history of the eventual domestication of wild plants and wild game, and the convergence of a number of survival skills derived from the knowledge of hunting and gathering, pastures, and the cultivation of crops.

A systematic study of ancient pollen deposited in lakes along the central Ganges plain suggests that the region was under the cover of extensive grasslands for the greater part of the last 15,000 years. Cultural pollens, suggesting human activity, indicate a significant population increase in the middle Ganges valley during the Mesolithic period and the systematic harvesting of edible wild seeds, including wild rice. It is unclear whether forest burning and the use of charred fallow was a standard feature of Mesolithic cultures in India.[93] However, by incremental clearance, planting, and herding, the early inhabitants of the Ganges plains would have begun to manipulate the forest and stake their fortunes to the edges of water bodies, including the Ganges itself. Evidence of such Mesolithic sites exists along extinct oxbow lakes, where they learned to fashion permanent shelters with plastered floors and to bury their dead with ceremony in cemeteries.

Lodged in the hills south of the Ganges valley, about 50 miles southeast of the city of Allahabad around the Belan River, are a group of sites known to archaeologists as the Vindhya Neolithic digs. Here, at Chopani Mando, Indian archaeologists have found charred remains of rice nestled in some burned clay dating roughly to the Early Neolithic period.[94] Further examples of impressions of grains and spikelets in pot shards in this region point to the cultivation of wild grasses. Other sites in clusters such as Chirand and Senuar have revealed troves of celts, scrapers, and arrowheads, along with red and red-black pottery. At Chirand, which would have coexisted with the Late Harappan cities of the Indus Valley at the confluence of the Ganges and the Ghaghara (about 7 miles from the district headquarters Chapra in the Saran district), we have evidence of ceramics made with levigated clay. Levigation is the advanced potter's technique of purifying clay by sedimentation, in which it is churned with water and left to stand. The coarser particles sink to the bottom of the trough, while the water and organic impurities rise to the top and are poured off. In the middle forms a layer of fine-textured clay. The potters of Chirand used such finely honed clay for their red and black-red ware and carinated pots and vases. They crafted specialized tools such as drills, spearheads, sockets, and neat punching tools from animal bones and antlers. They also made combs, bangles, and pendants. While they might not have fashioned large plowshares, it is likely that picks made out of large antlers served as the prime tool with which they broke clods of earth, and paddy husks stuck to burned clay pieces show that they had a steady diet of rice. These were the handiwork of the Neolithic denizens of the middle Ganges plains who lived in wattle-and-daub structures or circular huts supported by wooden posts and thatching. They used stone blades, ground stone axes, bone tools, and handmade pottery, and cultivated wheat, barley, rice, peas, lentils, and pulses. They also domesticated cattle and hunted buffalo, rhino, stag, and deer.

One of these sites, Koldihwa, dated approximately to the seventh millennium B.C.E., has revealed traces of the earliest version of the Indian species of rice, *Oryz sativa*.[95] Painted rock shelters have been found at Lekhania, near the Kaimur Hills, which form an abrupt massif in the vicinity of the old fort of Chunar where the Ganges makes a sharp bend. Here too, wild and cultivated strains of *Oryza rufipogon* and *Oryza sativa* have been found alongside the remains of wood and charcoal. The botanical record of nutrients gathered by these people suggests a broad spectrum of grasses and seeds gathered locally. These communities had access to plants such as goosefoot and purslane, millet-like grains, wild jujube, buckwheat, and mint. They would have also enjoyed the flowering *mahuā* tree (*Bassia latifolia*), whose crushed flowers may be cooked or consumed raw, and a potent liquor distilled from its seeds. The humble sesame seed also bespeaks an intriguing story of adaptation. We know that the earliest finds of sesame come from Harappa, most likely domesticated from wild varieties in the Indus Valley. Sesame cultivation became widespread in the Ganges basin in the second half of the second millennium B.C.E. We do not know whether this was a local find or whether it traveled from farther west.[96] Along with sesame and mustard, the settlers partook of foxtail grass, Job's tears, and wild millets. At some point both wild and domesticated varieties of rice emerged as a consistent dietary staple, supplemented by wheat and barley.

The gradual domestication of wild paddies marked the beginning of a long-term revolution in the making. Over the subsequent centuries the Ganges valley became home to some of the richest diversity in the production of rice, as both staple and luxury. However, the planting of rice in itself did not alter the profile of Late Neolithic river and lake dwellers. The story of agriculture in this regard might not necessarily follow the proverbial tipping point of their social organization. There is unbroken evidence of some form of plant husbandry from the Neolithic and Chalcolithic periods that extends all the way to early historical cultures that lie within the range of about 2000 B.C.E. to about 500 C.E. Scrub, marsh, and forest surrounded the immediate vicinity of the Neolithic dwellers of the Ganges basin, and the knowledge of the forest as a resource for nutrients, game, and medicine would have been crucial to their survival. In the long run, not only the expansion of agricultural staples such as rice and wheat but also the use of fire and iron would move the needle from dependency to the exploitation of natural resources.

The Advent of Iron

Numerous copper hoards have been found along the course of the Ganges, dating to the prehistoric period. Among these are celts, double axes, lance heads, swords, harpoons, bracelets, and human figurines. Bronze objects are also present alongside copper, many of them ornamental artifacts such as bracelets, anklets, and chokers. Copper and bronze are also associated with the ocher-colored pottery of the second millennium B.C.E. However, despite evidence of such a rich phase of Chalcolithic culture in the Ganges basin, we still do not know much about the wider use and efficacy of such metals and metallurgy. D. P. Agrawala, a pioneering scholar of ancient Indian technology, argues that there is not enough evidence to provide further insights into how bronze and copper might have been a part of a settled agrarian society. They were clearly scarce metals, difficult to mine and render, and accessible perhaps only to influential members of society. The weapons that have been found do not seem to have been intended for standard issue in combat but rather designed for ceremonial use.[97] The eventual mastery of iron-smelting technology would lead to a much more widespread change. In the long run, iron implements became common, everyday objects, and easy access to iron tools and weapons changed the physical and social landscape of the valley. In this long retrospect, it would seem that a combination of fertile soils, availability of water, iron deposits, and human enterprise created the agricultural surplus necessary to launch a spate of new cities, constituting what is known as the second urbanization in the Ganges valley. The age of iron is thus often seen as the basis for a new spurt of civilization: planned cities; increasing trade relations; efficient networks of trade routes; sharply defined social stratification and complexity; a more refined knowledge of the arts, literature, science, and medicine; and the birth of recorded human history arising from the obscurity of prehistory. At the same time, compelling as this story of progress may seem, it is also deceptively simple, which makes the appearance of iron seem like an inevitable step in the ladder of human progress.

We still have no clear idea where iron was first mined, smelted, and distributed in northern India. As late as 1958, the stalwart archaeologist Mortimer Wheeler, who generally believed that the resurgence of the civilization along the Ganges was a direct result of Persian conquest, suggested that the knowledge of the manufacture of iron was introduced in India during the fifth century B.C.E. because of Persian incursions into the Indus Valley.[98] This assertion, that the manufacture of iron was exogenous to India and came from more advanced civilizations of the Near East, has now been laid to rest. While textbooks on ancient India may still date the beginnings of iron smelting typically to about 700 to 600 B.C.E., newer archaeological finds matching traces of iron alongside specific styles of pottery, and sophisticated use of radiocarbon dating, have steadily pushed this date back. Some of the earliest iron smelted in northern India may date back to the second millennium B.C.E. From this period onward, iron tools become widespread across the floodplains of the Ganges and the Yamuna. Early iron objects have been located in the far northwest, in the Swat Valley of Pakistan, in what is known as the Gandhara Grave cultures of the eighth and ninth centuries B.C.E., which were possibly connected to settlements in Central Asia and Iran.[99] Recent excavations in the Middle Ganges valley in Uttar Pradesh, at Raja Nala-ka-tila, Malhar, Dadupur, and Lahuradewa (Lohradewa), dated and analyzed by archaeologist Rakesh Tewari, show that ironworking in India might have begun even earlier.[100] Radiocarbon dating of these iron implements, furnaces, tuyeres, and slag show a range of 1800 to 1000 B.C.E., which casts fresh doubts on whether ironworking was brought to the Indian subcontinent by outside migrants or developed within India independently and across multiple sites.

The last phase of the Bronze and Copper Age in India seems to have overlapped with the emergence of larger and more sustainable settlements. These communities and their ways of life began to change appreciably with the introduction of iron, which as the archaeological record suggests became common by the early second millennium B.C.E. The earliest evidence of this change comes from Pirak in the Indus Valley and from graves in the

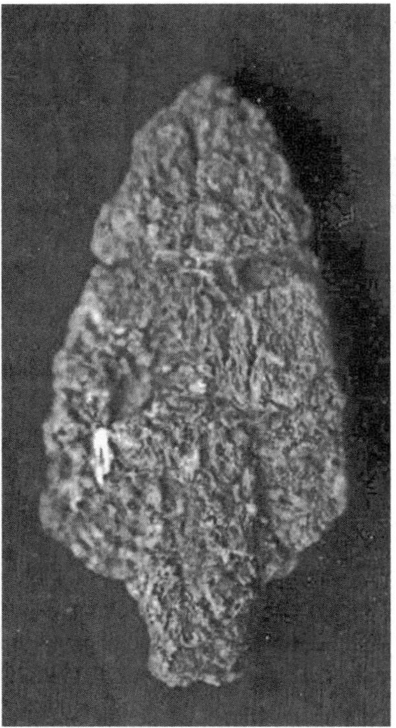

Corroded iron arrowhead, Dadupur, Uttar Pradesh, ca. sixteenth century B.C.E. Courtesy Dr. Rakesh Tewari.

region of Gandhara, followed by Punjab, Rajasthan, and the Ganga-Yamuna Doab. As noted earlier, the Iron Age has always been tied to the emergence of PGW, which has been found not only at Ahichchhatra and Hastinapur associated with the *Mahabharata*, but across a considerable swath of territory, from the now-extinct, desiccated channels of the Ghaggar in Bahawalpur and northern Rajasthan and from the Indus Valley to the Ganga-Yamuna Doab.[101] Excavations at Bhagwanpura in the Haryana district, 15 miles northeast of Kurukhshetra on the banks of the Ghaggar River, have yielded new evidence of PGW, suggesting an intervening phase between the last years of the old Harappan civilization and new settlements associated with the early Vedic period. Many of these sites have remnants of burned bricks and mud-walled houses, and there is evidence of the domestication of cattle, sheep, and pigs. At some point, it is surmised, the people who fabricated an early version of this pottery also became adept at smelting iron and breeding horses. A large part of this history is still unclear.

More than a thousand such sites have been found so far.[102] The major ones—Rupar, Kurukshetra, Panipat, Purana Qila, Mathura, Sonipat, Alamgirpur, to name just a few—all carry PGW shards, further supporting the assertion that the people who fashioned these pots and vessels also used iron tools. Although archaeologists have not been able to pinpoint the chronology and intensity of usage along the different occupation levels, early evidence of iron from preliminary excavations at Atranjikhera (1150 B.C.E.) and Jakhera, in the form of arrowheads, spearheads, axes with shaft holes, and tongs, along with the presence of wheat, barley, rice, cattle, pigs, and horses, points toward an expansion of agriculture and farming during this millennium. Over the course of this period, painted gray pottery is overlaid with black and red ware, dated roughly between the twelfth and ninth centuries B.C.E., and commonly associated with the later Vedic civilization.

The specialization evident in the manufacture of iron implements suggests the rise of new and complex forms of communal labor. The making of iron through a process known as direct reduction requires continuous work involving the separation of slag from melted pieces with the use of heavy hammers wielded with considerable force, precision, and dexterity.[103] Iron in a pure state melts at a temperature of 1,536°C, which would have been beyond the reach of most inhabitants of the valleys and forests of northern India. The early smelters had to heat the iron between layers of charcoal to a temperature close to 1,200°C, at which most of the slag or impurity could be drained. What was left was a spongy, malleable mass of iron known as the bloom, with some residual slag. This could be broken up and rendered further by reheating and hammering at temperatures above 250°C. During this early period, a prototype of Indian wootz (derived from the ancient Tamil *ukku*), a fine form of carburized steel, was likely also developed. Indian steel was cast by hand in a unique fashion, from bar iron melted in the presence of charcoal or wood and formed into narrow conical crucibles. These lumps of steel with a precise range of carbon content for hardness could be refashioned into some of the most durable and high-impact swords known to the ancient world. Over the course of centuries this form of raw Indian steel, known also as *ferrum indicum*, became highly prized in Persia and Afghanistan. It was also used in the manufacture of the legendary Damascus sword. Such specialization in the production of iron would have required a concerted tradition of family or group work. A large-scale manufacture of iron also implies the extensive use of wood charcoal derived from hardwood trees. Thus mining, smelting, and forging of iron seems to have taken place within or in the vicinity of the forest.

It is hard to pinpoint when and where iron tools began to have a decisive impact in the clearance of scrub and forest, making possible the extensive tillage of land and ultimately leading to the planting of extensive fields of rice, wheat, and barley. While the revolution in the smelting and use of iron would have taken centuries, many historians of ancient India have seen a decisive march of civilization foreshadowed in the story of the advent and spread of iron, especially in the period between 500 and 322 B.C.E., marked by significant changes in the basic economic organization of everyday life. The historian and mathematician D. D. Kosambi in his pioneering reflections on ancient Indian history proposed that there was a close link between the abundance of cheap metals and the widespread extension of agriculture. Kosambi was interested in a particular sequence of social change. The spread of agriculture, starting with the crude metal harrows to the finished iron plows that could break up tough soil into clods for planting, gradually replaced migrant tribal communities dependent on slash-and-burn methods of cultivation with a sedentary, village-dwelling population.[104] This, in turn, enabled the accumulation of a larger surplus capable of sustaining significant sections of the population that did not directly engage in agriculture but joined the ranks of priests and warriors, ultimately capable of supporting a ruling elite that acquired the power to command hereditary menial work. For Kosambi, permanently occupied fields were integral to the emergence of early landed property and a complex social hierarchy. A similar and more forceful argument about the transformative power of iron technology was advanced by another stalwart historian of ancient India, R. S. Sharma, who argued that the widespread adoption of iron smelting ultimately resulted in the cultivation not only of staples such as wheat and rice but also of relative luxuries such as sugar and cotton. Surplus production spurred trade, craft guilds, and towns. Sharma pointed to the history of the refinement of plowshares made by blacksmiths who perfected the use of bellows and the addition of carbon. He based some of his observations on a wide array of literary evidence, especially Sanskrit texts such as Panini's monumental study of grammar (*Astadhyayi*), which can arguably be dated to the fifth century B.C.E., or the Jatakas, or the life stories of the Buddha dating roughly to the same period. These texts refer to iron girdles put on wheels in horse- or bullock-drawn carts.[105] Not only did Sharma point out the connections between the use of iron, agrarian surplus, and the rise of urban economy in northern India, he went on to argue that the origins of Buddhism and Jainism as forms of belief and practice that challenged the traditional social order in the sixth century B.C.E. were direct responses to these changing material conditions.[106]

There is no one explanation of why particular forms of agrarian practice, ownership, and distribution of crop fields, taxation, or patterns of trade and exchange took shape in the Ganges valley. Advancement in the smelting of iron was one of many factors in this transformation. An increased demand for plows and weapons of war may also have made iron manufacturing a priority among warring kingdoms. Historians today do not single out the technology of iron production as a benchmark of the progress of Indian civilization. Metallurgy does not explain the changing complexity of a social or political order or the vibrancy of economic activity. What made life durable for kingdoms and city-states of the Ganges valley was not only the reclamation of land under tillage from forest and scrub but also the ability to mitigate the impact of droughts and floods. The uncertainties of monsoon that plague agriculture across the northern Indian countryside today were very much present during antiquity, and the use of water had to match the predictable regularity of crop cycles and the demands of labor for large-scale harvesting and planting. This is one of the reasons why the cultivation of both husked and unhusked rice and Indian six-row hulled barley became more widespread than wheat, which requires more water at regular intervals.[107] Such conjectures have become the staple of ancient Indian history because archaeology has not yet given us a comprehensive picture of what exactly led to the sharp rise in the number and distribution of warring kingdoms and city-states that became common during the period between the seventh and fourth centuries known as the "great settlements," or *mahājanapada*s. During this extended period, not only were major advances made in ceramics and the manufacture of iron plowshares, hoes, sickles, swords, wheels, and lances, but archaeologists have recovered the remnants of a number of unlined wells that indicate the attention given to irrigation at sites farther from the banks of the river. Cities built from this period to the first century C.E. began to invest in defensive walls protecting houses and granaries of fired bricks supplied with drainage. The increase in the number of seals and coins suggests a steadily rising traffic of overland and riparian trade.

Eventually, all along the Ganges valley, kingdoms and cities began to align themselves to natural strongholds and trading routes. They were able to weigh the needs of subsistence and surplus and sustain large standing armies, merchants, and artisan guilds. Toward the end of the Vedic period in northern India, these patterns of settled lifestyle, along with rising prosperity in trade and agriculture, undermined the traditional tribal order and the political power of the warring elite that had characterized older kingdoms such as the Kuru-Panchala.[108] New kingdoms rising in Kosala, Kashi, and Videha in the middle region of the Ganges valley were wealthy enough to reclaim the patronage of grand public rituals and attract Brahmins from farther west. These developments began to unravel the prescriptive sacrificial economies of the Vedic age, making room for the emergence of heterodox and new forms of animism. This was also the period of the rise of the two great pathways in the conception of life and the cosmos that diverged from the traditional beacon of Vedic teachings—Jainism and Buddhism. From the mythic and historic corpus of Jain and Buddhist texts we get a fragmentary but fascinating account of the evolution of new clans, kingdoms, and cities, refinements in military organization and political thinking, and new forms of patronage of writing, leisure, and the arts.

CHAPTER 4

RISE OF THE WARRING KINGDOMS

The *Maitrayana-Brahmana Upanishad* contains a remarkable episode about King Brihadratha's encounter with the great sage Sakayanya in the forest.[1] The king, having fulfilled his worldly duties, had wandered far from his kingdom in search of spiritual fulfillment. Alone with the elements, penitent and rapt in meditation, he was surprised by Sakayanya, who was pleased with his efforts and granted him a wish. The only blessing Brihadratha sought was a true knowledge of the self. Sakayanya offered him some thoughts on the impermanence of human affairs in a world of flux. The human body, he argued, was perishable. Needs, desires, and attachments were transitory. Life and death followed each other in endless succession. The only truth that reigned in the world was knowledge of the true self.

Buried in these words of wisdom about mortality, destruction, and the natural order of things is perhaps also a hint of a great social upheaval that had taken place in northern India. Sakayanya told the king about the passing of the age of great heroes and mighty kings, legendary archers, and rulers of great empires. Their glorious epoch was gone forever, and the signs of this passing were evident. Oceans were drying up, mountains were crumbling, and the gods of old were dying too. The pole star had been dislodged from its eternal perch in the sky. The human being was now "like a frog in a dry well": isolated, unsociable, and out of place. Many scholars of ancient Indian history, most memorably the stalwart historian and Indologist A. L. Basham, have referred to this episode to suggest the rise of speculative, pessimistic, materialistic thinking that attended the rising tide of Buddhist and Jain thought and a vindication of the atomistic philosophy of the Ajivika schools.[2] This sense of disenchantment with the traditional Vedic order of knowledge, centered on the rituals of oblation and offerings, may also have been a result of the breakup of the old tribes and chieftaincies and their absorption into new and larger kingdoms. The most eventful sites of rebuilding and change were shifting to the nether reaches of the Ganges valley, throwing into prominence regions that were on the eastern edges of the traditional warring tribes of the later Vedic period. What led to the rise of these kingdoms and city-states? How were the older tribes and clans realigned within this new order? How did they ensure a surplus production of crops and grains to feed their increasingly large and entrenched armies? This chapter takes up these questions to explore the political geography of the Ganges valley and its bearing on the rise of chiefdoms, oligarchies, and kingdoms, including the Nanda and Maurya Empires, which were supported by a burgeoning agricultural surplus and new forms of unfree labor.

The city-state in the ancient Near East in its early conception, the great geographer Lewis Mumford once argued, undermined the old symbiosis of the Neolithic village community, with its limited appetite for seasonal warfare.[3] The ability to command resources for bloodier and more sustained campaigns ultimately revised the efficacy of battles and consequently the governing ideas of both the city and the state. A similar development seems to have been taking place in northern India, with cattle raids giving way to organized confrontation between fighting forces mustered and equipped to defend citadels, granaries, and places of worship. One might say that there are seeds of the absolutist injunction of *bellum contra omni* in such principles of warfare designed to seize, subjugate, and assimilate people and wealth from neighboring tribes, which also advanced an understanding of property tied to the use of force and intimidation. Wars were fought not just for women, cattle, crops, or seeds but for prisoners of war and, increasingly, captives and slaves. Like the early dynastic city-states of Sumer and Akkad, which expanded their reach over the most productive and well-watered lands, the successful *mahājanapada*s of ancient India preyed on and exploited the resources of others, advancing a conception of wealth, power, and political order based on territory rather than the principles of kinship that made a virtue out of violence and exploitation. According to the historian D. D. Kosambi, during the course of the seventh and sixth centuries B.C.E. a "prolonged struggle for absolute power" took place across the northern plains.[4] This drawn-out competition between the warring states paved the way for the rise of great kingdoms of the lower Ganges plains such as Kosala and Magadha.

The word *janapada*, meaning "foothold," was originally associated with the stronghold of a tribe or clan. Later the word variously came to denote a country, state, or district. These settlements are directly associated with the second phase of urbanization in South Asia dating back to the middle of the first millennium B.C.E. Ancient India has often been compared with ancient Greece or China, where similar complexities of thought and social order were emerging during roughly the same period, which Karl Jaspers once called the axial age of sentient awareness that cut across the entire swath of human civilization. Whether this is true or not, one can detect an appreciable degree of existential introspection in northern India at this time, when certain aspects of philosophical and religious thinking came together, most likely catalyzed by the intensity of political and economic activity in the states, cities, trade, and markets. These winds of change affected most aspects of life along the middle Ganges valley between the present-day cities of Allahabad in the west and Bhagalpur in the east. It is difficult to conclude that these changes in philosophical outlook among certain sections of the population were simply a result of the extension of tillage or the fortification of cities. However, such changes seem to have coincided with heterodox, skeptical traditions of thought that had begun to confront the traditional Brahmin-dominated vision of a social and moral order. During this period, the fertile banks of the Ganges began to support major urban settlements such as Atranjikhera, Ahichchhatra, Hastinapur, Kaushambi, Mathura, Noh, Pataliputra, and Ujjain. Many of these cities were garrisoned and had defendable walls, sure signs of the emergence of large territorial states that were staking their claim over land under the plow and its produce.

A finer distinction between the village and the city lies in the difference between what appears as settled in contrast to the wild. This difference may also have been an expression of the way the inhabitants of the larger Ganges valley began to conceptualize the relationship between the human and natural worlds. Such thinking was also reflected in new social distinctions. Most historians of ancient India agree that a few hundred years before the birth of Buddha and Mahavira, a fundamental social transition took place in northern India in which clans, lineages, and tribes began to yield to new ruling councils and kings. This history of the refinement of the political order that opened up new ideas of territorial sovereignty, surplus wealth, and a gradual reshuffling of social hierarchy adhering to caste and kinship is still not entirely understood, and its relationship to fundamental changes in the material texture of life has not been fully tested by the archaeological or textual evidence.

The emergence of the *janapada* therefore may very well have been a result of a breakup of the later Vedic social order.[5] However, following the problems outlined in the last chapter about the history of use and the spread of iron implements, whether such transformation was directly indebted to a turn in technology is still not clear. Eventually iron tools enabled the clearing of forests on a much larger scale, including the removal of roots and stumps, and iron plowshares drawn by bullocks broke up harder and stonier ground with greater ease. One of the principal advocates of this long-term transformation has been R. S. Sharma, who initiated this debate more than fifty years ago. Sharma postulated a fundamental change in the human geography of the middle Ganges plains that led to the rise of new and denser urban settlements all the way to the Rajmahal Hills in the east, where the great river begins its bend southward.[6] He argued that this part of the river basin had some of the deepest alluvial deposits and attracted some of the heaviest rainfall during the monsoon months, supporting an abundance of thick vegetation. Settlement here meant a war against the forest, which could not have been fought without axes equipped with iron blades. Iron plowshares were also required to break up the hard, clay-laden soil for planting. In subsequent discussion of this idea of a technological tipping point, Sharma has repeatedly drawn attention to the surge in the production of NBPW and widespread references to iron plows in Buddhist and Jain literature.[7] A sudden spurt in the manufacture of pottery as a result of population density, which in turn could only be supported by an increase in the production of agrarian surplus—especially from the cultivation of rice paddies—seems eminently plausible.[8] While the manufacture and use of iron became increasingly common during this period in both agriculture and warfare, the timeline of changes in material culture of the great Gangetic *mahājanapada*s are not clearly aligned. As shown in the last chapter, the use of iron in this region can be traced back to more than five hundred years before this period. And yet, despite the presence of iron smelting in the middle and lower Ganges valley, archaeology has not really surprised us with new or overwhelming evidence about the use of iron in agriculture, especially in the shape of iron axes or plowshares.[9]

At the same time, reasonable doubts cast over the extent of iron smelting should not blind us to the relative stability achieved in sustained agriculture able to support the proliferation of cities. The emerging picture of the settlements along the middle sections of the Ganges basin is much more uneven and complex. Large agrarian groups settled in these parts long before the spread of iron, carrying forward patterns of subsistence harking back to the New Stone Age, or the ages of bronze and copper. This period saw a gradual obsolescence of what was essentially a combination of small-scale pastoral and agricultural economies that sustained the tribal organizations and the rise of an economic surplus defined by urban centers and monarchical kingdoms. In many of these formations the land was worked by a distinct and separate class of people at the bottom of society known generally as the *dāsa-bhṛtaka*—roughly speaking, a class of slaves and hired laborers.[10] However, as the economist Ester Boserup wrote many years ago, agrarian and pastoral lifestyles did not evolve predictably, nor did the extension of land under the plow necessarily lead to any revolutionary change in traditional tribal formations.[11] Boserup was also convinced that a hard-and-fast distinction cannot be made between cultivated and uncultivated land in ancient societies such as northern India. Most tribal groups tended forest fallows and had access to grassland pastures for cattle. The important question is when and how larger and more complex social formations broke out of small, stagnant populations that had achieved stable sustenance levels through an optimal mixture of hunting, foraging, herding, and tilling.[12] Not all early *janapada*s of antiquity abandoned tribal councils, clan oligarchies, or hieratic social arrangements. However, many witnessed key transformations in the specialization of labor and an intensification of the daily routine of agricultural work on unprecedented scale, enough to feed large groups of people devoted entirely toward priestly and military functions cloistered behind heavily defended facades. The frequency and scale of warfare was one indication of this change, which also entailed the search for captives and slaves who could be gainfully exploited as servile labor in the fields and mines.

The Lichchhavis of Vaishali

One of the great kingdoms of ancient India known as the Vrijji (*Vajji* in Prakrit) emerged out of a confederacy of tribes. One of these tribes was a city-dwelling clan of warriors known as the Lichchhavis. The territory of Vrijji was located toward the north end of the Ganges basin and extended up to the low-lying, marshy *terāi* regions of Nepal skirting the foothills of the Himalayas known as Madhesh, bounded on the west by the Gandaki River. Toward the east lay the forests lining the edges of the Koshi and Mahananda Rivers, which separated the Vrijjis from the great kingdom of the Mallas. The Lichchhavis, who ruled the countryside from the fabled city of Vaishali, were the most powerful members of the Vrijji confederacy. During the time of the Buddha they were considered as powerful as the kingdom of Magadha. The Pali canon and Jain sources suggest that Magadha and the Lichchhavis fought a number of wars testing one another's strength. Ultimately King Ajatashatru of Magadha was able to annex the historic city of Vaishali to his kingdom. However, this did not quite extinguish the warring Lichchhavi bloodlines or their political ambitions.

Founded by the mythical King Visala of the Ikshvaku lineage—the same as that of Prince Rama in the *Ramayana*—Vaishali had long served as the jealously defended citadel of the Lichchhavi res publica. It also became one of the most prosperous and renowned cities of ancient India and a draw for pilgrims, especially because of its proximity to the birthplace of Mahavira, the final emissary and founder of the Jain order, and a place that the great Buddha visited many times. Because of its frequent mention in Jain and Buddhist texts, we know quite a bit about the Lichchhavis and their capital city. Unlike many of the other settlements of the lower Ganges plains, the Lichchhavis conducted their political affairs in a fairly open and consultative manner through their councils of elders. The Vrijjis have been traditionally seen by scholars as consisting of a number of representative clans. The fifth century C.E. Theravada Buddhist scholar and critic Buddhaghosha's commentary *Sumangala Vilasini*, which is a standard interpretation of a major collection of the Buddha's teaching known as the *Digha Nikaya* [The Long Discourses], notes that the Lichchhavis were most likely one of the more important participants in the legal tribunals held in Vaishali, led by eight of the most prominent clans. Besides the Lichchhavis of Vaishali, the Vrijji federation also included the formidable Videhas of Mithila from the area comprising northern Bihar and Nepal, along with the Nayas of Kundapura and the Mallas of Pava and Kusinara. It is interesting to see that the Jain sources mention only the Lichchhavis and not the Vrijji confederacy, while the Vrijjis are frequently mentioned in the Buddhist literature. The Vrijji federation may have fully emerged after the arrival of the Lichchhavis, who originally hailed from Tibet and migrated to the foothills of Nepal, defeated the Videhas, and laid their claim to the city of Vaishali.

Buddhaghosha provides an interesting aside on the origin of the Lichchhavis.[13] Once upon a time the queen of Varanasi gave birth to a lump of flesh instead of a normal infant and tried to get rid of it in the swift currents of the Ganges. A hermit retrieved the basket with the newborn and looked after it for six weeks. Soon the lump began to turn into a boy and a girl. The boy was resplendent like gold, and the girl was fair as silver. The hermit's affection for the two children was just as strong as a mother's, and this maternal attachment turned his fingers into nurturing breasts from which milk flowed like "clear nectar" and sustained the two children. Their skins were brilliant and translucent, and the sage gave them the name of Lichchhavi, apparently derived from the words *līna chavī*, "gossamer skin," or perhaps *nichavī*, "skinless." This legend purportedly refers to the Lichchhavis being fair of skin and to their migration from the Himalayan Terai to the Ganges plains as a significant part of their genealogy. There has been some debate about this myth because the derivation of the term does not follow the usual rules of Sanskrit grammar. Some scholars have suggested that the tribe name might have been a Magadhi version of a Prakrit form of the Sanskrit word *ṛkṣavin* (*ṛkṣa* is one of the Sanskrit words for "bear"), or "people of the bear-clan."[14] Bears were abundant in the forests along the foothills north of the Ganges plains, and the Lichchhavis might have held on to a distant totemic reference to the land of their origin even as they evolved into a powerful and complex political order.[15]

The city of Vaishali is often mentioned in the epic *Ramayana* and Buddhist texts. Not only was it close to the birthplace of Lord Mahavira of the Jains and the capital of the Lichchhavis, it also hosted the Second Buddhist Council, held about a hundred years after the Buddha's demise. The site of the ancient city was first excavated for archaeological remains during 1903–1904 and 1913–1914, and seals engraved with the name of the city were found. It was subjected to more systematic spadework by the Archaeological Survey of India during the 1950s and again under the auspices of the K. P. Jayasawal Research Institute from 1957 to 1961. During these digs the tank Kharauna Pokhar was confirmed as the coronation tank whose waters had blessed the crowned heads of many Lichchhavi scions. The *Bhaddasala Jataka* describes a water tank in the city of Vaishali secured by a net made out of iron chains, where the family of monarchs got their water for the ceremony of sprinkling during the coronation of monarchs. These sacred waters were reserved for princes hailing exclusively from the Lichchhavi clans, who claimed this tank, described as the *abhiṣeka maṅgala pokkharaṇī*, or the "reservoir of auspicious coronation," as their peculiar and exclusive privilege.[16] A surrounding wall with rock and mortar plinths with many levels of occupation was also unearthed during these digs. Excavators found that between 150 B.C.E. and 100 C.E., rooms were constructed for archers with designated places to set down their quivers from which they could hurl deadly showers of arrows on potential invaders. The core of the citadel must have been settled from 500 B.C.E. onward, with a primary layer of black and red ware pottery, overlain by NBPW. An early phase of building dating to 350 to 150 B.C.E. corresponds roughly to the period after the Maurya takeover, when a mud rampart was put in place, later reinforced by bricks. Excavations of the large mound locally known as Raja Vishal Ka Garh ("Fort of Raja Vishala") have revealed the remains of ancient shrines, stupas, and habitations, including a fortified citadel. The city originally is said to have had three walls around it and was partially defended by the river as well. The Jataka stories describe the city of Vaishali as having formidable walls that were difficult to breach. Each was a league from the next, and one had to pass three gates with watchtowers to enter the inner quarters of the city.[17] The arrangement and scale of these ramparts remind us of the frequency and ferocity of warfare and the need for standing armies. They show how preparedness for battle was one of the key features that led to the fortification of the capitals of the major settlements, especially with the use of brick walls.[18] The fort and the town were often called by the same name, *durga*, which literally means in Sanskrit "that which is difficult to pass" and in later centuries became the default word for any castle or fort. The regularity of warfare proved to be an important catalyst for the rise of statecraft.

Contemporary literary sources say that the Lichchhavis erected a major earthen stupa over what they claimed as the relics of Buddha to the north and west of the city. The stupa became a landmark monument and was rebuilt four times over the centuries; the first restructuring was done in intricate brickwork during the Maurya period. A reliquary casket believed to contain corporeal remains of Buddha was found in an ancient breach inside the core of the structure. These remains were supposedly re-exhumed by the great Mauryan emperor Asoka for further redistribution among Buddhist followers and laity during the fourth century B.C.E.

The strategic and symbolic importance of the city of Vaishali during the fifth century is evident from the number of subsequent rebuilding projects. Over many years, archaeological digs at Vaishali have yielded rich hordes of coins, seals, terra-cotta figurines, ornaments, and other fragments suggesting continuous occupation. The original rampart overlying the pottery lair of NBPW was further reinforced with burned brick in the Shunga period; during Kushan times it was raised with earth and engirded by a moat. The Guptas of Magadha, who eventually made peace with the Lichchhavis through a dynastic marriage alliance, added barracks and other reinforcements to the site.

The discourses contained in the *Anguttara Nikaya*, one of the major collections of the teachings of the Buddha and his principal disciples, allow a glimpse of the nature of the tribes assembled under the alliance of the Vrijjis. When the Buddha lived for a while in the city of Vaishali at the Sarananda Shrine, a number of visitors from the clan of the Lichchhavis came to him for advice about war and peace with neighboring powers and the survival of their old political ways. In response, the Buddha shared with them the seven great principles of "nondecline"—or means of arresting political stasis—that might help them shore up their age-old oligarchy and tribal councils. As long as the constituent members of the Vrijji assembly took care to hold frequent meetings, the Buddha advised, they should continue to grow in strength. Such assembly, however, had to be held without rancor. As long as the affairs of the tribes were conducted in relative harmony, as long as the tribes kept to their established decrees and time-tested ancient principles, as long they venerated clan elders, as long as they carefully abstained from abducting girls and women by force, as long as they maintained and paid due obeisance to traditional shrines within and outside the city walls, and as long as they offered protection and shelter to the wandering Buddhist *arahant*s, who were spiritually worthy and thus deserving of support—the Buddha assured them—they would not come to harm.[19]

This well-known piece of counsel shows clearly that the Buddha, a prince of the Sakya clan, understood very well the shared nature of power within the old and venerable tribal units (*gaṇa*) and the consultative procedure of their decision making for war and peace. When he was residing in the territory of Rajagriha at the Gridhrakuta, or the Vulture Peak Hill, he was approached by the formidable king Ajatashatru Vaidehiputra, son of

Bimbisara of Magadha. Ajatashatru, unlike his father, did not support the Buddhist order. He had also declared his intent to wage war against the Vrijji confederacy. When he sent his chief minister, Vassakara, to ask the Buddha about what might be the way to confront the Vrijjis, the Buddha answered with his characteristic equanimity that the only way to defeat them was to sow the seeds of disunity among the tribes, and that this was a difficult thing to do.[20]

As many scholars have pointed out, the idea of the Buddhist monastic order, deliberately named the *saṃgha*, was in many ways indebted to the intertribal councils such as that of the Vrijji. The congregation of disciples after the Buddha's passing centered their organization on a council of elders (*sthavira* in Sanskrit, *thera* in Prakrit), recognizing the value of age, experience, restraint, and foresight. Within a couple of hundred years after the Buddha's passing, a much more concerted drive for wealth and resources had taken hold of the rulers of city-states and kingdoms along the Ganges valley. The concentration of resources in the grasp of the four major Mahajanapadas—Avanti, Vatsa, Kosala, and Magadha—and a more fierce and unyielding outlook toward wars of conquest and despoliation would partially determine the nature of life in both the city and the countryside of the northern Indian plains.

The Age of Rival Kingdoms

It is difficult to pinpoint which aspect of the overall transformation of the social order led to the rise of the warring states of antiquity along the northern Indian plains and the Ganges valley. Oligarchies, tribal elders, and monarchies were capable of sustaining states that engaged classes of soldiers and cultivators. Not only the ghost of Thomas Hobbes's *Behemoth*, but also the ruminations of Karl Marx and Friedrich Engels's *Anti-Dühring* on the gainful uses of violence, come to mind as we reconsider how early the city-state soared to new heights of military power in ancient India. During the age of the rise of the great *janapada*s, routine warfare, and therefore the routine fear of defeat, depredation, and enslavement, had become a necessary feature of life. In this regard Buddha's measured statements about an order of peaceful mendicants surviving in a world of kingdoms and states clashing over land, labor, and resources, acknowledging the futility and inevitability of such warfare, is similar to the speech of Hermocrates during the Athenian expedition to Sicily, in which he argued that it was basic human nature to extend one's dominion over those who were weak and powerless to resist. Much as in ancient Greece, walls and elevated perimeters in India were the decisive signs of the zeitgeist of the embattled city. Athenian Greeks, as Thucydides pointed out, understood the formation of the polis in much the same vein as a conscious political decision, evident in the term frequently referred to by Aristotle as *synoikismos* (συνοικισμός), or the act of "settling together"—a recognition of the physical concentration of the population in a single city for collective defense or an act of pure and forcible political unification. This is how Attica was gathered into larger units around 431 B.C.E., when Pericles sought to shield the growing population behind the city walls of Athens.[21]

Comparable changes, on a much larger scale, were becoming evident in northern India as cities and states multiplied in numbers, with frequent reversal of political fortunes. As Romila Thapar has argued persuasively, major transitions were taking place during this period, when many older

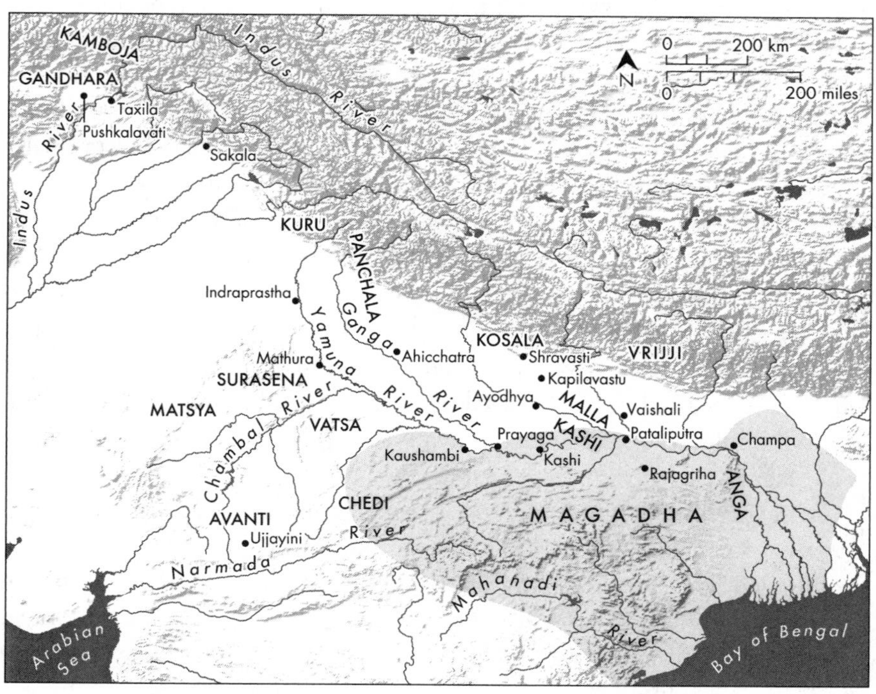

Early kingdoms of the Ganges plains. Map by M. Roy Cartography.

forms of ruling lineages with established genealogies of places and origins were evolving into dynastic sequences with more clearly delineated and exclusive claims of kingship, which in some cases led to larger monarchical groupings.[22]

In the middle of the Buddha's famous Long Discourses known as the *Digha Nikaya* is a version of the popular myth about the earth being inherited by the sons of Bharata.[23] According to this tale, at the center of all inhabited territories on earth once lay the realm of the great king Disampati, favored by Buddhist elders for his beneficence and virtue. After Disampati's death his son Renu, notorious for his flawed character and profligate habits, came to power. His accession made the Buddhist Samgha and the peers of the realm anxious. However, the young king turned to the Brahmin Jotipala for help, the one who had long been designated as the great spiritual guardian (*mahā govindo*), along with six of the oldest nobles of the realm, and he asked them about the extent of the cities and kingdoms around him. Who can divide this vast territory (*mahā paṭhaviṃ*) that widens as one travels north and narrows like the yoked head of a wheeled cart (*sakaṭamukhaṃ*) toward the south? He was told that seven equally powerful kingdoms (*janapado*) belonging to specific tribes controlled this entire realm. Among these, Mithila belonged to the Videhas, Champa to the Angas, and Varanasi to the people of Kashi.[24] While enmeshed in allegory and fable, the Buddhist scriptures give us a rough geographical sketch and a fair idea of the rivalry between the major kingdoms of the middle and lower Ganges valley. This is also supported by a passing mention in one of the last commentaries on the *Atharva Veda*, the *Gopatha Brahmana*, which states plainly that in the end four major political contenders were vying with each other for power and attention: Avanti, Vatsa, Kosala, and Magadha.[25] They had emerged triumphant from older wars between neighboring kingdoms, whose long memories persisted in mythical tales. Some of these conflicts were legendary in their day—for example, the rivalry and eventual truce between Anga and Magadha or the fight between Kashi and Kosala.

A degree of complexity in the understanding of political affairs must have become necessary for survival in this period of rapid change. The heightened scale and intensity of armed conflict beyond cattle raids and border skirmishes put pressure on traditional tribal councils as well as on the norms of hereditary kingship. New laws for religious rites and social codes had to be invented, indicating a growing concern about the differentiation of labor and the stability of the social order. These changes did not occur at the same time in every settlement along the northern Indian plains. However, certain passages attributed to the grammarian Panini, composer of one of the earliest known linguistic treatises in the world who probably lived during the fourth century B.C.E., give us some ideas about new hierarchies being carved out by groups wielding power, which would be later described in much greater detail and force by Kautilya in his celebrated political treatise *Arthashastra*. In the context of basic political organization we can discern actors of the state: rulers, ministers, priests, council members, administrative officers, and members of the king's inner household. We also see the prevalence of a number of distinct taxes, including those imposed on agricultural produce and capitation, also apparent from the layout and planning of cities, forts, walls, and the lay of the cultivable land, especially after 550 B.C.E.[26] Both oligarchs and kings during this time were busy manufacturing punch-marked coins, evident in the silver currency of all the major city-states such as Kashi, Kosala, and Magadha.

In one of the oldest Buddhist Jataka tales, two kings reigning over Varanasi and Kosala wanted to find subjects who would help them diagnose faults in their approach to governance.[27] The two kingdoms were comparable in territory, army, wealth, and social order of caste, tribe, and family. Their courtiers would not dare tell them the truth about their shortcomings, so the two monarchs went around asking people in the urban quarters outside the palace walls, around the city gates, and in the adjoining countryside. At the end of this tryst, the king of Kosala concluded that evil could be overcome only by evil, while the king of Varanasi realized that anger can only be conquered by calm, wickedness by goodness, parsimony by beneficence, and lies by truth. The story is a parable about contemporary Buddhist ideals of kingship with little basis in actual history, but it tells us something about these kingdoms and their capital cities, and the fact that different classes of subjects lived within and outside their main ramparts.

When the Buddha resided for a while among the Magadhan people of Matula, he delivered to them the *Cakkavatti Sihanada Suttanta* [Lion Roar on the Turning of the Wheel Sutra].[28] He forewarned an audience of monks about the inevitable karmic cycles of prosperity and misfortune through which all human affairs must pass over the course of time. Herein lay the expectations and duties of the great wheel-turning monarchs to come (*cakravartin*).[29] Over successive generations, the Buddha held, human society would inevitably suffer the gradual deterioration of moral values, and the life span of future progeny would be severely stunted. At the end of the cycle they would turn on each other with swords in hand, like savages. However, this anarchy would also end one day. The inhabited earth, Jambudvipa, would be powerful and prosperous once again, thick with people as the jungle is thick with reeds and rushes. The great city of Varanasi would become known as the legendary Ketumati, powerful, prosperous, crowded, and awash in goods and provisions. The Buddha predicted that the city of Ketumati would be at the apex of 84,000 new cities all over India. And there would emerge King Sankha, a new wheel-turning, justice-dispensing, righteous monarch, a conqueror of the four quarters of the world, who would restore moral law, not just by the dint of the sword but also by the strength of virtue, and his reign would make way for a new, fully awakened and exalted Metteya (Maitreya) to emerge. What is remarkable about these passages, apart from a wonderful insight into the Buddha's view of the classical karmic order, is the conception of human prosperity in terms of urban renewal.[30] While so much of early Buddhist pilgrimage is associated with secluded, forested, and idyllic *vihāra*s and the withdrawing, contemplative ideal of monastic life, it is easy to forget that the moral political of order of Buddhism was resolutely centered on the cities and city-states of the period of the first great urban resurgence of the Ganges plains.

Buddhist sources from this period are replete with references to guilds, manufactories, and the flow of everyday commercial life. The ideal city of Ketumati, modeled on contemporary Varanasi, is projected not only as a royal capital, fortified and secure from external attacks, but bustling with people, busy and well-fed. Passages in the *Sangiti Suttanta* of the *Digha Nikaya* speak admiringly of the muslins and dyes of Varanasi, with their vivid colors of yellow, red, and indigo and their intricate floral patterns.[31] Varanasi textiles appear also in the famous *Mahaparinibbana Sutta*, where the Buddha discourses about nirvana, the final exit from the coils of mortal suffering.[32] Early Buddhist doctrines reflect the view that the prosperity and refinement of kingdoms and cities are tied to trade and manufacture and also to the cultivation of the land that clothes and feeds them. The *Majjhima Nikaya* states categorically that just as some cities and districts are cultivated and some are not, some human beings are cultivated and some are not.[33]

The frequent references to artisanal production, commodities, and exchange in the early Theravada teachings are telling indications of how the rise of Buddhist *saṃgha*s in northern India and their dedicated support of the public veneration of relics and gifts began to undermine the traditional, sumptuary Vedic order of animal sacrifice based on a strict and hieratic conception of social order with warriors and priests at the top. The *saṃgha*s not

Ruins of the Mulagandhakuti Vihara near the Dhameka stupa, Sarnath, Uttar Pradesh. Photograph by author.

only replaced over time the prerequisites of blood and oblations to gods and ancestors mediated by Brahmins and protected by warrior clans, they also helped reconstitute a new order of Buddhist monks, pilgrims, and laity, supported by a burgeoning economy of gifts, donations, and monetary transactions. At the heart of this bustling network of commerce were the cities and trade routes of the middle and lower Ganges plains, the evidence of which can be simply gleaned from the rise in the number of stupas, *vihāra*s, and monasteries.

Some of the oldest rock-cut temples of ancient India are nestled in the Barabar Hills, close to the Dharwar outcrop near Rajgir (old Rajagriha), one of the first great cities of ancient India and the onetime capital of the great *mahājanapada* of Kosala. The Dharwar outcrop has been recognized since antiquity for its rich veins of iron ore and deposits of copper. Immediate access to these mines gave Kosala a decisive advantage over its rivals. Metallurgy was important, not only for the advanced weaponry required in the age of the warring states but also for the manufacture and circulation of coins.[34] Silver mines were particularly prized because silver provided the base for a more durable metal currency. Over the two centuries following the lifetime of the Buddha, the major inhabited tracts of the middle

Ganges basin were inundated with the flow of metal coins, especially silver.[35] The discourses attributed to the great Central Asian (Indo-Greek) king Minander known as the *Milinda Panho* [The Questions of Minander], which are generally dated to the second century B.C.E., contain references to a vast array of coins marking old and new dynasties circulating along the middle and upper Ganges plains.[36] The diversity and extent of this currency indicate not only the vibrancy of commercial life but also the attempt of rival kingdoms and republics to put their stamp on long-distance traffic. Evidence from Kapilavastu, the center of the Sakya clan, and Rajagriha, the capital of Kosala, suggests that large numbers of people were employed in the crafts of bow making and fletching, dyeing and weaving of textiles, pottery, ivory-working, glass-making, and iron manufacture. A few of the Buddhist Jataka stories describe plowshares being manufactured and sold in the city. Brick making from 300 B.C.E. onward, R. S. Sharma points out, is a clear indication of surplus production that would sustain a large section of the population living in urban quarters.[37]

The rise of identifiable monarchical kingdoms and city-states from a myriad of old warlike clans took place over several centuries. Even during the heyday of the Mahajanapadas, the distinction between tribal heads and kings was not entirely clear, especially if we consider the warrior lineages such as the Mallas, who claimed to be a cluster of five hundred kings, or for that matter the Yaudheyas and the Lichchhavis, who boasted that they came from the land of a thousand kings. It is said of the Videhas that they were once organized around a monarchy but came together as a *gaṇa* during the sixth century B.C.E. Later Buddhist texts such as the *Ekapanna Jataka* mention that in the city of Vaishali there were always 7,707 kings stationed to run the kingdom with a contingent of deputies, military commanders, and treasurers.[38]

Only a few monarchical kingdoms eventually conquered and absorbed smaller tribal polities. However, conflicts between kingdoms that pitted new monarchies against traditional warrior clans were already breaking out during the lifetime of the Buddha. The *Bhaddasala Jataka* gives us a candid account of the struggle that unfolded between the Sakya clans around the region of Kapilavastu and the kingdom of Kosala, after King Prasenjit (Prakrit Pasendi) of Kosala, in a carefully calculated political move, proposed to marry into the lineage of the Buddha.[39] The story narrates that the Sakyas, too proud to marry one of their daughters into the Kosala royal line, arranged for a slave girl, Vasabhakkhattiya, to take the place of the bride.

After she had borne the king a son and a daughter, King Prasenjit found out that he had been deceived. It is said that the Buddha himself intervened to save Vasabhakkhattiya and her children, arguing that the father's social standing overrode the queen's humble origins. Although the Buddha might have forestalled a violent outcome, when the son of the slave mother stepped into his father's inheritance, he laid the Sakyas to waste to avenge his father's humiliation. The story also illustrates the predicament of the clansmen of the Sakyan kingdom after they learned of marital propositions brought to them by the envoys of the mighty king of Kosala. If they accepted the proposal, they feared that the age-old sanctity and name of their clan might fall to pieces, because their blood lineage would no longer be considered pure. On the other hand, if they went to war and suffered defeat at the hands of Prasenjit's army, they would be annexed, and their ways, customs, and clan structure would not survive.[40] Something similar happened later, when Magadha in its early ascendancy advanced on the Lichchhavis and after a long struggle eventually incorporated them within their polity.[41] At the same time, many of these old tribes and clans maintained their traditional standards and identities even after they had been incorporated into larger kingdoms. Many raised their heads in defiance once again after the mighty empire of Magadha fell apart, and some of them survived all the way into the Shunga and Kushan periods through the second century C.E.

The ancient Indian *gaṇa-saṃgha*s were neither oligarchies nor democracies in any simple sense. Power was vested in the hands of an aristocratic class led by warrior lineages. Chieftains functioned largely according to the principle of *primus inter pares*. They were important as leaders of the community but also beholden to the council of elders. The heads of many of these warlike Kshatriya lineages were given the generic term *rājā*. Kings, clans, and tribal councils coexisted in ancient northern India for more than a thousand years. Their internal contradictions ultimately led to the ascendancy of decidedly dynastic, monarchical kingdoms. Early terms for a monarch such as *sārvvabhauma*, "owner of all the lands," or *chakravartin*, "turner of the wheel," give us a sense of these early drives for territorial autonomy and claims over the moral and social orders of Buddhist India.

Mud and Brick Ramparts

The great city of Rajgir, known as Rajagriha ("Abode of Kings") of Indian antiquity lies in a valley enveloped by lush hills that are part of the northernmost limits of the Gaya range skirting the southernmost flank of the Ganges plains. A small river, Banganga, runs through the city. The old capital can be identified from the ruins lodged in the legendary peaks of Gridhrakuta, Vaibhara, Vipula, and Udayagiri that are closely associated with the wanderings and teachings of the Buddha. The fabled grove of Yastivana (Jethavana in Pali sources) lies within the boundaries of this ancient pilgrimage, where King Bimbisara found the Buddha as a young mendicant and, overwhelmed by his grace, fell at his feet and asked for his blessings. It was in these hills during the rainy season that the Buddha spent some of his most memorable moments of contemplation, discoursing with disciples and doing the rounds with begging bowl in hand inside the city gates of Rajagriha. It was here also at the mouth of the cave at Saptaparni that the First Buddhist Council was held after the master's nirvana.

The present-day town lies just outside this valley. There are ancient ramparts from the period after the rise of Buddhism, and experts surmise that the city was relocated sometime during the middle of the first millennium B.C.E. Two natural passes serve as gates between the elevations toward the north and the south. It is not a stretch to imagine why a promising city would emerge in this area, fortified as it is by the natural lay of the land. Rajagriha eventually became the capital of the imperial state of Magadha and held its position until the fifth century B.C.E., when a new capital was erected in the city of Pataliputra.

After 1950, some of the first systematic excavations of the city were undertaken by the Archaeological Survey of India at the foot of the Vaibhara Hill, which is located at the outer edge of the walls of the old citadel. The city stands on deposits of clay covered by pebbles, and its earliest layers of occupation show an unknown type of burial: pits with oval-shaped bottoms and additional funnels dug deeper into the earth. Usually packed with clay, these makeshift receptacles contain fragments of bones and crematory ashes dating back to the fifth century B.C.E. or earlier. We do not know for sure whether it was during King Bimbisara's reign that the great city of Rajagriha was founded. Legend says that it was plagued by fires and repeated attacks from the petty chiefs of Vaishali.[42] Rajagriha proved to be too provincial and isolated in the long run for the burgeoning territorial ambitions of the Magadhan Empire. Shishunaga, the last king of the Videha line, removed the capital to Vaishali, and his son Kalasoka Mahananda began the construction of the great metropolis of Pataliputra on the banks of the Ganges. Although Ajatashatru was supposed to have rebuilt Rajagriha, there is little

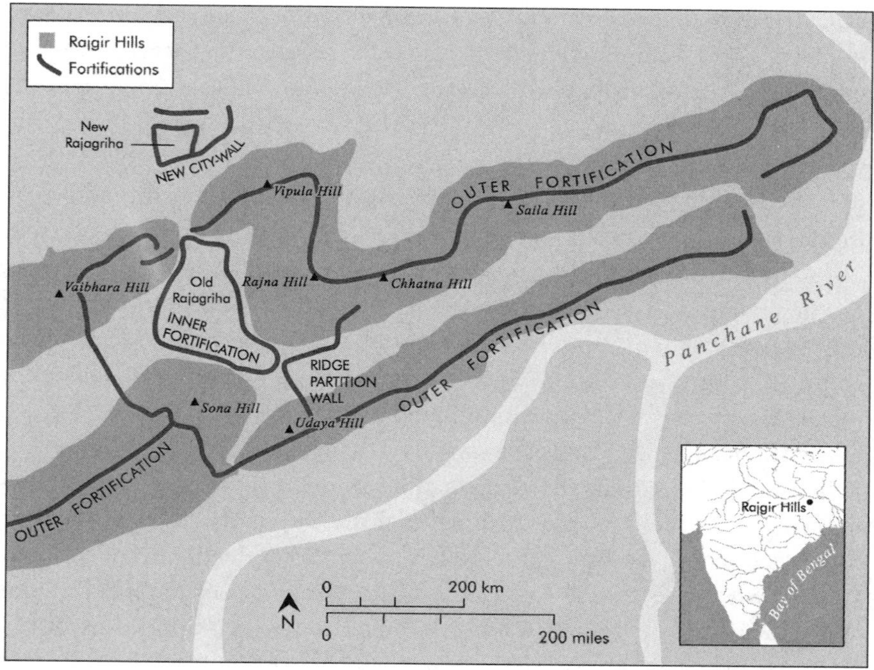

Schematic diagram of the Rajgir Hills (ancient Rajagriha) showing remains of walls and fortifications. Drawn by M. Roy Cartography after M. Kureishi and A. Ghosh, *Rajgir*, Delhi: Archaeological Survey of India, 1958.

evidence that a fully fortified city existed in the valley of the hills before this time.[43] The hilltop fortification of dry masonry was in essence more a symbol than effective defense, and only the entrance to the city was initially fortified.[44] A. M. Broadley, an amateur archaeology enthusiast and an assistant magistrate in Bihar, writing as far back as 1872, found that the walls of the city and its gates were still traceable about a half a mile from the foot of the mountain, directly facing the north entrance of the valley.[45] Subsequently, remains of walls eleven or twelve feet high were found, with parts made out of compacted clay, indicating that the city at some point prepared for a large-scale military onslaught.[46] However, it was not just the garrisons but rather the busy trading routes traversing the valley that put Rajagriha at the head of a complex network of settlements and markets.

Rajagriha was by no means the only city of the Ganges valley in this period braced for recurring conflict. Contemporaries often compared its hilly ramparts to the other great and impregnable city of Kaushambi on the eastern bank of the Yamuna River, built under the aegis of the Vatsas, one of the great *janapada*s of ancient India west of Magadha and Kashi. Kaushambi's geographical importance lay in the fact that other than Varanasi, this was the other great terminus for trading routes between the northern plains and the Deccan south. The Vatsas appear time and again in contemporary literary texts owing to the legendary exploits of Udayana, one of their early kings. Udayana was not the first king of Kaushambi and the Vatsa kingdom, but he managed to get to the throne even though he was behind seventeen other claimants. The Buddhist sources tell us that he knew secret magic commands for controlling elephants, which was a great military advantage during the period of the Warring Kingdoms. Perhaps the most well-known tale attributed to Udayana is about the attempt of neighboring King Pajjana from the state of Avanti to abduct him and wrest the powerful formula with which he manipulated the most frightful elephants. In the manner of a Trojan horse, King Pajjana built a hollow wooden elephant that could be worked with ropes and levers from within by soldiers hidden inside. Once the elephant was let loose in Vatsa, someone spotted it and sent word to Udayana, who like a true hunter went after it. This is how the king was tricked and caught, then brought to the palace of Avanti and locked up. Languishing in captivity, Udayana decided to share his magic formulas. He agreed to reveal his secret to the king's daughter, who came to visit him in his chambers. However, she was smitten with his charm and fell in love. The captive and the princess plotted a joint escape, which they accomplished on the back of a royal elephant, scattering gold coins to distract onlookers. This way, the crafty elephant-charmer Udayana was able to spirit the daughter of Vatsas to his kingdom of Kaushambi, where she would become one of his major wives. Buddhist sources also mention how the Buddha visited the capital of Vatsa several times to deliver some of his most important sermons. It is said that through his entreaty King Udayana became a great patron of the *saṃgha*, and the city of Kaushambi became one of the most important centers of Buddhism. This was an important turning point in the history of a place well known in an earlier period for its Vedic practice of sacrifices, and excavations have revealed the remains of large altars, surrounded by bones of humans, buffalo, horses, elephants, and goats.[47]

The nondescript village of Kosam still stands, not far from the present-day city of Allahabad. It was hailed by the veteran archaeologist Alexander Cunningham as the site of the ancient and renowned town of Kaushambi when he conducted early surveys there in 1861. Kaushambi was a major urban center on a key trade route on the Yamuna River, between Mathura in the west and Pataliputra in the east. It is one of the oldest urban sites in northern India, perhaps as old as Varanasi itself. The city is mentioned in the *Aitareya Brahmana*s, which are dated to the eighth century B.C.E. Kaushambi came into prominence after the extensive excavations conducted near Allahabad by G. R. Sharma, first in 1949 and then again in 1959–1960. Sharma located mounds spread across four square miles; parts of a high wall were found, indicating early experiments in fortification. The city was designed with major roads coming from each direction, and every road was secured by high watchtowers.[48] Kaushambi was a well-known urban site as early as the late fourth century B.C.E., and evidence of many states and dynasties is buried in its earthen layers.

At least fifteen layers have been identified that give us some indication of settlements over more than seven hundred years, the oldest dating as far back as 1000 to 800 B.C.E.[49] The first traces of human occupation are evident in the typical PGW pottery shards along with holes dug deep into the ground, suggesting the early use of wooden posts and clay to construct houses. Early settlers built walls of solid clay blocks. In the layers where we find NBPW are traces of large pits dug for waste and sewage and reinforced wells for rainwater and groundwater. At a level dated to around 400 B.C.E. we find a massive earthen rampart erected to protect the city from outside attack. The same people built geometrically aligned streets and thoroughfares, fortified earthen ramparts, and houses crafted of timber, clay, straw, and stone. Bits of charcoal and terra-cotta suggest kilns for bricks and large construction projects, and the presence of beads points to a vigorous and ongoing trade with neighboring kingdoms.[50]

The capital of Anga, a kingdom ultimately subordinated by Magadha, was the famed city of Champa, situated in the present-day state of Bihar close to the city of Bhagalpur, which curves along the banks of the Ganges. It was built where the Champa River once met the Ganges, forming a natural boundary between the rival kingdoms of Anga and Magadha. Its formal name was Champapura, "City of the Frangipani Flower," and it was protected by a fortress and an encircling wall. It was also famous as a riverside mart whence boats laden with merchandise, including its legendary perfumes, left for various ports. Jain texts such as the *Campakasresthi Katha* [The Story of the Merchant of Champa] and the Buddhist treatise *Sumangala Vilasini* suggest that the city was famous for its trade and commerce. Near its vicinity was a water tank called Gaggara dug by a former queen (possibly of the same name) that was always surrounded by groves of flowering frangipani trees (*campaka vanam*), which were harvested and pressed by perfume sellers.[51] The *Campeya Jataka* story narrated by the Buddha to his disciples at Jethavana describes the internecine rivalry between the kingdoms of Anga and Magadha during the early days of the Warring Kingdoms.[52] During one of these numerous wars, when the king of Magadha was fleeing the battlefield in the face of defeat, pursued by the warriors of Anga, he plunged with his horse into the Champa in desperation. He did not die, however. The great serpent king of the Champeya kingdom who dwelled in his underwater palace took pity on him and saved his life by sheltering him in the subterranean depths. In return the Magadhans offered a bejeweled pavilion each year to honor the great monarch of the riparian deep. With the help of the serpent king and his special prowess, the king of Magadha was able to defeat the kingdom of Anga and unite the two territories under one formidable power. This story in many ways is an indirect acknowledgment of the serpent icon as part of a wide variety of animistic practices that found their way into popular and demotic forms of Buddhism.

Looking through the prism of the Buddhist texts, we can reach back in time and see the patterns of the large-scale transformation taking place in the demographic and ecological landscape of the greater Ganges plains. One indication of this was the rise of large, fortified cities as capitals of dominant kingdoms. However, it was not just these urban concentrations but the networks to which they were attached that bring into focus the crisscrossing pattern of overland and riverine trading routes.[53] Settlements such as Kara, Sringaverapura, Jhusi, Bhita, and others near Kaushambi on the banks of the Ganges provide clues to the extent and intensity of trade and production. The mound of Jhusi, for example, rising from the cliffs alongside the river opposite present-day Allahabad, was an ancient settlement around the site of the holy confluence of Prayag dating back to the eighth century B.C.E.[54] Excavation at the nearby site of Sringaverapura alongside the Ganges has yielded the remains of the great tank complex, which existed at least as early as the second century B.C.E., roughly corresponding to the Kushan period.[55] It sits on one of the most ancient fords on the river, and the western approach contains remains of perhaps even older fortifications, with traces of occupation within as well as outside the city walls. These excavations show that the countries of the middle and lower Ganges plains were connected to the western Ganga-Yamuna Doab area as well as southern trade routes leading through the forests of the Vindhya Hills onward to the countries in the peninsular plateau. Trade routes along and across the Ganges valley fed the rising economic demands of a patrician society of kings, warriors, Brahmin priests, and Buddhist monks. Mercantile wealth left its mark on the eastern cities of Champa, Pataliputra, and Buxar (known as Vedagarbhapuri) on the banks of the Ganges. These cities were connected to a vast hinterland comprising the Rajmahal Hills, where the Ganges enters Bengal, as well as the hilly tracts of the Chota Nagpur Plateau hinterland, which supplied mineral and timber.[56]

Beginning in the second half of the sixth century B.C.E., the urban settlements of the upper and middle Ganges plains began to encroach on the dense forest canopy of the immediate environs of the river. There is evidence of forest clearance from this period forward. In the kingdom of Vatsa, for instance, while there is evidence of a forested area between the confluence of the Ganges and the Yamuna and the city of Kaushambi, we also find villages situated far away from the river, which indicates a network of roads. The scale and ambition of urban construction must have also required a more intense search for resources. Kaushambi is estimated to have occupied at least ten hectares during the early phase of its building, between 1000 and 800 B.C.E., and as the capital of the Vatsa kingdom, it grew to about five times that size.[57] Both Rajagriha and Kaushambi supported populations upward of 40,000 at the height of their prominence.[58] Much of this expansion was based on the extensive use of timber. There is evidence of lumber being gathered from near and far for use as a building material, identifiable as belonging to certain species of trees identified for the durability and strength of their wood, the most common being the pipal (*Ficus religiosa*), bel or wood-apple (*Aegle marmelos*), udumbara (*Ficus rasemosa*), teak (*Tectona grandis*), and neem (*Azadirachta indica*). Elephants were most likely used in the felling and transportation of logs. During the early digs at Kaushambi, G. R. Sharma found a square coin with the impression of a tree enclosed by a railing on one side and a hill and an elephant on the other, which was identified later as a coin belonging to the Anga Dynasty during the period just before the rise of the Magadha Empire.[59]

Within a century of the end of the Buddha's life, the entire Ganges plains had come alive with the traffic of monks and pilgrims. Early Buddhist orders began to receive the support and patronage of major political regimes of northern India, and they in turn facilitated trade and commercial exchange. In one of his discourses the Buddha had explained to a rich Setthi (moneylender) of Rajagriha that one could express one's devotion to the order through works of architecture: *vihāras* as shelter for monks, cloisters with turned up-eaves (*addhayoga*), towers, stone houses with flat roofs, or crypts. It is reputed that this Setthi singlehandedly patronized sixty buildings.[60] The Buddha had enjoined that monks should strive to live a life of austerity, abjuring the comforts of house or hearth. They were asked to follow the path of the ocher robe and the begging bowl, finding shelter wherever they could: in the woods, at the foot of a tree, on a hill, in a grotto, in a mountain cave, in a cemetery, in a forest, in the open air, and on the occasional bed of straw.

Soon after the Buddha's death, however, kings and merchants rushed to get hold of his body, casket, and possessions as relics. Buddha's followers divided his remains and effects into eight portions. On the urn itself a shrine was built. Even the coals from the great crematory fire of the Buddha's *parinirvāna* were lovingly carried away by the Mauryas of Pippalivana. At this time the first eight great stupas were built in the major cities of Rajagriha, Vaishali, Kapilavastu, Allakappa, Ramagrama, Vethadipa, Pava, and Kusinara.[61] Buddhism ushered in a spate of building activity in ancient India. Buddhist texts such as the *Mahabhagga* contain a wealth of information about these early monasteries and forested retreats, and also the building of more elaborate structures such as hospices (*āvāsas*) and commemorative towers (*saṃghārāmas*).

The concentration of wealth and patronage in the cities of the Ganges plains was also tied to significant changes in the rise and distribution of the agrarian surplus. It is said that once, during the middle of the rainy season, the Buddha returned from the city of Shravasti in central India to visit the southern hills of Rajagriha, where he observed the Brahmin Bharadvaja directing agricultural laborers in his fields at the village of Ekanala.[62] The Buddha told Bharadvaja that much like the cultivators, he too sowed, plowed, reaped his harvest, and filled his belly. Bharadvaja was surprised by this remark because he did not see a plow, a plowshare, a goad, or oxen with the Buddha. Hearing this, the Buddha described his path of enlightenment to the cultivation of the mind. "Faith is the seed I sow," he pronounced, "modesty is the plow-shaft, the mind is the tie of the yoke, mindfulness is my plowshare and goad."[63] The true spirit was akin to a team of cultivators and their bullocks leading the mind to safety. The bountiful rain-drenched fields and the coming harvest of the kingdom of Magadha thus provided for the Buddha the ready, natural metaphors for spiritual fulfillment.

Early Buddhist literature contains extensive discussion of villages (*grāma*s, Sanskrit; *gāma*s, Prakrit) and agricultural fields. The same word, *gāma*, used for caravan halts was also used to describe a settled agricultural community. Villages were often led by kinship groups dominated by traditional Brahmin families. The grammarian Panini provides us with words such as *ghoṣa*, "herdsmen's camps," and *kheta*, "plots."[64] The term *gahapati*, which simply meant "one who lives in a house," also began to gain wider currency during this time.[65] This term also implied "householder" or "master of the house," distinguishing men of substance other than kings, priests, and warriors.[66] The title *gahapati* has been associated with commercial activities and was often used to refer to donors of caves and patrons of buildings in the Buddhist canonical literature.[67] While some scholars have seen this term used primarily for property owners and taxpayers, others have seen it in conjunction with the associated terms *kuṭumbin* and *kuṭumbika*, an early description of independent peasant kin groups who had rights to land and tillage.[68] We also find frequent references to *nigama*s, or market towns, in the Buddhist Jatakas, which also mention old families, with their kin groups and retainers, adopting and patronizing Buddhism in the villages. The terms for *city*, *village*, and *market* from this period are often interchangeable and lack specificity, which makes it difficult to define what constituted a proper city. Cities were not always seats of state power.[69] The definitions of villages and cities were rapidly changing during this period, especially when viewed through the lens of Buddhist pilgrimage, which revolved around the six cities that received the Buddha's remains: Champa, Kashi, Shravasti, Kaushambi, Rajagriha, and Saketa. Most of these centers were located in and around the floodplains of the Ganges.

The Second Coming of Magadha

The probable boundaries of the early kingdom of Magadha were the Ganges to the north, its tributary the Son to the West, the old country of Anga to the east, and the dense forests of present-day Chota Nagpur to the south. Across the Ganges to the north was the country of Videha. The rise of Magadha as an empire was the result of a long and bloody history of violence and intrigue, during which a succession of dynasties tried to lay claim to its resources. Magadha had long been a rival of the kingdoms of Kashi and Kosala and had been consolidated under Bimbisara, a patrician of modest origins hailing from the Haryanka clan, who was anointed king at age fifteen. His first capital was Girivraja, which has been identified with Rajagriha, and he entered a marriage alliance with the sister of Prasenjit, ruler of Kosala, a kingdom that nevertheless remained a political rival. The Sinhalese Buddhist text *Mahavamsa* suggests that King Bimbisara, who ruled from the great city of Rajagriha, had a hard time defending it from fires and repeated attacks from the Lichchhavi chieftains of Vaishali.

Bimbisara had been king for an unusually long period of fifty-two years when he was imprisoned and killed by his own son Ajatashatru, who also turned against the influential Buddhist order that his father had patronized. Known for his ruthless political ambition, Ajatashatru finally defeated Kosala and incorporated it within his kingdom. Kosala had already taken over Kashi and the city of Varanasi, and now the entire region around Varanasi was part of the expanding Magadhan polity, with its core region located between Patna and Gaya in the present-day Indian state of Bihar. Ajatashatru also entered into a bloody war with the Lichchhavi confederacy in a dispute over spoils. Lichchhavis and their tribal oligarchy posed a challenge, but eventually Ajatashatru's Brahmin chief minister Vassakara (much as the Buddha had once feared) succeeded in sowing seeds of dissension within their ranks and weakened their resolve. In the end Magadha succeeded in humbling the Lichchhavis, laying claim to the legendary city of Vaishali.[70]

Ajatashatru died around 461 B.C.E. and was succeeded by a string of ineffective and parricidal rulers under whom Magadha slipped from its prominence, until the disgruntled notables of the kingdom elevated one of the ministers, Shishunaga, to the throne, naming a new ruling dynasty after him.[71] Shishunaga's father was a Lichchhavi prince of Vaishali and his mother a mistress. In effect the last king of the Videhas, he temporarily removed the capital of Magadha to Vaishali, raised a formidable army, and finally defeated Aryaka, the ruler of the Pradyota Dynasty of Avanti, a major threat to Magadha, and annexed its territories and the wealthy city of Kaushambi. By the fourth century B.C.E. the fortified settlement Pataligrama on the Ganges, which had been initially built to keep the raiding Lichchhavi chiefs and other members of the Vrijji alliance in check, was anointed as the new capital of Magadha, replacing Rajagriha. Now Magadha not only ruled the Ganges basin but also held territories as far afield as Rajasthan and Punjab. It had become a part of what the *Aitareya Brahmanas* prophesied: the region that providence had anointed as the crucible of expanding empires.

In retrospect, Magadha seems to have enjoyed certain decisive advantages over its rival kingdoms. It held a key strategic location between the upper and lower reaches of the fluvial plains of the Ganges. It had an impregnable citadel surrounded by the hills of Rajagriha, and it had built another heavily fortified encampment at the confluence of two major rivers in Pataliputra, which also benefitted from the rising wave of trade and commerce borne by river and overland routes. In its day Magadha controlled most of the major trade routes that ran across northern India. It also had access to some of the most fertile soils of the entire Ganges valley. It had steadily built its resources of metals and mines and had an exclusive hold of iron deposits.[72] As the governor of Anga, young Ajatashatru fought one of his early battles with the Lichchhavis of Vaishali over the rights to a mine.[73] The imperial experiment in Magadha might have also taken off because its greatest rulers came from outside the Brahmin-dominated heartland of northern India and rose to power by the use of strategy and force. In this regard the caste order always made space for the powerful at the top of the social ladder, as it made space for the destitute at the bottom.[74] Magadha was able to recruit and reward people in its administration of the countryside and erect a long-lasting bureaucratic arrangement that was based not only on cities but also on the agrarian countryside, where even a village headman could take a meaningful part. Outside observers often spoke with genuine admiration for the system of justice, roads, irrigation, and hospitality in the kingdom of Magadha.

This was a time of change, opportunity, and political adventurism, and the environs of Magadha attracted fortune seekers such as the brothers of the Nanda clan of the frontier provinces (*pachchanta nagaram*), who reputedly came from a lowly social rank. Their leader, Mahapadma (also known as Ugrasena), founder of the Nanda Dynasty, ran with a band of roving outlaws at an early age and was introduced to a life of looting and fighting. He rose up the ranks as a mercenary, began leading raids on neighboring kingdoms, and eventually succeeded in usurping the kingdom of Magadha. Buddhist texts such as the *Mahabodhivamsa* simply state that former robbers had now become the new kings, and although Buddhist sources are reticent in stating how exactly he deposed the Shishunaga king of Magadha, later dynastic chronicles such as Banabhatta's *Harshacharita*, written in the seventh century C.E., state that he was killed by a knife thrust in his throat near the capital city, Pataliputra.[75] These violent attempts to secure power and wealth have long been associated with the rise of Magadha as a ruthless territorial power, and they provide a ripe context for the purpose and function of the state outlined in the *Arthashastra,* one of the world's earliest treatises on the ends and means of sovereignty and power. It is said that one of the Nanda monarchs, appropriately named Dhana ("riches") Nanda, was so obsessed with the acquisition of gold that he began levying taxes on trees and stones.[76] People believed that he had a huge treasure trove buried under a rock somewhere in the bed of the Ganges.[77] This was the same king to whose court the great scholar from the western city of Taxila, Chanakya, also known as Kautilya (of the *kuṭila* clan, or *gotra*), author of the *Arthashastra,* sought refuge and patronage.

One Buddhist source states that Chanakya was known for his expertise in the three Vedas and incantations and the art of political strategy and intrigue, and also for his physical deformities, including an unsightly complexion and misshapen legs.[78] Chanakya was hopeful that the Nanda king, who had recently turned to more responsible governance and public charity, would find a place for him in his court. Instead, the king, repulsed by his appearance, insulted him and threw him out.[79] Fearing further reprisal, Chanakya fled disguised as an Ajivika recluse but swore to destroy the dynasty, family, and entire progeny of the rulers of Magadha. Not only did he succeed in this mission by putting his student and protégé Chandragupta Maurya on the throne, he also put into effect the blueprint for one of the most formidable empires of antiquity.

A View from Macedonia

The political education of the young adventurer Chandragupta Maurya is best understood in the context of Alexander of Macedon's invasion and the turbulence of the expatriate Greek raids along the northwestern frontiers of India. This is the period during which the trans-Gangetic political landscape comes into view in Greek and Latin classical accounts. Fragments of contemporary histories have been preserved in the accounts of writers such as Strabo, Justin, Pliny, Plutarch, and Ptolemy the Geographer. They give us some of the first outside impressions of the eastern parts of the Indian subcontinent far from the theater of Alexander's conquests across the western Himalayan frontier kingdoms. The word that appears for the eastern reaches of the Ganges plains in the Hellenistic and later Latin accounts is *Prasii,* a Greek version of the Sanskrit *prācya,* or "east," and *Gangaridai,* a

Greco-Latin version of the Sanskrit word *gaṅgāhṛda*, "Ganges at its heart," or the Gangetic heartland.[80]

The following account is given by Quintus Curtius Rufus, the Roman historian, who was perhaps the only author to write a biography of Alexander of Macedon in Latin, *Historiarum Alexandri Macedonis*.[81] In the monsoon season of the year 326 B.C.E. an exhausted Alexander, with a mutiny brewing among the rank and file of his troops, asked about the conditions of India beyond the Hyphasis River (Beas) they had just crossed. Alexander had been advised by Phegeus (possibly the Sanskrit Bhagala), a local prince who ruled the land immediately next to the Beas valley, who gave him an impressionistic and largely mistaken account of a march of twelve days through the desert to a great plain where one would have to cross a river more than 32 stadia (over 3 miles) wide. This river was what the Greeks called "Ganges." Farther downstream lay the lands of the Prasii and the Gangaridai, whose king, the formidable Xandrames, had posted military garrisons along the main highway leading to his kingdom.[82] This mighty eastern monarch was reputed to have an army of 20,000 cavalry and the same number of infantry, along with 2,000 chariots and 4,000 giant caparisoned war elephants. Alarmed at the prospect of approaching this powerful kingdom, Alexander apparently asked King Porus of Indus to verify the report. Porus endorsed Phegeus's account. He confirmed the military strength of this great eastern king but added that he was a lowly son of a barber who had somehow succeeded in gaining access to the inner court, seducing the queen with his charm and good looks, assassinating the reigning monarch, and murdering every viable heir. Xandrames was his son, an unpopular but much-feared king. Curtius Rufus seems insistent on the point that Xandrames was a "man from the lowest class" who had earned the hatred and contempt of the people, which might have contributed to his downfall and the eventual usurpation of his kingdom by another adventurer.[83] It is said that upon hearing this, Alexander began to worry about crossing the mighty Ganges, given the vagaries of the Indian rainy season, the size of the army of the eastern monarch, and the number of his war elephants.[84]

We have to consider that both Curtius and also the Greek historian Arrian of Nicomedia were rendering fragments of history that were removed far away in time and suffused with the heroic legends of Alexander. It is expected that these histories would seek to elevate the status of the world conqueror by embellishing the broken spirit of the Macedonian army and by showing the depth of Alexander's courage when he was confronted with the daunting task of a further campaign into the heart of unknown India.[85] Alexander, as we know, decided in the end not to venture beyond the Hyphasis and challenge the young king Sandracottus (Chandragupta), who ruled over the eastern valley from Palimbothra or Palibothra (Pataliputra), the largest city on the Ganges.[86] Arrian, in his account of the campaigns of Alexander, records that there were Greeks in Alexander's retinue who knew the lay of the land beyond the last tributaries of the Indus.[87] The people of the eastern Indian empire were rich and productive. They had good farmers and fine soldiers and lived under an orderly and efficient social system. The governments in that region were mostly aristocratic but not unfair or oppressive. The reliability of the reports that reached Alexander has been seriously doubted by more recent scholars, such as A. B. Bosworth.[88] While the early history of the Macedonian encounter in India remains murky, it is crucial to the reconstruction of the story of the Maurya Empire, of which only fragments and secondhand accounts have survived.

In the aftermath of Alexander's expedition and premature death, the Greeks added another region to their map of the world and a new chapter to their knowledge of the land through which the mighty Ganges flowed. This is difficult to piece together because most of the original writings from that time are lost, and we have only fragments preserved in later commentaries. The first Greek accounts of India most likely came from two figures associated with Alexander's navy in India: Nearchus, a naturalized Macedonian from Crete, who was a commander, and Onescritus, who was in charge of navigation. While these earliest accounts may have been based on actual Indian sources and experience, they became grist for the mill of marvelous and exotic histories that were centered on the celebrated figure of Alexander the Conqueror. The most popular histories of this genre, such as those of Cleitarchus of Alexandria or Calisthenes of Olynthus, are full of colorful embroidery and must be treated with caution.[89]

The Greek view of the world as a picture, or *mappa mundi*, emerged from the idea of *oikoumenê* (οἰκουμένη), that is, the entire known and inhabited world as a singularity with unknown, shadowy edges. Hectataeus of Miletus, who lived during the sixth century B.C.E., was the geographer who first divided the inhabited world into two basic parts: Europe and Asia.[90] To this cosmology Herodotus introduced his fantastic accounts of India, which quickly assumed the status of one of the more important antipodean realms in the Greek imagination. In Strabo's great map of the world, for example,

Indians occupied one of the world's four parts in the direction of the four Aeolian corners, with Ethiopians, Celts, and Scythians inhabiting the other three.[91] Strabo was also convinced that it was only because of Alexander's Indian campaigns that parts of distant Asia came under Greek purview. The Greeks were also fascinated with the great rivers of the world. Strabo recalls a letter written from India by Craterus, commander of the Macedonian phalanx and the left-wing infantry in the battle of Issus, writing to his mother in Macedon, which mentions how he saw the River Ganges for the first time. Strabo argues that the Ganges was the largest of the rivers known in the three continents, larger than the Danube, the Indus, and even the Nile.[92] Arrian, echoing observers in the ranks of Alexander's army, reasoned that if all rivers, large or small, in various parts of the world were capable of depositing mud and silt brought from down from the uplands where they rise, then there was no reason to doubt that India was also *one* giant alluvial plain. Rivers in the west would not, he marveled, "for sheer volume of water, compare with any single one of the rivers of India—to say nothing of the greatest, the Ganges, with which not even the Egyptian Nile or the European river of the Danube is to be mentioned in the same breath."[93] The most read and discussed authority on the river was, of course, Megasthenes, who traveled as an envoy to the legendary kingdom situated near its delta. Megasthenes noted that many rivers drained into the Ganges. He measured its width as 30 stadia, flowing north to south, emptying its waters into the ocean that formed the eastern boundary of the nation of Gangaridai. This country had the largest elephants that could be trained and used in battle with lethal effect. This, according to Megasthenes, was the principal reason their country had never been conquered by any foreign kings. And it was this reason above all that gave Alexander pause as he contemplated his onward march.[94]

India entered Hellenistic geomancy as a distant, fantastic land of luxuries, kings, and elephants, which had a lasting discursive effect. Grant Parker points out that Megasthenes's *Indika* and subsequent commentaries based on it indulge in a premeditated discourse about the distance, dimension, and difference of India as a place and Indians as people. A wealth of geographical details had been amassed by the end of the fourth century B.C.E. that dispelled the murky accounts of Herodotus of India as terra incognita, a vast unexplored desert with strange gold-digging ants. However, as Parker points out, the image of India in Greco-Roman antiquity was one of plenitude and abundance.[95] India was a geographical cornucopia of exceptions:

elephants, monkeys, snakes, spices, crops—a limitless source of gold and silver. In this view Indus was the last known river, and thus the outer limit of the Persian and Indo-Grecian world. The Ganges was the unknown river that flowed to the fabulous city of "Palibothra" (Pataliputra) and its swarming fleet of elephants. The knowledge of the land of India between Taxila, a key frontier province of the Maurya Empire in the valley of the Indus, and Pataliputra, its capital on the banks of the Ganges, remained for a long time a source of wonder and unverified assertions.[96]

As Alexander's campaigns in India were coming to a close, dramatic changes were about to take place on the political stage of Magadha. There are no firsthand accounts of the bloody encounters that took place in the vicinity of Pataliputra. The Greek authors mention in passing that the Nandas had been overthrown by Chandragupta. There are veiled references in the Puranas that Chandragupta, with the help of his maverick councilor Kautilya (also known as Chanakya), secured people from the oppressive hand of the Nandas (*uddhāriṣyati tān sarvvān*).[97] According to the later Buddhist text the *Milinda Panho*, a mighty struggle was waged between the Nandas and the Mauryas in which thousands of elephants, horses, charioteers, and infantry were slaughtered, and the headless corpses danced in a frenzy on the bloody battlefields. We do not know exactly how Chandragupta succeeded in toppling the Nandas, but it took several attempts. The strategy seems to have been that after securing a base in the Punjab, where he either subdued kingdoms and tribes or made alliances with them, he waged a war of attrition from the frontiers of the kingdom of Magadha. His initial army was crafted out of smaller chieftains, mercenaries, and kingless peoples (*arāṭṭa*s), which may well have included highland brigands, and people of the forests (*āṭavika*s). He also amassed a formidable corps of elephants. By the year 312 B.C.E. his territories stretched to the Narmada River in the west and the Vindhya Hills in the south.

Chandragupta had begun his campaigns just as Alexander and his disgruntled troops turned away from India, leaving behind generals and soldiers of his retinue such as Seleucus and Antigonus, who would soon carve out their own independent territories on the northwestern frontiers, which later came to be known as the Bactrian or Indo-Grecian satraps that recruited a bulk of their soldiers from Persia. The evidence is scant for Chandragupta's rise to prominence. It is believed that the Greek invasions gave him an ideal opportunity to strike. Accounts such as that of Plutarch mention that he had the opportunity to meet Alexander in person; others suggest that he carefully studied the Macedonians at war.[98] He is also reputed to have sought the assistance of King Porus, who had embraced the suzerainty of Alexander and fought on the Greek side. After Chandragupta secured his hold on the kingdom of Magadha, he fought with the Macedonian satraps to seal the northwestern part of his empire. One of these rulers was Seleucus (Seleukos), a general who had served in Alexander's army. After Alexander's death, Seleucus had marched into Babylon and seized most of Bactria. Seleucus secured his Indian flank by signing a lasting pact with Chandragupta, ceding several of his eastern provinces to the Maurya Empire. Among these was Gandhara, roughly the territory between the Kunar and Indus Rivers, in addition to Arachosia.[99] This was a time of intrigue and great rivalry among the Greek satraps across the Hindu Kush. Sibyrtius, the ruler of Arachosia and Gedrosia, had just been slain, which is one of the reasons why Seleucus might have decided to declare a truce with Chandragupta; if the account of Megasthenes is to be believed, this involved not only the reception of Bactrian envoys at the court of Pataliputra but also the introduction of Greek women into the Mauryan seraglio in exchange for a gift of five hundred war elephants for Seleucus's army.[100] The significance of this gesture has been debated by many historians. The popular account of Justin (Justinius), based on the *Philippic History* by the Romanized Gaul Pompeius Trogus, suggests that when Seleucus made peace with the Mauryas, Chandragupta's forces had already crossed the Indus River.[101] Several versions of this event, including those of Appian and Strabo, suggest that Seleucus surrendered his easternmost satraps of Gandhara; Parapamisadae, west of present-day Kabul; eastern Gedrosia, roughly in southern Pakistan; and possibly also Aria, Arachosia, and other territories as far as present-day Herat, Afghanistan. The Greeks also entered into a nuptial arrangement with the Mauryas that may not have been dynastic betrothal, *epigamia* (ἐπιγαμία), but rather an informal gesture, *kedos* (κῆδος), in which a female relative of Seleucus was given in marriage to the Maurya court.[102]

The Maurya Pax Indica

Chandragupta Maurya's empire at its height extended from its stronghold around the middle and lower Ganges plains to the far valley of the Indus and beyond. It may have stretched as far as the coastal parts of present-day Andhra Pradesh to the south. One of the reasons why Chandragupta could afford to set up garrisons in faraway cities such as Ujjain and Taxila beyond the Punjab, or Pundranagara in the Bengal delta, was the ability of his state to tax the agrarian bounty and bustling commerce of the Ganges plains. We can form a plausible picture of the Maurya administration from the account of Megasthenes, the Greek envoy in Pataliputra, from the *Arthashastra* and from the famous edicts of Emperor Asoka. One key aspect was the balance between autarky and the recognition of local and regional customs.[103] Besides the directly ruled provinces of the empire, there were also subordinate territorial powers that enjoyed a degree of autonomy under the Mauryan imperial canopy.[104]

Chandragupta built a royal thoroughfare leading to the gates of Pataliputra, defended by well-equipped troops with horses, chariots, and war elephants. He styled himself as an *ekarāṭa*, or the unmatched sovereign who had uprooted all the traditional warrior lines (*sarva kṣatrāntaka*), although this was more bluster than fact. The idea seems to have taken root that in these times of great political instability, subjects needed a strong and unyielding sovereign. In the *Arthashastra*, Kautilya states categorically that a true king is a firm holder of the scepter (*danḍadhara*). Without the *danḍa*, which in Sanskrit is a word with a range of meanings—"scepter," "sentence," "punishment"—human society became prey to the "law of the fishes," or *matsyanyāya*, when big political powers devour smaller ones. Only the king's supreme and fearsome authority keeps the weak strong and the strong weak.[105]

Many scholars over the years have questioned the veracity of the *Arthashastra* and its author, Kautilya. Extant versions of the text survive from a much later date, and they all exhibit the signs of accretion to varying degrees. Nevertheless, the stark consistency in its arguments speaks to a consistent, authoritative voice with an exceptional political acuity that would have been hard to replicate.[106] What is more important for our purpose here is that the *Arthashastra* as a text was conceived in view of the rise of imperial Magadha beginning with the Haryanka and Shishunaga Dynasties and ending with the Nandas and the Mauryas, a long political experiment in which the Indian monarchical state outrivaled other forms of political organization, including the old republican oligarchies. There is a "relentless logic with which the implications of state policy are worked out," writes K. A. Nilakantha Shastri in his masterful survey of the Maurya conception of the state.[107] The key to this pursuit of dominance was a rising awareness of the interdependence of power and wealth, driven partly by the surge of money economy that also helped the proliferation of mercenaries, militias, and armed camps. Both Megasthenes and Diodorus following him describe how coinage and bullion—gold, silver, iron, and copper—were flooding the cities and markets of the great northern Indian plains at the time. In many ways this provides the precise context for the disposition of the *Arthashastra* toward the instinctive pursuits of power and domination but also its tone of restraint and moderation in the exercise of force. Also, the text tends to pivot toward more traditional, bookish forms of Brahminical orthodoxy.[108]

In many respects Kautilya presents the outline of a model imperial city-state. The kingdom is secure only if the capital is secure. Hence the capital must be fortified, along with the frontier towns. Exquisite details are given about palace layout, moats, secret exits, and the personal safety of the king secured by the careful planning and inspection of the palace compound: audience halls, council chambers, and the station of the guards.[109] Kautilya says that fortified towns, especially at the frontier, must not be islands of control surrounded by an indifferent body of people who do not share the same allegiance toward the empire. The loyalty of subjects at the distant outposts is vital to the interests of the state, as important as the support of people within the kingdom. Kautilya is wary of hidden sources of disaffection among subjects owing to the king's personal failings. Ignorance of this is tantamount to negligence of the duties of governance, which must anticipate and uproot discontent. The *Arthashastra* frequently mentions the absolute reliance of the state on the covert acquisition of information through the deployment of spies and agents. It takes great care in detailing the various informants in the countryside that are eyes and ears of the state, often disguised as disciples of religious orders, monks, merchants, nuns, prostitutes, and ascetics.[110]

In the political landscape that emerges from Kautilya's text, despite its theory of undiluted sovereign authority, many cities and territories within the empire seem to enjoy a certain residuum of autonomy. The *Arthashastra* mentions a number of different *saṃgha*s, a term used loosely to describe tribes and nations. Thus the frontier territories of the people inhabiting the northwestern borders of the Maurya Empire, such as the Yona, Kamboja, Gandhara, or Pitinikana, could not be taxed. The text is wary of such volatile political groups because it is preoccupied with the perceived threat of *vyāsana*.[111] This word is difficult to translate because it has a range of meanings—"addiction," "passion," "vice," "misfortune," "ruin," and "disaster," owing principally to the fault of the character and distemper of the sovereign. Kautilya uses it in a sense close to the Latin *infortunium*, natural disasters or calamities caused by human error or evil, as in the case of fire and arson having the same result. Among such *vyāsana* are external enemies, succession struggles, rebellious tribes, and highway brigands. There is frequent mention of jungle tribes, fractious and defiant, who behave like independent kings, rising up in periodic rebellion.[112] Years later, Chandragupta's grandson, Emperor Asoka, addressed similar concerns—although in a different style—in some of his key edicts. The autonomy of smaller states, especially along the frontier, as well as the forest people far from cities and villages, seemed to have been equally important to the peace and security of the empire.

The *Arthashastra* has a clear and detailed view of the agrarian heartland of the empire, which it considers the foundation of the king's treasure. Imperial land must be productive. There should be no swamps, rocky ground, saline soil, desert, or rugged terrain where unruly or seditious groups can hide from the troops. Kautilya asks the king not to distribute land held in common to nobles, because the unchecked power of the higher-born was inimical to industry. An idyllic vision of the land along the valleys of the Yamuna and Ganges emerges in this account: a landscape of arable land, with timber, forests full of elephants, and pastures teeming with cattle. Care must be taken to secure pastures from wild animals and prevent overgrazing. Domestic animals should be cared for in a humane way. Fines are stipulated for dairies that do not allow cows to nurse their calves.[113] Agricultural fields must not depend only on rainwater; there should be tanks for rainwater and wells. Fields and towns should also be connected by good roads and waterways, so that manufactures from the country can flow to markets and cities to support a high level of taxation. The text also warns about the destructive effect of floods on rural life. Kautilya prescribes that when rivers are in spate, the villagers who live near the banks must be moved to higher ground; they should be ready with planks, bamboo, and boats. Dried gourds, skin bags, trunks, canoes, boats, and ropes should be used to rescue people who are in danger of being washed away. According to Kautilya, although floods and forest fires were bad, droughts were worse, for they destroyed livelihoods. To propitiate the elements he prescribes that offerings be given to the Ganges and to the old rain god Varuna. He also gives precise steps to be taken during famines: public distribution of seeds and food to relieve the afflicted, inspection and repair of irrigation canals, erection of royal stock houses for grains, and, if need be, provisions for the physical evacuation of the suffering population.

This view of the supervision of the king's land reveals the depth of the Maurya administration in the Indian countryside. The title for the governor of a province, *samāharta*, derived from the term *āharaṇa* (Sanskrit root *hṛi*), which means "bringing together" or "collecting." The *samāharta*, aided by the *sthānika*s and local *gopa*s, were entrusted with collecting the emperor's share of the produce, known as the *bhāga*, which was one sixth of the total produce. Megasthenes says that the emperor collected even more, almost one fourth of the produce in places, with additional tributes. Villages were administered by *grāmabhojaka* (also *grāmabhṛtaka*), appellations that are connected to verbs for consumption, enjoyment, and service, which emphasize the value placed on the usufruct of the soil in the countryside.[114] Comparing these details, it becomes clear that the central inspectors, *āyukta*s, worked closely with village heads and elders, which also means that the local *gopa*s and *sthānika*s were men on the spot. Peasants were valued in the kingdom. We hear of the high status of village elders and a certain degree of leniency and restraint in the gathering of urban and rural taxes, which varied according to the fertility of the soil.[115] Villages were typically left alone during times of war and conflict. Megasthenes famously saw tillers going about their daily toil unperturbed by heated battles taking place nearby and the clash of arms within their earshot.[116]

Such instances of paternalism show how closely aligned land and labor had become to the inner workings of the state. Megasthenes mentions the presence of the *agronomoi* and the close control of cultivated and irrigated land.[117] There were superintendents of rivers, measurers of land, inspectors of canals and sluices, and forest guards to keep tabs on the hunters, gatherers, and poachers.[118] Officers known as *rājukas* were appointed to measure the amount of taxes to be collected. Although the word appears to have been derived from the word *rāja*, a more plausible explanation might be that they knew the art of holding the rope (*rajju*) to measure the land, hence *rājuka*, a variation of *rajjuka*.[119] Pastoralists and hunters were expected to pay tribute in cattle and kind. Artisans and traders paid their taxes in money. Soldiers were paid well enough so that they could afford keepers and trainers for their horses and elephants and camp attendants, and leaders in the army could keep an active and effective supply of spies.

Reading the fragments from Megasthenes's *Indika* alongside the *Arthashastra* reveals a vibrant and dynamic image of the Maurya Empire centered on the Ganges plains, whose outer flanks are marked by mountain passes, caravan trails, and littoral entrepôts, connected to roads leading to markets, towns, and cities, with shelters for traders, travelers, and pilgrims at regular intervals. The *Arthashastra*, which places a good deal of value on the importance of financial solvency in affairs of the state, also attends to the relative advantages of different types of trade routes and provides lists of imports borne along these routes across the Ganges valley, with precious goods such as elephants, horses, perfumes, skins, tusks, gold, and silver.[120] It speaks of commerce flourishing across the Himalayan frontier as far north as Gandhara.[121] These are also networks of currency and credit. Remnants of coins, *kārṣāpaṇa*s of silver or copper, along with gold *niṣka* and *suvarṇa*, indicate the intensity of this long-distance trade. The lists of fines in smaller currency (*paṇas*) for acts of civic negligence indicate the reach of monetary circulation at the time.[122]

Much of this early administrative order remained relatively unchanged for the reign of Bindusara, known as Amitraghata, the "Destroyer of Enemies," and his illustrious son Asoka. Bindusara extended the limits of his kingdom farther south across the natural barrier of the Vindhya Mountains. It is said that he sent many envoys to the Indo-Grecian courts and became fond of Greek wines, figs, and philosophy.[123] According to the Buddhist account of Taranatha, he humbled the rulers of sixteen major city-states.[124] He was also a great patron of the ascetic and rationalist philosophers, the Ajivikas, whose early doctrines had much to contribute to the materialism and logic of core Theravada philosophy. This set the stage in many ways for the widespread adoption of Buddhist ethical teachings by his son Asoka, who emerged as one of the most powerful and memorable emperors and brought a large part of the subcontinent under the banner of his sprawling empire.

Much has been written about the conversion of Asoka to the path of nonviolence and the tolerant, ethical stance he adopted after the end of his bloodiest war, which resulted in the conquest of Kalinga from 262 to 261 B.C.E. Asoka spent the latter part of his reign addressing his subjects with edicts engraved on rocks and pillars conveying words of amity and peace alongside public injunctions. Asoka's edicts have been found as far west as the Girnar Hills on the Kathiawar Peninsula in Gujarat to Sopara near present-day Mumbai, and from Raichaur and Chitaldurg on the southern Indian plateau to Dhauli and Jaugada as far east as coastal Odisha.[125]

Asoka was able to portray a larger-than-life image of his sovereignty. He seems to have had a keen appreciation of the power of royal pronouncements set in the common language of his lay subjects. About thirty-five inscriptions have been found to date written in the languages of the region where they were carved: Aramaic, Greek, and Prakrit. Asoka's administration was in good order, his empire had unbounded access to agricultural surplus and mineral wealth, and a lion's share of the crop-bearing land was directly or indirectly in the hands of the state.[126] The bulk of this revenue surplus came from the great alluvial plain of the Ganges, on the lower reaches of which stood the fabled city of Pataliputra at the meeting point of four other rivers, including the Son and the Punpun. During this time the Maurya Empire extended into the great deltaic portions of Bengal, laying claim to Samatata and the great overseas trading outpost of Tamralipti. While most of his rock-face inscriptions are located at the frontiers of his empire, a succinct idea of the core of his territorial power can be gleaned from the placement of the five major pillar inscriptions: the Delhi-Topra inscription from the area of present-day Haryana; the Delhi-Mirath inscription from present-day Meerut; three inscriptions from Champaran Bihar (Lauriya-Araraj, Lauriya-Nandangarh, and Rampurva); and the Allahabad-Kosam inscription, originally from the territory of Kaushambi.[127] A map of the location of these pillars would roughly correspond to the layout of the Ganges floodplains. This is not an accident. Parsing some of the details of Asoka's pronouncements, one gets a clear sense of this landscape of riverside cities and towns, the extension of land under tillage, and large swaths of forested tract still inhabited by hunting and gathering tribes. Asoka says in Pillar Edict V:

> Twenty-six years after my coronation I have declared that the following animals were not to be killed. Parrots, mynas, *aruṇas*, ruddy geese, wild geese, the *nāndīmukha*, pigeons, bats, queen ants, terrapins, boneless fish, the *vedaveyaka*, the *pupuṭa* [dolphins] of the Ganges, skates, tortoises, and porcupines, squirrels, twelve-antler stags, bulls which have been set free, household animals and vermin, rhinoceroses, white pigeons, domestic pigeons, and all quadrupeds which are not useful or edible.[128]

In the same edict, Asoka declares that forests "must not be burned without a reason or in order to kill living creatures," and on fifty-six days of the year creatures living in "elephant forests or the fishermen's preserves" must not be killed. A straightforward explanation of these proclamations may be sought in a keen sense of overarching paternalism in which all subjects and all manner of living things were placed under the watchful care of the emperor. Asoka frequently addresses his subjects as "progeny," which in Prakrit is somewhat synonymous with "subject" (*sava munise me pajā*). On the other

Entrance to the Lomas Rishi Cave, Barabar Hills, Bihar. © The British Library Board 11/14/17. Photograph, Shelfmark: Photo 1003/(44b).

hand, this is also the vision of what a kingdom ought to be: not simply a constitution of human subjects but a custodial domain dedicated to the preservation of age-old ways of life and sustenance (*porāṇā pakiti*) such as "elephant forests," fishermen, and fish, a relationship that we may usefully translate as symbiotic and ecological. It is thus not a surprise that Asoka sought to dispatch emergency help for the welfare of animals and medical aid to various corners of the kingdom in the same breath.[129] In such obligations expressed in official decrees, there seems to have been some common ground between the idea of an eminent imperial domain articulated in the *Arthashastra* (*cakaravartīkṣetra*) and the Buddhist idea of the wheel-turning, righteous, and compassionate monarch.[130]

The geography and territorial reach of Asoka's extensive kingdom is also evident in the contemporary understanding and practice of Buddhist pilgrimage. In the fifteenth year of his reign, Asoka paid public homage by erecting a memorial pillar to one of the Buddhas known as Kanakamuni, who had preceded Siddhartha Gautama, in what today would be the village of Nigalisagar in southern Nepal. Legend has it that this was the actual site of Buddha's *parinirvāna,* as was later testified in the account of the Chinese pilgrim Xuanzang.[131] Some sources suggest that Asoka was taken by his preceptor Upagupta to Buddha's birthplace at Lumbini and to Bodhimula Cave near Rajgir in the Barabar Hills. An expansive network of Buddhist monasteries, shelters, caves, groves, and grottoes was coming alive across the Ganges basin in this period, supported by a rising volume of trade and commercial exchange. The Maurya Empire laid the foundation for one of the oldest and most vibrant material cultures throughout the Ganges valley that lasted for centuries after Asoka's descendants had all but faded from historical memory.

GUARDIANS OF THE MIDDLE COUNTRY

A few oval-shaped intaglio lapis lazuli seals have been recovered from the region known as Gandhara near present-day Kandahar, Afghanistan, which were highly treasured as talismans around five hundred years after the passing of the Buddha.[1] At the center of one of these seals is a carefully carved impression of the footprints of the master (*buddhapada*). Footprints of the Buddha were often carved into stone or wood as objects of devotion in northern India and across the trans-Himalayan tracts because the early monastic orders did not support the adoration of graven images for a long time and clearly not before the end of the first millennium C.E. The *buddhapada*, along with other relics of the living Buddha, was the focus of many of the cults associated with the Buddha that had spread across the Ganges plains and beyond. Such signs in contemporary parlance would have been called *pāribhogika*s, literally "sites of joy," which meant places that were touched by the people and objects dating back to the time of the master and his first disciples. Purported remains of the Buddha's body had become a kind of reliquary currency across the subcontinent, commonly referred to as *uddeśaka*, meant to illustrate his tangible traces even after his nirvana.

Footprint of the Buddha from the Pala period, ca. eleventh century C.E., Bodh Gaya, Bihar. Courtesy Indian Museum, Kolkata.

Buddhism spread, flourished, and recombined into various cults and subgroups during the period between the end of the Mauryas and the rise of the Gupta Empire in Magadha. A plethora of new memorials and relics spread across northern India, far exceeding the number of pilgrimages reputed to have retained the original physical remains of the Buddha (*śārīrika*) derived from his actual body (*śarīra*). Scholars such as Susan and John Huntington have pointed out the enormous importance of such sites and objects associated with pilgrimage in early Buddhism as an outward expression of popular devotional practice.[2] The journey of every pilgrim was inspired by the idea of the Buddha as a wandering, traveling, and teaching mendicant (*bhikkhu*). The faithful longed to follow in his footsteps, retracing the map of his original spiritual journey toward enlightenment. A proliferation of sites claimed possession of the physical remains of the Buddha: bits of clothing, fragments of his begging bowl, or the impression of the soles of his feet. Such items were part of an extended landscape of remembrances around which networks of travel and pilgrimage had sprung up in the Ganges valley.[3] The great stupas of this period and their elaborate narrative panels donated by families, guilds, villages, and towns, as John Walters has suggested, are testament to this great surge of popular devotion and the enduring appeal of the stories of Buddha's life and his previous births.[4] Stupas were also the manifest embodiment of the Buddha, sanctified by his relics and those of his principal followers—visible reminders of his physical presence (*rūpakāyā*), almost as important as his doctrine and teachings. These narratives and emblems, especially during the post-Asoka period, helped forge a link between the present and the hallowed past, situating the stupa in what Walters describes as a "cross-section of cosmic time and space."[5] This chapter traces how Buddhism in its many incarnations—ascetic, monastic, and popular—inherited some of the oldest animist and spirit traditions around the veneration of trees, stones, and totems that had long flourished among people of the middle and the lower Ganges valley. Many such sites became part of an extended reliquary of the Buddha and his disciples.

Embedded throughout the vast body of accumulated legends, including stories of the Buddha's previous lifetimes (*jātaka*s) and those surrounding his birth, enlightenment, and nirvana, are rudiments derived from archaic practices of animist worship dedicated to natural elements such as water, trees, and serpents. Many of these stories are set in an idyllic city, most often alluding to Varanasi, and a river much like the Ganges, surrounded by lush forests. One such story is recorded in the *Mahaparinibbana Sutta* of the *Digha Nikaya* (Long Discourses) written in Pali about the miracle of the Buddha crossing the Ganges during the time of the first kingdoms. Ministers of the principality of Magadha, Sunidha and Vassakara, who were at the time overseeing the defenses of the city of Pataligrama against a perceived threat from the Vajji confederacy, decided to follow the Buddha, who was traveling north. They wanted to name the gate through which the Buddha departed the city after the master and also the place where he crossed the Ganges. At the time the Ganges was in full flow, brimming with water, so that even crows could dip their beaks and drink from the surface. The people accompanying the Buddha went in search of boats, and others went in search of rafts. The Buddha did not wait for them. Just like a human arm being extended, he appeared in the blink of an eye on the other side of the river. From there he observed people trying to get their boats ready and struggling to tie their rafts, and, reflecting on the moment, he said that in life, people who were mindful and mentally prepared to cross to the other shore were the ones who had been truly blessed with inner wisdom.[6]

The Ganges and its banks appear time and again in stories about Buddha's journey from this life to the next, or in the case of the previous births of the bodhisattva, from one mortal incarnation to the next. The *Mahavastu* compilations, dating back roughly to the second century B.C.E., narrate the story of the four bridges made out of boats built across the Ganges for the

Buddha's crossing: the first dedicated by King Bimbisara of Magadha, the second built by the people living inside the city precincts of Visala, the third made by people living outside the city of Visala, and the fourth put together by the serpent-kings residing in the river, Kambala and Asvatara. The story goes that the Buddha, looking at all these bridges and unwilling to disappoint any of his devotees, performed the miracle of crossing all the bridges at the same time. In recognition of this deed, King Bimbisara made a gift to the Buddha of five hundred parasols, and so did all the people of Vaishali. The *nāga* kings Kambala and Asvatara did the same. To the amazement of all witnesses, the Buddha seemed to appear simultaneously under each and every parasol.[7]

Not just the major reliquaries but many secondary sites and memorials associated with the life and legend of the Buddha were frequented by pilgrims, travelers, and members of the laity. A testimony to this phenomenon are the elaborate and exquisite carvings and sculptures, only a few of which have survived from the great Buddhist monuments dating back to the Maurya period. One of these sites is the great stupa standing on the hill of Sanchi near the ancient city of Vidisha at the meeting of the Bes and Betwa Rivers, straddling the two most important trade routes connecting central India to the Ganges valley. Made out of brick and polished sandstone quarried from the Kaimur Hills in the Chunar district of Bihar, the structure as it stands was largely completed during the period of the Shunga Dynasty in northern India, ca. 100 B.C.E., although gates and buildings were added to the main site as late as the ninth century C.E. On the gateways and the railings of the Sanchi stupa are some of the most elaborate depictions of the many lives of the Buddha, along with natural and celestial icons from the *jātaka* tales. Among these motifs are guardian spirits (*yakṣa*s and *yakṣī*s), centaurs, heavenly musicians and dancers (*apsarā*s), serpent-kings, serpents (*nāga*s) and *makara*s (water creatures in the shape of crocodiles or dolphins), ducks, cranes, parrots, peacocks and peahens, winged lions, elephants, tortoises, rhinos, humped bulls, antelopes, squirrels, and griffins.[8] There are also a number of distinctive vegetal motifs such as garlands, creepers, grapes, custard apples, and lotus flowers.

The artisans and craftsmen of the Maurya and Shunga periods were known for their intricate wood and ivory carvings, and specimens of their fine work are exhibited in the stone portals of Sanchi. Typical examples are the carvings depicting the "Lotus Tree of Life" most commonly associated with the presence of the Buddha, with elaborate stalks, buds, and rosaces.

On the south pillar of the western gate are scenes from the *Mahakapi Jataka*, which is the story of the bodhisattva born as a monkey living in the forests around the banks of the River Ganges, who saved his people from the evil invaders of Varanasi by suspending his body between two giant trees across the river, creating a living bridge.[9] Similar impressions have been found in the Bharhut stupa, originally discovered as ruins near Jabbalpur in central India by the veteran British archaeologist Alexander Cunningham during the 1870s. The impressions are now on display at the Indian Museum in Kolkata, dated to the first quarter of the second century B.C.E. during the Shunga Dynasty and later.[10] During Cunningham's time these were the oldest available illustrations of the Buddhist Jataka narratives that had been brought to light. The carvings of Sanchi, Bharhut, or Amaravati in southern India, taken together, give us a remarkably intimate portrait of the imaginative life of northern Indian Buddhism, along with its view of the River Ganges and its wooded banks. Through the graven muse of master craftsmen, an entire landscape comes alive in carved stone. These carvings give us an idea of the geographical image of the valley among common followers, almost five hundred years after the time of Buddha and his first disciples. In many ways, the verdant plains of the Ganges with its flora and fauna provide an ancient backdrop for the reliquaries, pilgrimages, and monasteries developed over the centuries, and an inspiration for the early naturalism evident in a unique vocabulary of visual icons that we see in contemporary Jain and Buddhist art and architecture. Along with actual historical figures of the Buddha and his disciples, or the great Jain emissaries who preceded Mahavira, we also find protohuman creatures and mythical denizens in a mélange of myth and history—reminders of primeval forms of nature worship.

Ancient Guardians

A large number of terra-cotta figures in the Allahabad Museum are guardian deities conceived during the Maurya Empire and after. They were collected over many years from various parts of the northern Indian plains, from Mathura west of the Ganges-Yamuna Doab to Tamluk in eastern, deltaic Bengal. Most of the artifacts can be traced to urban centers such as Mathura, Rajghat near Varanasi, and Sankisa.[11] A few are large and imposing, but most are smaller, orotund, crouching, and deliberately grotesque figures of dwarfs. These are various representative forms of the *yakṣa*, typi-

Yaksa figure from the Bharhut stupa, ca. 300–200 B.C.E., Indian Museum, Kolkata. Photograph by author.

cally tree spirits, who were also regarded as the custodians of sacred places. Although associated largely with trees and water, they were venerated in many other settings: mountains, lakes, rivers, caves, tanks, cremation grounds, homes, and even city gates. The figure of the *yakṣa* has been part of the extended pantheon of northern Indian culture since Vedic times. Its origin, as Ananda Coomaraswamy pointed out in his eminent treatise on the subject, is difficult to pinpoint because a number of minor deities and guardians, especially in Buddhist literature, are seen as kindred spirits that once inhabited the world in the guise of humans that might one day rise again in human form. Even during the period in which Buddhism spread and flourished in India, Coomaraswamy noted, *yakṣa*s were treated with both fear and respect, as entities "wonderful, mysterious, supernatural, unknown," endowed with magical powers.[12]

The railings and gateways of Bharhut show how widespread the *yakṣa* icon had become between the third century B.C.E., when the structure was first commissioned, and the second century B.C.E., when the last friezes and sculptures were added. The guardian spirits of Bharhut were venerated by pilgrims who visited the stupa to offer prayers and perform the ceremonial walk around the structure. Cunningham described them as the "guardian demigods" of Bharhut: Kuvera, the king of the *yakṣa*s, represented as the defender of the great northern quarter, and the great Yaksa Virudhaka, the sentinel of the southern quarter.[13] Many sites blessed by the presence and teachings of the Buddha were also the sites of *yakṣa* worship. In one of his last discourses presented to the residents of the city of Vaishali, the Buddha is supposed to have confirmed that the city and its people would prosper as long as they honored, respected, and made regular offerings to all the major *yakṣa* shrines nearby. It is also said that the Buddha directed the setting up of particular stupas in these auspicious places guarded by these primordial entities. A story is told about another vengeful demigod, Yaksa Gardabha, whose wrath fell periodically upon the inhabitants of the city of Mathura.[14] They turned to the Buddha for help. The Buddha was able to restore peace to the community by requesting that a monastery be built in the honor of the *yakṣa*. Over time, this became a popular spot for worship and offerings, and it paved the way for the rise of a multitude of *yakṣa* memorials in the region. While this story may be seen as largely apocryphal, it shows a direct association between these early Buddhist monastic complexes and sites of *yakṣa* worship. It gives an example of how smaller local communities found a place within the emergent landscape of Buddhist piety and votive practices. By the same token, while local guardian deities were made a part of the monastic complex, they were also placed at the margins of the larger confraternity of *bhikkhu*s and *arhat*s.[15] After the decline of the Maurya Empire, as some Buddhist texts suggest, many cities were assigned to specific *yakṣa*s and *yakṣa*-cults.[16]

A similar account of the rise and proliferation of *yakṣa* worship can be traced in early Jain texts. *Yakṣa* images are also found in latter-day Jain temples, and we can infer that they are derived from similar sources that inspired the Buddhist Theravada worship of guardian spirits. They suggest that these cults once reigned across the Ganges valley during the rise of the great kingdoms. Sacred natural sites (*caitya*s and *devāyatana*s) dedicated to the *yakṣa*s Purnabhadra and Manibhadra are some of the examples of older cults that survived within the early Jain and Buddhist traditions.[17] Many caves and tumuli that later became Jain pilgrim sites were once sites of *yakṣa* worship. There is little evidence to suggest that Jain liturgy actively sought to assimilate such cults within its teachings. However, guardian deities appear routinely in Jain temples dating to the sixth century C.E. and later.

Alongside male *yakṣa*s, there were also powerful female *yakṣī* figures. The Buddha was known to have honored the great *yakṣinī* Hariti, who was famous for her supernatural powers and her role as the guardian of children. The text *Mulasarvastivada Vinaya*, which recent scholars date to the pre-Kushan era (ca. fourth to fifth century C.E.) speaks of an important patron guardian of Rajagriha, Abhirati. According to legend, she was once a simple herdswoman from the city of Rajagriha.[18] Once when she was pregnant, she had to travel to the marketplace to sell buttermilk, where a drunken celebration was going on, and she was forced to dance and suffered a miscarriage. Cast into a world of grief from the loss of her unborn child, she had the good fortune to meet one of the early solitary Buddhas, who helped her in her quest to be born again, this time as the mother of five hundred children. In this incarnation, she took the name of Hariti and wielded power over the lives of all humans, especially children.

Alice Getty, in her detailed study of the evolution of Buddhist iconography in northern India, recounts a slightly different version of the same

Hariti and Panchika, Kushan period, ca. second century C.E., Jamalgarhi, northern Pakistan. Courtesy Indian Museum, Kolkata.

story, told in the Buddhist text *Samyuktavastu*.[19] When the Buddha dwelled in the *vihāra* of the bamboo grove in Rajagriha, there was a *yakṣa* called Satagiri who was known as the protector of the region. His sister Abhirati was betrothed to the son of another great protector of the Gandhara region, Yaksa Pancala. One day Abhirati divulged that she was overcome with the desire to kill and devour the children of Rajagriha, a confession for which she was reprimanded by her brother. Alarmed by her state of mind, Satagiri hastened to marry her to Pancika, Pancala's son. Their marriage was consummated, and soon she became the mother of five hundred children. And yet, children were still disappearing in Rajagriha, where aggrieved and desperate parents gathered and appealed to their king. Having heard this alarming news, the king decided to throw a great ritual feast and make offerings to the guardian Yaksa of Rajagriha. After this propitiation, Yaksa Pancala began to appear in people's dreams, alerting them to who the culprit was and telling them that they must go to the Buddha and ask for his help. From that day on Abhirati became known as the infamous Hariti, the terrible "seizer of children."

The people of Rajagriha banded together and traveled to the *vihāra* where the Buddha was residing. Prostrating themselves in front of the great teacher, they begged him to stop the Yaksini Hariti from stealing and killing their newborns. The Buddha listened silently. The next morning he went to the city as usual with his begging bowl, but afterward he paid a visit to Hariti's house. The she-demon was on the prowl and not to be found, but her son, Priyankara, was there, and the Buddha cleverly hid him in his begging bowl. When Hariti returned she found her son missing. Beside herself with worry and grief, frantically searching for her son everywhere, Hariti sought out the greatest *yakṣa* of all, Kuvera (here named as Vaisravana)—the lord of wealth—for his help. Kuvera asked her to put her faith in the Buddha. When she came back to her house she found the Buddha waiting for her there, bathed in a splendid, unearthly light. Her heart filled with joy. She fell to the floor in front of the Buddha and begged to see her son. The Buddha made her promise that she would follow his precepts and help guard the good people of Rajagriha. She and her children also promised to abjure human flesh and agreed to take food from the monastery instead. Other versions of the story suggest that to teach her a lesson, the Buddha took all of her children to the monastery. Searching for her children, she became frantic and gave up food. Finally, the Buddha told her that the people of Rajagraha loved their children just as much as she loved her own. After she saw the light and submitted to the Buddha, she and her progeny took refuge in the Samgha.

A significant detail in this narrative is this eventual permission given to Hariti and her children to enter the sanctum of the monastery. Hariti was endowed with unusual and terrifying powers, and only after the Buddha's intervention were these powers redirected toward the good and harnessed to protect monks, followers, and portals of monasteries. Coomaraswamy found this aspect of the legend "more like an explanation or justification of a cult than a true account of its origin . . . an edifying sanction for an ancient animistic cult too strong to be subverted."[20] The worship of Hariti, elevated to the status of queen *yakṣinī*, may have continued in the form of a virtually independent goddess cult within the everyday practices of monastic Buddhism. Otherwise we would not hear stories of how the Buddha laid down the rules for apportioning the food offered to her at every monastic gathering. Through these practices, lay devotees brought bits and pieces of their traditional beliefs into the Buddhist fold, raising Hariti to the position of a mother goddess surrounded by children.

Hariti is still worshipped in parts of northern India and Nepal by parents who have lost their children in childbirth or from diseases such as smallpox. Some commentators trace this and similar figures to an older, possibly Vedic deity known as Jataharini. The *Kashyapa Samhita*, dated to the seventh century C.E., enumerates at least fifty ancient "child-snatchers," or *jātahāriṇī*s, who were feared as malignant spirits that invade the body of pregnant women, often choosing between different castes and types, and destroy their fetuses. They are derived from the goddess of opulence, Revati, who in Puranic myths once fought the original demons during the creation of the world but could not exterminate them entirely, precisely because they began to assume the form of humans. In ancient Indian medical texts, *jātahāriṇī* is simply a female icon of miscarriage and infant mortality.[21] The widespread and varied nature of such local cults and their place in Buddhist legend and iconography is testimony to the groundswell of popular Buddhism across the cities and villages of northern India. Hariti is not only imagined as an *upāsaka*, an ardent worshipper of the Buddha, but a new deity in her own right, with shrines dedicated far beyond the city of Rajagriha, at many major Buddhist monasteries located along the major routes of pilgrimage and trade.[22] The Chinese pilgrim Xuanzang saw altars dedicated to Hariti in every monastery, her image placed in a statue or painted near the doors or on porches typically leading to the refectory. She was depicted as seated or standing with a child at her breast or on her hip, holding up the fruit of the pomegranate as a sign of fecundity, a reminder that she was once blessed by the Buddha, who cured her bloodlust by offering her the red pomegranate kernels instead of actual human flesh and blood.

Male *yakṣa* figures were widely accepted across northern India during this time, evident from the abundance of stout, grimacing dwarfs on the top of pillars or crouching at their bases. They can be seen in the remains of the Bharhut stupa or on the railings of Sanchi, and they adorn the architectural ruins of the oldest urban settlements of northern India in Junagarh, Kaushambi, Mathura, Bhita, Tamralipti, and Chandraketugarh. It seems that the popularity of *yakṣa* cults grew rapidly with the spread of Buddhism during the Gupta and Shunga periods. In this manner new life was breathed into *yakṣa*s, some of the oldest surviving icons in India, associated with tree worship, fertility, and fecundity, dating back to Vedic or even earlier times.[23] One of the most familiar Yaksa figures from the period of early Buddhism in India is the valiant *yakṣa* Vajrapani, who later became one of the most beloved bodhisattvas among the followers of later Mahayana Buddhism.

Yaksa motif from the Bharhut stupa, ca. 300–200 B.C.E., Indian Museum, Kolkata. Photograph by author.

Popular Buddhism absorbed within its fold a wide array of fertility cults centered on trees and plants. Female *yakṣīs* in both the Jain and Buddhist traditions typically represent the essence or the vital fluids of a tree. *Yakṣas* also are represented as part of plants, trees, and flowers. A motif of the Bodhi Tree supported by a *yakṣa* holding a lotus in his hands became a common subject of sculpture for the decorative facades of the stupas.[24] Similar vegetal motifs appear in early medieval sculpture, such as those produced by the Mathura School of the Gupta period, with long stalks of the lotus plant bursting forth from mouth and navel, a symbol of plenitude and the elemental force of all creation.[25] They are typically presented alongside overflowing pots (*pūrṇakumbha*) and the *śrīvatsa*—a figural, nonhuman representation of the lotus goddess associated with lakes, rivers, and oceans. These are signs of good fortune, abundance, and perpetuity. The carved pillars and medallions from the Bharhut ruins dating to the Maurya and Shunga periods carry earlier versions of these female fertility figures emerging from either the lotus plant or lushly fruited trees.[26] Placed under them are flowering plants disgorged in profusion from the mouth or the navel of squatting dwarfs.[27] Female gatekeeper figures are often shown accompanied by auspicious vessels of plenty (*pūrṇaghaṭa*), which in later times would become directly associated with the veneration of the goddesses representing fecundity and good fortune, such as Shri and Lakshmi. They stand witness to the natural and effortless ways in which such feminine cult figures, like tree maidens, or *vṛkṣikās*, or tree-embracing *dohada*s desirous of child—figures tied to the rituals and ceremonies of pregnancy—were included in everyday aspects of popular Buddhist devotional practices dating back to antiquity.[28]

Serpent Cults

*Nāga*s, which are serpentine, aquatic figures in both reptilian and human forms, feature prominently in the graven representations of the Buddha's life stories. The Mahayana Buddhist *Lalitavistara Sutra*, which is a compilation of older sources, records the episode of the birth of the Buddha to Queen Maya in Lumbini, where the two great snakes Nanda and Upananda, serpents from the waist down, suspended in midair, bathed the newborn in streams of hot and cold water. It is said that centuries later, in remembrance of this event, Emperor Asoka built a commemorative stupa at the very spot.[29] The same two serpent-spirits are said to have stood witness to the miracle that took place in the city of Shravasti, where an image of the

Buddha appeared sitting on a golden lotus, replicated multiple times in the same instant.[30] It is said that the primordial serpent Kalika became aware of his imminent enlightenment from the sound of Buddha's footsteps. It is also said that at the time of his spiritual rebirth, after undertaking a very long fast, the Buddha visited the legendary river of the serpents, the Nairanjana, where a *nāga* daughter brought him a bejeweled throne. Lord Vishnu's mascot, the eagle-headed and winged Garuda, a natural enemy of the snakes, swooped down to seize it but did not succeed. Garuda wanted to fly back with it to the heavens as a gift to Indra, the king of gods, who had ordered a festival in heaven to celebrate the acquisition of the subterranean and exquisite Naga throne.[31] Denied his prize, Garuda had to beg for it. This legend is carved on the pillars of Amaravati, where Nagaraja, the king of serpents, is shown with a seven-headed hood next to the Buddha's feet. The Buddha is shown as crossing the fabled river of Nairanjana after his realization of ultimate wisdom under the Great Bodhi Tree. It is an auspicious moment. Birds fly in a circle above his head. This river is most likely the river today known as Phalgu, which runs through the district of Gaya, and is still one of the most frequented pilgrimage spots for both Hindus and Buddhists in India.

Not all serpents that appear in these stories are good or auspicious. After his first sermon at the deer park near Varanasi, the Buddha traveled to a place called Uruvilva. People warned him that an extremely venomous and savage snake resided there, who also knew black magic.[32] The Buddha proceeded to test its power, agreeing to stay in a hut where this creature was seen lurking. The hut was soon set on fire by the serpent, but the Buddha was able to repulse the flames spewed by the snake with his own fire. In the end the malevolent reptile was tamed and lay in harmless coils in the Buddha's begging bowl. It is also said that this was the test of strength that finally convinced the brothers Kasyapa of Uruvilva that the Buddha possessed certain magical powers. This scene too is represented in a bas-relief on the eastern gateway of the Sanchi stupa. There is no representation of the Buddha, but his seat is clearly visible, as well as the fire and the three Kasyapa brothers with matted hair, wearing the tree-bark loincloth typical of reclusive disciples.

In an early landmark essay, Lowell Bloss noted the connections between the oldest Buddhist *cetiyas* (*caitya*, Sanskrit)—built around relics, memorials, and trees—and popular animist cults that predated them. This is one of the reasons why deities of the forest or water are so prominently featured in the Jataka tales. The stories of the life cycle of the Buddha thus record some of the oldest forms of folk religion in India.[33] The same people who were attracted to the cult of the Buddha, especially the larger laity outside the formal Theravada orders, continued to seek benedictions from the spirits and forces of the darker netherworld, especially those that had been held in reverence for the promise of water, including relief from the devastation of floods and the blessings of rain for fields and crops. In many ways early Buddhism owed its popularity to these older folk traditions. They help explain the miracles attributed to the Buddha, especially the taming of evil *nāga*s who threatened to bring floods and destroy crops, and also the legends of serpents guarding or watching over the Buddha during his peregrinations after the attainment of enlightenment. Sometimes these are actual snakes, sometimes serpent-kings in human or quasi-human form, such as in the legend of the great tree of Muchalinda and its resident serpent-king *nāgarāja*, who protected the Buddha from the wind and rain by coiling around him seven times and spreading his giant hood over his head. The legend describes the Buddha sitting cross-legged under the Muchalinda Tree in uninterrupted meditation for seven days, while a great thundercloud appeared and a storm raged on earth, bringing floods and darkness. Watching this, the serpent-king came out of his abode to protect the Buddha from both the heat and the cold and also from flies, gnats, and other insects. When the storm passed, the serpent-king appeared at the foot of the Buddha in the guise of a young man with hands folded in supplication. The Buddha then said to him that happy were the people who exercised "restraint toward all things that have life."[34]

These stories, marked by the appearance of serpents and serpent-kings, point to the proliferation of Buddhist sites directly associated with older cults dedicated to trees and water bodies such as tanks, wells, streams, and rivers. As the pioneering scholar of Buddhist doctrine T. W. Rhys Davids once pointed out, Buddhism simply added to its list of veneration ancient places of worship mentioned in the epic tradition. Many of these ancient cults reappear under new guise in the Buddhist Jataka tales in the form of trees, stone altars, pools, or streams. Some of these icons appear in the sculptures around Buddhist *caitya*s and stupas, where carved and votive railings were put up specifically to acknowledge such borrowed facets of popular worship.[35] The unknown authors of the *Maha Samaya* [The Great Concourse] exhibit an intimate knowledge of this landscape of cults and relics. They list the most important primal spirits of the planet Earth and the great mountains before the time of the Four Great Kings, who were the

Seated Buddha sheltered by the Naga Muchalinda, Indian Museum, Kolkata. Photograph by author.

early guardians of the four quarters: east, south, west, and north. The first of these was Vessavana (Vaisravana) Kuvera, a deity who watched over all guardian spirits. Next came the Gandharvas, heavenly musicians, minders of pregnancy, birth, and the newborn. Then came the Nagas, serpent sirens, who, propitiated in the right manner, always came to the help of devotees. While they took the shape of cobras in the presence of humans, they

Veneration of the Buddha with Naga serpents in attendance, Bharhut stupa, ca. 300–200 B.C.E., Indian Museum, Kolkata. Photograph by author.

dwelled underwater in hidden, luxuriant realms, guardians of treasures and precious gems. Some Jain accounts state that when the twenty-third emissary Parshvanatha, who preceded the Mahavira, was a young prince living near Varanasi, he saved a pair of snakes hiding in the kindling about to be incinerated in a sacrificial fire.[36] These snakes were reborn as a *yakṣa* and a *yakṣī*: Dharanedra and his consort Padmavati. It is said that of all the great Jain preceptors, sites of worship dedicated to Parshvanatha were the most numerous because of their close association with snakes and other guardian spirits.

Nagas have persisted in folklore and superstitions across India. They are featured in bas-reliefs of early Hindu temples, and even more ancient mounds and tumuli—quasi-human figures with cobra hoods rising from behind their head and scaly coils from their waist downward. At the height of the popularity of Theravada Buddhism in the Ganges valley, these mythic denizens of the aquatic deep were raised to an iconic art form. At Firozpur,

Nagaraja, the serpent king, Bihar, ca. ninth century C.E.
Photograph by author.

less than two miles west of Gulgaon, in the present-day state of Madhya Pradesh in central India, not too far from the Buddhist stupa of Sanchi, is a group of three serpent sculptures: a freestanding *nāgarāja*, a female *nāginī*, and a carved lotiform bell-capital surmounted by two *nāgarāja*s and two *nāginī*s. These have been dated roughly to the early fifth century C.E., judging from their similarity to well-known images of the same period from Udayagiri, Nagaur, Eran, and Mathura. These Gupta-era sculptures show clearly the continued popularity of serpent deities and serpent worship almost a thousand years after the time of the Buddha.

The figures of the *nāga*s and *nāginī*s are reminiscent of the images from nearby Sanchi. Not only do they show a degree of continuity in the popularity of motifs tied to the history of Buddhism, they also show how these images underwent subtle changes following the popularity of Vaishnava devotional practices during the Gupta period. Moreover, the iconography of serpents and serpent-kings became crucial to the legitimacy and lineage of the kings of the Naga clans and oligarchies from the region of Vidisha—especially dynasts who forged political and matrimonial alliances with the imperial Guptas, most notably Chandragupta II (380–415 C.E.), who married the Naga princess Kuberanaga.[37]

Venerable Trees

Many Jain and Buddhist *caitya*s, caves and mounds, started simply as sites of worship of trees or entire groves (*vanacetiya*).[38] Trees were the abodes of traditional gods and wooded spirits (*vṛkṣadevatā*) and anthropomorphic *yakṣa*s.[39] There were smaller hemispherical tumuli resembling later stupas, enclosed areas with railings, pillars, and flag posts (*dhvajastambha*) decorated with animal motifs. The most typical decorations on these votive pillars would have been the eagle-headed Gadura, mascot of Vishnu; the Vrisha bull, mascot of Shiva; and Makara, the dolphinlike snouted water creature, a mascot of the river deity Ganga and other celestial beings such as the Kandarpas. In some cases animals and mascots at the top of the pillars made way for tree motifs, the wishing tree (*kalpadrūma*) or the palm tree with a crowning cluster of leaves. The text *Prayaga Mahatmya*, which is an expanded version of passages originally found in the *Matsya Purana*, mentions three eternal guardians of the sacred pilgrimage of Prayag: Vishnu as the four-limbed idol *venīmādhava* (*venī* as in a braid or confluence), Brahma as the shalmali, or silk cotton tree (*Bombax ceiba*), and Shiva as the *akṣyavaṭa*, or the everlasting banyan tree (*Ficus sp.*).[40] The veneration of such trees is a testament to the persistence of certain forms of animistic devotion stretching far back into Indian antiquity.

As noted earlier, Siddhartha Gautama attained his enlightenment under the great Bodhi Tree near present-day Bodh Gaya by the banks of the River Nairanjana—most likely the contemporary Phalgu—one of the tributaries of the Ganges.[41] At the foot of this great tree he came to realize the chain

Worship of the Bodhi tree representing Krakuchchhanda Buddha, Sunga period, ca. second century B.C.E., Bharhut, Madhya Pradesh. Courtesy Indian Museum, Kolkata.

of causation that sustained the cycle of human consciousness and suffering in this world. In recognition of this fact, early Buddhism identified not only the *bodhidrūma*, or the Bodhi Tree (*Ficus religiosa*)—also known as a peepal or the *aśvatthva* tree in north India—as the tree of enlightenment, but all ancient trees as significant mythical and historical landmarks. There have been many attempts to identify the botanical species of the original tree, which seems to have been recorded differently in different sources. Ideally, every Buddha who descended to earth in different epochs chose his particular tree of wisdom, leading to a further proliferation of trees and tree veneration associated with the spread of Buddhism and Buddhist pilgrimages in India.[42] Thus the *śirīṣa* (*Albizzia lebbeck*); the *udumbara* (*Ficus glomerata*), well known for its medicinal properties and a supposedly rare flower said to bloom only once every three thousand years; and the *nyāgrodha* (*Ficus indica*) are all known as Bodhi trees, *bodhi* being a generic adjective denoting enlightenment. Not just the tree of enlightenment but many other trees also marked the various stages of the Buddha's spiritual quest.

After his enlightenment the Buddha went in search of the Ajapala Tree belonging to nearby goatherds.[43] Here once again, he sat in meditation cross-legged for seven days until he was visited by a Brahmin who challenged him, and the Buddha explained to him what the true qualities of *brāhmaṇa* were supposed to be. From there he journeyed to the Muchalinda Tree mentioned earlier, under which he also spent days in rapt contemplation, oblivious of the natural elements including a fierce storm. This was where the serpent-king protected him by coiling around his body and sheltering his head with his hood.[44] From there he went to the Rajayatana Tree for meditation.[45] Here he received the two merchants Tapussa and Bhallika, who had traveled on the road from faraway Odisha (Ukkala) and from whom he accepted honey and rice as gifts in stone bowls. After that the Buddha went back to the Ajapala banyan tree, where he was visited by none other than the great Brahma Sahampati himself, who reasoned with him not to withdraw forever from human society but to come back as a teacher to enlighten and save the fallen people of Magadha.[46]

The image of a seeker sitting at the foot of a tree is one of the most appealing figures in Buddhist teachings of this early period. With the consolidation of the Buddhist orders in India, as monks turned from alms and itinerancy to monastic life, renewed doubts were cast over the true intent of sheltered or secluded living. One term that retained its place for a long time among the prescribed Buddhist austerities was *vṛkṣamūlika* (literally, the "root of the tree"), living and contemplating under the shelter of a memorable tree. This was a technical term used by anchorites and the practicing *bhikkhu* confraternity. Living in a forest, sitting under a tree, and residing outdoors, without shelter, exposed to the elements, were all parts of the general body of observances under the head of *dhutaṅga* (attributes of scrupulousness) or *dhutaguṇa* (attributes of "shaking off") that were taught by the Buddha himself to his closest disciples such as Mahakasyapa, Pindolabharadvaja, and Upagupta.[47] This is one reason why certain shade-bearing trees appear time and again in Buddhist teachings: aspen, rose-apple, banyan, fig.

Emperor Asoka was led by Sthavira Upagupta, his preceptor, to the Bodhi Tree, where the Buddha vanquished the forces of Mara. In recognition of the secluded spot, the emperor gave a hundred thousand gold coins (*suvarṇas*) for the upkeep of the sacred tree and donated the resources for the building of a memorial *caitya* nearby. Asoka also made similar donations to the places where Prince Siddhartha was born, where he first spoke about the wheel of law, where he attained nirvana, and where he rested during his many travels. But the spot of the Bodhi Tree was the one that touched him most profoundly.[48] It is said that Asoka was so taken with the welfare of the tree that he donated his favorite queen Tisyaraksita's jewels for its upkeep. This aroused much jealousy, because the queen saw the tree as her rival—so much so that she summoned a woman from the *mātaṅga* caste, known for their sorcery, to put a spell on the tree. After the curse took effect, the tree indeed began to shed leaves and dry up. It is said that the emperor fainted when he heard this, and seeing his distress, Tisyaraksita asked the *mātaṅga* woman to take back the curse, and the tree was restored by digging a furrow around the trunk and pouring a thousand vessels of milk around its roots. Over time, this treatment seemed to have restored the fabled tree to its original state.

There should be little doubt that the sight of the Mauryan emperor Asoka led by his preceptor on a journey through the hallowed ground of Buddhist soteriology was a definitive public statement of his devotion. Asoka seems to have been committed to the preservation of old sites and monuments. The Pali canon, as presented in the *Mahavamsa*, states that the emperor not only foreswore to protect the tree of enlightenment but appointed eighteen representatives from his own dynasty and extended family for support. Asoka also chose eight prominent families from his royal entourage: eight reputable Brahmin families, eight representatives of the major guilds, and the chiefs of the eight major clans, including cowherd, hyena, and shrike-tailed sparrow. He also asked weavers, potters, and the mythical guardians of relics, monuments, and treasures: the Yaksas, Yaksis, and Nagas. He donated eight auspicious vessels of gold and silver for the welfare of the tree and the pilgrims. Asoka was also keen to bring a part of this tree as a gift to the people of Sri Lanka, along with the message of Buddhism. Perhaps it was a cutting or a sapling that was destined for the island, carefully brought aboard a ship on the Ganges that would eventually make its way across the delta, down the Bay of Bengal, and across the Palk Strait. The gift accompanied his daughter Princess Sanghamitra and eleven nuns who accompanied her, as well as his son Mahendra. The scene is set evocatively in the *Mahavamsa,* which describes in sweeping terms the significance of the moment as the tree is being prepared for its public send-off.[49] Not only did people line up at the harbor, but *nāga*s and gods also came to watch the tree being lifted onto the ship. The emperor himself strode out into the water until he was immersed up to his neck to make sure that the tree had been placed safely. He requested that his dear friend King Devanampiyatissa of Lanka pay homage to this tree along with all the subjects in his kingdom. The *Mahavamsa* embroiders the episode further, stating that lotuses began to bloom around the ship to the sound of unseen musical instruments. Great serpents of the sea, covetous of this ancient and legendary tree of wisdom, tried unsuccessfully to seize the ship, but because it was blessed, they were not able to harm it. Not only was the tree successfully replanted in Sri Lanka, but if contemporary chroniclers are to be believed, it flourished for centuries at the monastery town of Anuradhapura and was being venerated by King Silakala and King Mahanaga during the sixth century C.E.[50]

A Landscape of Relics

Buddhist travelers in India braved long distances and hardships of the open road. They reenacted parts of Buddha's own journey as a wandering mendicant: his quest for enlightenment, and his extensive travels punctuated by sermons on the road, sometimes in the company of steadfast disciples who accompanied him on foot from city to city, village to village, through the forests and valleys of the Ganges plains. As Victor and Edith Turner once noted, the ultimate appeal of pilgrimages lies in the fact that they appear to transcend the bounds of ordinary human time. They are sites where miracles took place, are still taking place, or might take place again in the near future.[51] The journey is expected to bring supplicants face to face with the possibility of the threshold of something that is believed to be timeless. Buddhist practice in northern India after Siddhartha Gautama's lifetime, along with the three gems and five precepts of the true path, placed great value on the homage to the relics of the Buddha. Buddhism not only assimilated ancient objects and spirits of veneration such as stones, trees, snakes, and guardian spirits, it added a myriad of new places and auspicious dates to a devotee's calendar of planetary conjunctions and phases of the moon.

Individual monks were committed to a life with few possessions: robe, bowl, brush, needle and thread—essential items for a lifetime of wandering.

While the frequency of travel and pilgrimage heightened, so did the number of *caityas*, *vihāras*, shelters, and hospices that received and redistributed significant amounts of wealth.[52] No doubt the level of donations and patronage the various sects and orders received during the period of the late Maurya Empire could not be sustained after the reign of Asoka, but lay patronage continued unabated all across India and rose further after the decline of the Mauryas and under the auspices of the Brahminical Shungas and the steppe empire of the Kushans, whose frontiers touched the westernmost edges of the northern Indian plains.[53] The cult of stupas, tumuli, and memorials led over time to the building and rebuilding of larger and more sumptuous stupas, along with spots marking the worship of Buddhas and bodhisattvas who had alighted on earth before Siddhartha. By the time of the decline of the Maurya Empire they had spread all across India, far beyond the original map of reliquaries dedicated to the clothing, shadow, bowl, or footprints of the historic Buddha.

During the fifth century C.E., when Buddhaghosha, the great scholar of Theravada, was perfecting his commentaries on Buddhist doctrine, the sects had proliferated beyond the original eighteen divisions, such as the Andhakas, Uttarapathakas, or Theravadins, listed in the Sri Lankan Pali texts like the *Dipavamsa*.[54] The history of the development of these later orders, especially the hardening of the different interpretations of doctrine and practice between the first council at Rajagriha after the demise of the Buddha and the second council at Vaishali a hundred years later, still remains sketchy. As Rhys Davids once mused, we can still question how far the word *schism* might be useful to talk about the splintering of Theravada Buddhism practiced in northern India. The split of the eighteen different sects, as noted in the Sri Lankan chronicles, was relatively peaceable and bloodless, what the venerable Chinese pilgrim Xuanzang described as conflicting schools of philosophy dedicated to the same ultimate goal.[55] In Buddhaghosha's commentary on the *Kathavatthu* we can detect the discomfiture of doctrinal controversy among practitioners of the old school, having to do with their adherence to particular interpretations of the nature of self-realization of the Buddha that illuminated the path to fulfillment and the explication of what were known as the basic *niyāmas*—the word *niyāma* translated as "assurance," "fixity," or "certitude."[56]

The proliferation of orders also meant a significant growth in the number of sites that were associated with the veneration not only of the life, journey, and historical traces of the Buddha, but also of his principal disciples and their followers. The public adoration of relics had become a key feature in the spread of early Buddhism in the Ganges valley. Wealthy and powerful patrons eagerly embraced the cult of relics and shrines to express the depth of their piety. The rise of the major cluster of reliquaries from the Sanchi in central India are the earliest known examples of this surge in public interest. The worship of secondary relics spread quickly across the northern Indian plains, where cities and towns had sprung up near along the course of the Ganges, especially in the area of Magadha.

The *Mahaparinibbana Sutta,* explaining the spiral of death and rebirth, records a conversation between the Buddha and his disciple Ananda in which the he tells Ananda that his body should be handled after death much like that of a true Cakravartin, the turner of the wheel.[57] His disciples should do what the wisest among nobles, Brahmins, or heads of houses would normally do. One had no dominion left over one's corporeal remains after death. The disposal of the body was now simply a matter of established custom. They must not worry unnecessarily about the anointment or its preparation for the funeral. Early Buddhism, it seems, attempted to refrain from any gesture that would place an undue emphasis on the veneration of the Buddha's body. The eternal body of his teaching was supposed to be the true legacy of the master. It is difficult, however, to determine whether these sanctions against using the mortal remains of a deceased master as objects of veneration were obeyed in the long run. Many historians of early Buddhism in India have suggested that what the early texts have to say about the place of monks in society and their relationship with the laity is merely prescriptive and not always a reliable indicator of what took place in everyday practice. Gregory Schopen argues that monks and the laity participated jointly in the remembrance and the preservation of the memorabilia of the great master and his early disciples during the early centuries of the institutionalization of Buddhism in India.[58] Evidence from inscriptions related to donors, as well as the extensive archaeology and history of monastic architecture, show that monks participated fully in the cult of stupas and relics as early as 150 B.C.E.[59] Many older *caitya*s had already been assimilated into the monastic complex, and similarly, the stupa was also integrated into the cenobite life, relying on the active participation of monks in the worship of mortuary relics. The belief that a certain trace of the master was alive in his bodily remains preserved in the stupas can be detected in the early texts, and the later Pali canon likely began to edit some of these features because of the knotty existential problems they posed to the Theravada doctrinal order, which was threatened by the rapid growth of popular devotion and the worship of graven objects.

Legend has it that after the Buddha's cremation, his remains were distributed among the eight great kingdoms of the Ganges valley, each portion celebrated by a major stupa. The Mauryan emperor Asoka is said to have exhumed and redistributed these across a multitude of new sites.[60] Asoka's public acts of piety also contributed to the rapid growth of pilgrimage sites. His handsome gifts to the two caves in the Barabar Hills with polished interiors in southern Bihar were early examples of such acts of veneration. After he had presided over the Third Buddhist Council, he undertook extensive travels to all the major sites, escorted by a large army. His stops included Sarnath, near Varanasi, where the Buddha first began to teach his doctrine; Shravasti, where he lived; and the places of his birth and enlightenment. Asoka made huge donations for the upkeep of the stupas of the Buddha's most important disciples, such as Sariputta, Maudgalayana (Prakrit, Moggalana), and Ananda. These new reliquaries incorporated sites of much older worship by *yakṣa*s and *nāga*s, who had now become the official guardians of the relics. A legend recorded in the travel account of Chinese pilgrim Faxian says that when Asoka desired to reopen eight great stupas and redistribute their contents, the last one, Drona stupa, right by the side of the River Ganges at Ramagrama, was guarded by a venerable old *nāga* who, seeing the emperor about to exhume the great master's remains, assumed human shape and guided the emperor to the pool where he lived with the old vessels and ritual objects. The *nāga* said to the emperor: "you can dismantle this site only if you think you can guard and worship the relics better than I have over all these years." The legend states that Asoka decided not to destroy the site, leaving the relics intact. The story that one of the most powerful emperors to have ruled India was actually persuaded by a snake-guardian to let go of the Buddha's remains reveals something about the resilience of popular animism in Gangetic northern India.

Asoka's efforts to regulate the sacred spaces and objects throughout his kingdom show how important the building, dedication, and patronage of stupas had become for monks and the laity alike.[61] The earliest stupas had been simple earthen mounds built to house the relics of Buddha and his main followers.[62] These largely primitive structures would not have survived for many centuries without substantial rebuilding efforts. Nearly all of Buddha's cremated remains were redistributed at the end of the Maurya period. The repair and addition to old sites became an established feature of kingly and elite patronage. Most of these structures were so extensively rebuilt over subsequent centuries that it is often hard to figure out what the original sites might have looked like. Bodh Gaya, the site of Buddha's enlightenment, is a prime example; it saw continuous building activity and resettlement from the third century B.C.E. to the seventh century C.E.—the succession of endowments inspired by its reputation as a site blessed by the living presence of the Buddha.[63] The stupa of Sanchi and the Bhilsa Topes of central India are also examples of both individual and collective efforts at reclamation. Inscriptions found at the apsidal temple of Satdhara dating back to the Maurya period, located about five miles west of Sanchi near old rock shelters and prehistoric platform terraces, reveal a long history of collective donations.[64] Structures continued to be added to this site all the way to the ninth and tenth centuries C.E.

Such rebuilding was further encouraged by the popular idea that the life force, or *prāṇa,* of the blessed Sakyamuni Buddha had survived in these remains and that relics contained his spirit and that of his early disciples such as Hemavata, Mahamogalana, and Sariputta. Sariputta, whose relics are said to be preserved in the urns at the Bilsa Tope in the country of Vidisha, was not only a famous student of the Buddha but one who had vowed to provide monks with food, clothes, shelter, and medicine. These examples also show that Buddhist pilgrimage continued unabated during the period of expansion of the great steppe empires: the Indo-Greeks, the Shakas, and the Kushans. It is generally believed that there was a resurgence of Brahminical establishments in northern India during the rule of the Shungas (ca. 100 to 400 C.E.), although the idea that the founder of the Pushyamitra Dynasty actually persecuted Buddhist monks and destroyed monasteries has been questioned by recent scholars.[65] Even if such a "revival" took place—although historians are not in agreement—it did not necessarily come at the expense of the dominance of the Buddhist orders.[66] Popular forms of Buddhist practice outside the strictures and rules of the monastic order had struck roots into the cultural subsoil of India, and they changed the landscape of both liturgy and piety.[67] The *saṅgha* now represented much more than just the followers and disciples of Buddha who had retreated from society and local communities. Royal patronage and the support of merchants and guilds over many centuries gave the Buddhist orders of northern India a more worldly occupation and a central place in everyday material culture.

Worship at the major stupas was not just meant for the population at large but was required of Buddhist monks as well. At the most important sites—Bharhut, Sanchi, Taxila, and Ratnagiri—there are clusters of small satellite stupas alongside the large ones where human bone ash can be traced in the form of mortuary offerings. Even fragments of sacred texts have been found at Ratnagiri. Such secondary and tertiary layers of relics have also been found in Bodh Gaya, confirming Alexander Cunningham's view that generations of monks sought the privilege of having their corporeal remains placed as close as possible to those of the Buddha and his earlier disciples. Veneration of relics and mortuary practices continued apace in northern India after the breakup of the Maurya Empire. The Shunga king Dhanabhuti Vachiputa donated decorative gates (*toraṇa*) and stone buildings to major stupas.[68] Later Shungas patronized the old capitals Vidisha and Pataliputra, which had now become centers of Buddhist monastic life, education, and pilgrimage. Similar patronage seems to have been extended during the time of the short-lived Kanva Dynasty, founded by Vasudeva in 75 B.C.E., which ruled over the remnants of the Maurya Empire in eastern reaches of the Ganges valley. After this period a new epoch of Mahayana Buddhism began in the northwest under the auspices of the Indo-Greek trans-Himalayan regimes.

Remembering the Dead

Many factors may be counted toward the emergence of a landscape connecting relics, reliquaries, monasteries, pilgrimages, and markets. The aspirations of kingly and elite patronage dovetailed with the need to support and protect overland and riverborne trade routes for peddlers, caravans, and boats that connected the largest and the most prosperous cities associated with the rise of the great monastic orders of latter-day Buddhism in India: Rajagriha, Shravasti, Vaishali, and Kaushambi.[69] We have, however, little evidence before the early centuries of the Christian era to reimagine what the early monasteries looked like and in what capacities they functioned. The extant liturgy suggests the makeshift nature of early monastic habitats, typically humble shelters for wandering monks, sometimes caves, sometimes simply retreats to escape the ravages of the rainy season (*vessana*). However, during this period the *saṃgha* became more prosperous, sedentary, and conversant with the material culture of the laity. It is not just the repatriation of the monastic cloister to the world of everyday culture that deserves our attention but the fact that monasteries had become indispensible to trading and commercial networks, bridging the divide between lay and prelate, ordained austerity and popular piety.[70]

Recent archaeological evidence gives us a clearer picture of the activities of monasteries from the third century B.C.E. onward in the Ganges valley after the decline of the Maurya Empire. This was when a host of minor stupas and other tumuli emerged around the widespread practice of relic worship and changed the character of the monastic complex, with the dwellings of saints and monks built into the same space as shrines, *catiya*s, and groves. To a degree, this was the legacy of the measures taken by Emperor Asoka during the last years of his reign, when his precepts and officers of *dhamma* helped elevate the status of the Buddhist mendicant order to a trans-Indian, imperial institution. In time, however, the support of the laity and the guilds surpassed the scale of imperial gifts that the Mauryas were once able to bestow. Clearly, the need for a physical reminder of the presence of the Buddha and his legendary pupils animated the ubiquitous practice of erecting earthen or stone memorials to the dead. This urge to honor the dead changed the structure of lay and kingly patronage, leading to the building of grander and more towering stupas, endowed lavishly and intended to dominate the landscape. During the second phase of this architectural resurgence, stupas and *vihāra*s were built in tandem with new and larger monastic centers across the northern countryside. The gigantic stone platforms of Sanchi and Bhilsa, which formed the base for tall structures built around grand vistas, or the plinths laid for rectangular halls and courtyards, similar to the ruins that stand near the great Dhameka stupa at Sarnath, indicate a new language of architecture and temporal authority. This was when stupas were being erected on hilltops for fear of pillage and theft but also for a commanding view of the surrounding landscape in sites such as Andher. This was also when the largest courtyard monasteries were built, often in competition with the new surge of temples being built for Hindu deities and Brahmin establishments.[71]

Richard Gombrich suggests that popular aspects of Buddhist piety were closely attuned to the rising volume of trade and urbanization across the greater Ganges plains.[72] Buddhism gained wide popularity in northern India when an extensive redistribution of surplus production was taking place along the agrarian heartland, sustaining larger kingdoms and their armies, as well as a surge of newly found mercantile capital. The early Buddhist texts contain many references to long-distance trade along overland routes.

They also speak of the standardization of weights and measures, along with a constant refrain of matters relating to debt, credit, profit, and loss. There was an affinity here between the pietistic ideals of merit and virtue and the notions of credit and trust in commercial exchange, which gives us some clues about the participation of monasteries in everyday commercial transactions. As Lars Fogelin has suggested, both Jainism and Buddhism gave traders, merchants, and craftsmen plenty of opportunity to associate with people across the divide of clan and caste, and by the second century C.E., mercantile groups and artisanal guilds by their sheer numbers had loosened the hold of the old social order based on caste hierarchy.[73] Guilds now profited from both landholding and money lending, and they often became the leading patrons of monasteries and pilgrimage centers. Lay members of society could acquire merit by donating to monasteries. There are extensive records of such gifts made by bankers and merchants (*śreṣṭhin*, Sanskrit; *seṭhin*, Prakrit), guild masters, masters of caravans (*sārthavāha*s), and even monks and nuns.[74] Merit accrued also from donations for the ceremonial dedication of relics at major stupas.[75] A typical offering for a reliquary deposit was the "seven jewels" (*saptaratna*): gold, silver, crystal, beryl, carnelian, coral, and pearls. Such developments took place over long stretches of time, even centuries, but they clearly transformed prevalent attitudes and practices relating to relics and memorials to the departed.

Popular Buddhism in India did not entirely dispose with age-old funerary rituals but adapted long-established rites toward its own views of death and the end of suffering. Cremation and the disposal of charred bones and other bodily remains (*asthi*) in funerary urns, placed in aboveground tumuli, are discussed in the *Satapatha Brahmanas*, as are burial mounds (*śmaśāna*) reminiscent of ancient megaliths. Such funerary memorials can be traced back to ancient times, and evidence suggests that the practice, while not always endorsed in the Brahminical tradition, was common enough to be mentioned in the Vedic literature.[76] In fact, many different kinds of monuments were dedicated to the dead, of which *eḍuka*s or *aiḍuka*s (charnel houses) were the most common. This tradition was revived during the long period of ascendancy of Buddhism and ultimately led to what has been described by historians of architecture as the "memorial temple."[77] Ancient urn burials, prototypes of early Buddhist reliquaries, have been found in places like Ujjain in central India dating to the period of the early Mahajanapadas. The *Mahabharata* has references to memorial mounds to honor the dead along with the more common practice of consigning crematory ashes to the waters of sacred rivers.[78] With the passage of time, some of the rites dealing with the visceral remains of the dead such as inhumation and exposure were deemed as primitive. However, the tradition of dedicating monuments to the departed continued, which is important to keep in mind while considering the evidence of a variety of mortuary remains.[79]

Beyond the world of scholastic learning and monasticism, Buddhism changed many facets of everyday culture. Buddhist monks and followers took leading roles in medicine, credit, education, and the management of agricultural resources. Most of all, they dealt with the dead, taking over as specialists in the realm of funerary and mortuary practice.[80] Along with the distribution, display, and tending of the bones, ashes, and casket fragments of the elders of the faith, Buddhist monastic establishments also began to handle the mortuary remains of prominent members of the laity. This led to significant overlap between the Brahminical and Buddhist mortuary traditions, including disposal of ashes from cremations in the currents of the Ganges.[81]

This changing landscape of reliquaries and memorials helps us reconstitute the related experiences of pilgrimage and travel in northern India beyond the period of the Shunga and Kanva Dynasties, all the way to the time of the Gupta Empire, especially along the banks of the Ganges and its tributaries, where a myriad of sites welcomed visitors and pilgrims all year round for oblations and offerings. Some of the most prominent stupas and *vihāras* were located in agreeable natural surroundings, actively maintained by kings, guilds, and monasteries. The rising tide of Mahayana Buddhism, with its embrace of the image of the Buddha, added another element of appeal to this sacred geography. Indeed, the form, layout, and architecture of the later stupas show that they were quite attentive to the needs of the rising number of pilgrims who visited in large groups for the circumambulation of the main structures and a glimpse of the image of the Buddha. Later installations of the figure of Buddha at older sites such as Kaushambi show the rising importance of image worship along with obeisance to the relics in the popular practice of piety and veneration. Buddha images, for example, as Fogelin points out, were put up on all the main pilgrim paths from the surrounding countryside leading to the Dhameka stupa at Sarnath.[82]

The Middle Kingdom

In a collection of poems written in the Pali language known as the *Therigatha* written by women who followed in the footsteps of the Buddha, we find verses that describe the trails of mendicants between the cities and kingdoms of the northern Indian plains. These are intimate reflections of women who found the courage to renounce their preordained social roles as wives and mothers to join the order.[83] Notable among them is Bhadda Kundalakesa, of the "coiled hair," who was born during the lifetime of the Buddha into a wealthy merchant banker family in Rajagriha.[84] A compassionate young woman, she once tried to save a robber being led to his execution. The ungrateful man, taking advantage of her kindness, tried to murder her and take her belongings. To save her life, she killed him, throwing him from the top of a cliff. Taking any human life was a sin, and she could not go back to her family and undergo the prescribed rituals of penitence, choosing instead the path of a Jain ascetic. She tore out her hair and went through severe routines of self-mortification. And yet she could not find solace in the teachings she received. Then she met Buddha's great disciple Sariputta and engaged him in a debate, but she could not defeat him in argument. Sariputta took her to see the Buddha, and the master ordained her as soon as she approached him. She took up the robe and begging bowl of the initiate, living on meager alms, now an aging and gaunt figure with shorn hair, dusty feet, and one tattered robe. She went on to lead the hard life of a itinerant *therī*, not a nun cloistered in the *saṃgha*. In a touching poem, she remembered her days of wandering.

> I cut my hair and wore the dust,
> and I wandered in my one robe,
> finding fault where there was none,
> and finding no fault where there was.
> . . .
> I have wandered throughout
> Anga and Magadha
> Vajji, Kasi, and Kosala;
> Fifty-five years with no debt,
> I have enjoyed the alms of these kingdoms[85]

She does not mention the cities and pilgrimages we have described in this chapter but acknowledges the five original kingdoms (*mahājanapadas*) of the Ganges valley. Over the centuries that followed, as Buddhism rose, prospered, gained widespread social acceptance, and split into different doctrines and branches, this entire landscape was transformed.

The suffering of the Buddha and the austerities embraced by his peripatetic followers eventually passed into the realm of myth. Buddhist texts of the later period reflect an awareness of material advancement alongside their concern for spiritual well-being. They seem to appreciate the labor that makes the cultivated triumph over the fallow, the civilized over the uncivilized, and settled tracts over uninhabited wilderness. *Vihāra*s and *saṃghārāma*s had always been located at spots carefully chosen, within the view of verdant parks, well-tended groves, and rivers and lakes. However, wilderness was now truly relegated to the margins of this urbane, well-tended material order. Mahayana Buddhism, with its well-endowed shrines and well-appointed monasteries, was moving away from the traditionally ordained path of the self-suffering seeker through uninhabited forests, rivers in their monsoon spate, thickets of thorn and scrub, and inaccessible mountain steeps.[86]

A succinct idea of settled and prosperous domains is implicit in the ancient Sanskrit expression of "middle country," or *madhyadeśa*, which during this period became more commonly known as *majjhimadeśa* in Prakrit. This was the territorial core of Buddhism that the Mauryas embraced and that subsequent kingdoms such as the Shakas and the Kushans of the steppes, or the Shungas and Kanvas of the valley, attempted to reclaim. A rudimentary sketch of this geography can be discerned in both Buddhist and Brahminical texts. One of Baudhayana's famous *Dharmasutra*s, which also contains some of the best-known passages on moral codes of conduct in early Hindu liturgy, describes this territory as the region lying to the east of the tract where the River Saraswati disappears into the ground, west of the great Kalaka Forest, south of the Himalayas, and north of the Pariyatra Mountains. This was the land of the *ārya*s, where rituals and social conduct were appropriately observed, located somewhere between the rivers Ganges and Yamuna, stretching westward "as far as the black antelope roams." If Baudhayana's descriptions are reliable, then the edges of his *āryāvarta* would have included Avanti, Anga, and Magadha.[87] This outline is clearly inscribed in Patanjali's *Yogasutra* during the second century B.C.E. as the land of the distinguished *āryas*.[88] Patanjali suggests that the eastern boundary of this swath of territory ended with the forested belt of Kajangala, east of the kingdom of Anga, which we can identify as the Rajamahal Hills, where the Ganges takes the southward bend toward Bengal. Later texts such as the *Manusmriti* and the *Kavyamimamsa* (on poesy and criticism) extend this geographical concept farther to the east, including the great city of Varanasi and perhaps even the upper reaches of Bengal.[89] In Buddhist texts this middle country of the *ārya*s is further refined to include the territories of Anga and Magadha. The *Vinaya Pitaka* mentions the River Salalavati (Saraswati) forming the western limits of that which is in the middle (*majjhe*)—a core territory where early Buddhism had left its most significant cultural imprint.[90]

This middle country was also transected by long-established routes of trade that connected the major sites of pilgrimage. As noted earlier in this chapter, the first recorded wanderings were of the Buddha himself. The routes he traveled during his quest for wisdom, or later during his mission as a teacher and mendicant, are noted dutifully in all the Pitakas. During his lifetime the Buddha passed many times through the cities of Rajgir, Nalanda, Pataliputra and the villages of Kotigama, Ambagama, Jambugama, Bhoganagara, Pava, Kusinara, Kapilavastu, and Setavya.[91] These cities were located on some of the oldest and most important routes of trade.[92] As the sinews of commerce and pilgrimage intertwined and proliferated, the major roads passing from Rajagriha to Shravasti, capital of Kosala to the north of the river Ganges, gained in importance. Four main branches connected the city of Vaishali to Shravasti, one of which passed through the venerated city Varanasi on the river, present-day Banaras. Some of the most frequented roads connected Veranja to Vaishali, Kaushambi to Shravasti, Kashi to Champa, Anupiya in the country of the Mallas to Rajagriha, Champa to Rajagriha, Gaya to Banaras, and Gaya to Rajagriha. Many of these ancient cities were located near the main channels of the Ganges or its principal tributaries. The coming prosperity of these cities and towns had been revealed to the Buddha, if the conversations recorded between Buddha and to his chief disciple Ananda are to be believed. When Ananda asked the Buddha why he wanted to choose Kusinara (Kushinagar in the present-day state of Uttar Pradesh in northern India) as the site of his nirvana, when there were many other, larger cities that could be blessed by his passing on, the Buddha assured him that Kusinara was important because it was situated on the great highway of trade between Shravasti and Rajgir.

It is important to emphasize that the dimensions of this middle country are so frequently invoked in contemporary literature because it maps so neatly on the area watered by the Ganges and its tributaries. This region had always been connected through commercial traffic to the Punjab corridor and the Himalayan foothills in the northwest, to the elevated plateau of the south, and through the great delta in the east to the ports along the Bay of Bengal. With the rise of Magadha and the Maurya Empire with its core territory located toward the east along the lower Ganges valley, the political importance of the older cities of *āryāvarta* in the western parts declined. However, the great trading routes of the western plains, recounted in the *Ramayana* and the *Mahabharata*, seems to have been active over many centuries. These ancient routes connected the mythical kingdom of Ayodhya to the western Punjab and the great caravan trails descending from the Hindu

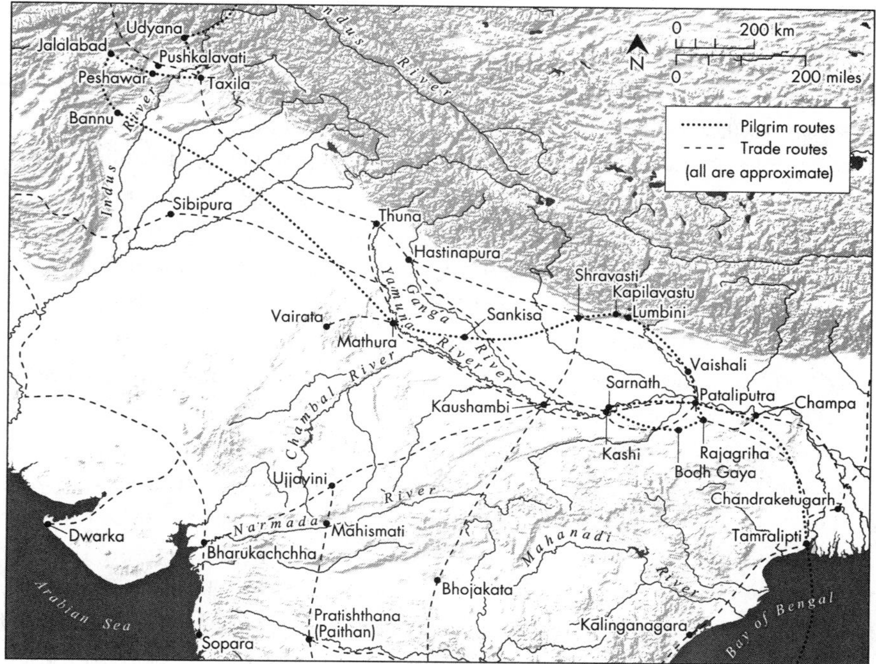

Major routes of trade and pilgrimage in ancient northern India.
Map by M. Roy Cartography.

Kush mountains, from Peshawar on the Afghan border through Lahore to Jalandhar in the Punjab, and from there to Ferozepur, Bhatinda, and Delhi.[93] One route came down to Rohtak, forking at the open fields of Panipat. A second went toward the southeast through the cities of Thaneswar, Indraprastha, Mathura, Sankisa, and Kanauj; another route following the right bank of the Ganges ran through the confluence of Prayag near Allahabad and from there went all the way to Varanasi.

The *Sutta Nipata*, one of the early texts of the Pali canon that extols the virtues of honorable trade and mindful industriousness as respectable paths to the acquisition of wealth, also mentions certain key urban centers that directed the flow of trade across the middle country.[94] It records how the great teacher Bavari sent sixteen of his pupils to go and seek out the Buddha all the way from Pratishthana near the River Godavari to find out whether his claims to enlightenment were indeed true. It also describes the route they took and the major cities they passed along the way. The most important among these were the cities of central India such as Mahismati, Ujjayini,

Gonaddha, and Vidisha located in the kingdom of Avanti; the city of Kaushambi on the banks of the River Yamuna; Saketa, close to the ancient site of Ayodhya by the River Sarayu; and Shravasti on the basin of the River Gomati. It also mentions the major halting places on the high road between the cities of Sharavasti and Rajagriha: Setavya; Kapilavastu, birthplace of the Buddha; Kusinara; Pava; and the major city of the Malla confederacy, Bhoganagara. These cities were located near the waterways along the Ganges that connected the region of Rajgir to the region of Kaushambi in the west.[95] They were all associated with Buddha's peregrination and gained further prosperity and renown over the succeeding centuries. For example, the old trade route from the city of Veranja, named after the Brahmin who came to question the Buddha, to the cities of Prayag and Kanauj on the Ganges, and Mathura on the River Yamuna, became fabled for their guilds, merchants, and commerce. It is also hardly surprising that all the key strategic crossing points where the Ganges was forded—whether it be Prayag at the meeting point of the Ganges and the Yamuna or Pataliputra at the meeting point of the Ganges and its major tributary Son, or crossings on the southern bank of the Ganges at places like Kalpi in the west and Chunar in east—were all bolstered by the rising tide of riverborne trade that brought goods and traders of the middle country to the great trading gateways of the west.

Through these long-distance routes of trade and commerce, cutting across multiple kingdoms and terrains, Buddhism spread northward across the Himalayas. Wandering mendicants and preceptors carried the message and teachings of the Buddha along high-altitude caravan trails frequented largely by merchants and nomads. Mobility and simplicity were the hallmark of such transmission.[96] This is one of the reasons why metaphors and analogies of roads and wheels abound in contemporary accounts, with epithets of the Buddha as a leader of caravans (*sārthavāha*), along with images of bodhisattvas as traveling merchants. Thus *saṃghārāma*s and *vihāra*s were the lifelines of homeless, itinerant preachers traveling on distant and difficult routes. The building of the monasteries in the vicinity of major trading routes, as Erik Zürcher's extensive studies suggest, helped bridge daunting distances, which would eventually lead to the rise of Buddhist societies in China.[97] Here, the simplicity of precepts, doctrine, and guidelines seems to have been the key, especially where contacts between distant monasteries were sporadic and the responsibility of spreading the word lay with individual preceptors.

The significance of such intimate relationships between pilgrimage, urban life, and trade, while indisputable, is not always easy to measure, especially as early Buddhism placed such great emphasis on the discarding of worldly ties and possessions. During this period, however, the austere worldview of the traditional ocher-robed ascetics was largely transformed by the sumptuary needs and practices of lay worshippers and patrons. Apart from the inferences we can draw from the vast body of Buddhist canon, archaeological evidence tends to support the argument that important urban centers associated with the spread of the great monastic orders—at Bodh Gaya, Vaishali, or Rajgir—were also the most vibrant conduits of trade across the Ganges valley.[98] Much of this resulted from the rising tide of lay support for Buddhist institutions, which led to an extensive economy of gift exchange, where there was little conflict between mercantile profit, display of wealth, and pious donations.[99] In this context, the most significant ritual practices of popular Buddhism took place not just around the most prominent stupas but around a multitude of lesser structures that mushroomed across the valley. At every site of a major stupa we find a host of lesser votive and funerary tumuli. Many of these are called *kula*s, a final resting place for bones and ashes of notables brought from elsewhere. Much like the Buddha's sacred relics, the bodily remains of multiple generations of devout practitioners and elders were distributed throughout these sites, where both royalty and laity paid their homage in prayer and offering.[100] The word used often to describe such relics in the Theravada tradition is *dhātu*—"substance," sealed in votive urns alongside precious memorabilia. As an inscription from second-century-C.E. Sanchi attests, harming or desecrating a stupa was equivalent to defiling the memory of a person of the highest rank, while rebuilding and adding to extant structures were seen as eminent acts of virtue and kindness.

Caravan trains and pilgrim bands continued along the major routes through the forests and scrubs of the Gangetic plains long after the fall of the Maurya Empire. During the rise of the Gupta Empire, as we shall see in the next chapter, close ties had developed between the Buddhist monastic orders and the various trading and financial groups. Similar patterns of mutual sustenance can be found between mercantile groups, Jain monastic orders, and Hindu temples. Long- and middle-distance trade in the valley, from this long-term perspective, cannot be studied separately from the evolving map of the middle country that we have traced at great length in this chapter. This overall geographical conception loomed larger than the specific networks of trade, commerce, and travel, or the territorial claims of contemporary empires and kingdoms, encompassing a landscape that

Buddhism had helped create over the *longue durée*. We have seen in this chapter how Buddhism in its ascetic, monastic, and popular incarnations brought within its fold some of the oldest and most vibrant animist and spirit traditions of the copses, grottoes, and riversides of the Ganges valley. Added to these were the relics of the Buddha, his disciples, and their followers, preserved in myriads of large and small mortuary monuments, creating a dense patchwork of sacred sites and contributing to the overall sanctity of the River Ganges and its environs.

Landscapes as phenomena, as Christopher Tilley has pointed out in his discussion of historical archaeology, fall somewhere between the natural and human.[101] They represent an alchemy of places, distances, and time that ultimately reside in collective memory. More than particular cities or monuments, a landscape is woven around sites endowed with experiential significance and accumulated reminiscence—historic, sacred, ancestral. Buddhist pilgrimage, in many ways, was intended to recapitulate the contours of such a preordained landscape. Latter-day Buddhism, with its cults of reliquaries and ancestral sites, with its lavish endowments of residential monasteries and sumptuous stupas, not only was able to sustain such a landscape of remembrances but bequeathed a rich geographical imaginary for the inhabitants of the Ganges valley. As Denis Cosgrove in his classic study of symbolic landscape once identified, the relationship between landscape—tied to aesthetics, experience and sentiment, and geography—dedicated to mapping, order, and the control of space, is notoriously difficult to figure out.[102] Nevertheless, while no one empire or kingdom—Shungas, Kanvas, Kushans, Shakas, Nagas, Maghas, or Murundas—succeeded in establishing undisputed mastery over this stretch of territory, they were all invested in safeguarding reliquaries, monasteries, pilgrimages, and the major routes of trade.

The commercial prospects of pilgrimage and travel, or the massive endowments for monasteries and hospices, would have been unimaginable to the early disciples of the Buddha, who left their homes, families, and cities for the hardships of the open road, traveling through the wild and inhospitable terrain between distant cities and villages, armed with nothing but their walking staves and bowls of alms. The stories of their journey, and their sufferance in the face of adversity, were forever imprinted in history and memory, but the country through which they had once traveled had been transformed by the surge of religious and social movements they initiated in the name of the Buddha and his message.

CHAPTER 6

THE GODDESS OF FORTUNE

The last great scion in the illustrious line of Gupta emperors, who ruled most of northern India from his ancestral kingdom of Magadha in the eastern Ganges valley during the fifth century C.E., was a battle-hardened prince named Skandagupta. He has gone down in legend as one of the last valiant guardians of Magadha, after whose reign the empire began to fall apart. Skandagupta claimed to have shored up the fortunes of the empire that had suffered badly at the hands of rebels and internal enemies during his father's regime. In one of his most ambitious proclamations—the Bhitari pillar inscription found on the banks of the Ganges at Ghazipur, fifty miles east of Banaras—his exploits are rendered in a few striking Sanskrit passages.[1] The inscription states that early in his reign, returning to his kingdom after a string of victorious campaigns, he stood triumphantly in front of his mother, a woman who had suffered terribly at the hands of a rival claimant to the throne, Purugupta, who most likely put her in confinement. Much the way Lord Krishna had once stood before his mother, Devaki, Skandagupta announced that the enemies of the kingdom had been slain. He had stamped out the bloody rebellion of the Pushyamitra tribes of Malwa and set the marauding Epthalite Huns on the run in the northwest, having inflicted on them a crushing military defeat. Like an earth-shaking, terrible whirlpool (*dharā kampita bhīmāvartta*)—the engraving proclaims—his army fell upon the barbarian Huns (*hūṇa*s) with the deafening roar of the mighty River Ganges (*gāṅga-dhvani*).[2]

The Epthalite Huns of Central Asia during this period had begun to overrun the northwestern frontier kingdoms. Having routed the Sasanian Empire in the mid-fifth century and secured a stronghold in Afghanistan, they edged forward toward northwest India. It is likely that Skandagupta's forces, distinguished by some of the finest archers in the country, confronted and defeated the Huns somewhere north of the Ganges plains. The victory over the Huns was a memorable achievement, and it kept them at bay for at least another decade. At the end of Skandagupta's reign, the Huns, led by Toramana, took Punjab and pushed farther southwest toward Malwa, where the Gupta Empire had been particularly vulnerable. The date of Toramana's coins place his Indian campaign circa 460 C.E., overlapping the last years of Skandagupta's reign. The Guptas ruled the whole of northern India below the Punjab, from Gujarat to the foothills of Nepal, but after Skandagupta's death they were unable to ward off the Hun advance. By the early sixth century C.E. the Huns swept across the Ganges plains, which is evident from the abundance of coins in this period featuring the head of Toramana facing in a direction opposite to the usual relief of kings found on Gupta coins.

The Bhitari pillar of Skandagupta was erected in the heartland of Gupta territory, the floodplains of the Ganges between Varanasi and Magadha. The fact that the royal panegyrists chose the torrential Ganges as the most compelling simile for imperial prowess they could summon against the fearsome Hun invaders tells us something about the status of the river as an imperial emblem. In this chapter we explore how this precedent came about during the early years of Gupta ascendancy. As noted before, the root of the word *Ganges* is derived from the verb *gaṃ* ("to go"), denoting motion, flow, direction, and force. The Guptas seized upon this riparian image along with other icons of imperial authority such as Vishnu and his mascot Garuda, the king of birds and the destroyer of venomous snakes. During the Gupta era sculptors fashioned the Ganges in the exquisite image of a goddess flanked by fantastic aquatic creatures representing power, prosperity, and plenitude. She now became synonymous with the imperial territory through which it flowed and on which it lavished fertility and abundance. Such a rendition is quite unlike the usual mythological invocation of the river as a sacred body of water in the traditional sources such as the Vedas, Puranas, Aranyakas, or Brahmanas. This imperial distinction was bequeathed by the Guptas to a multitude of regimes and dynasties that emulated them across India. In order to test the depth of this legacy, we have to explain why the founders of the Gupta Empire succeeded in exercising such a powerful hold over the political, cultural, and historical imagination of the subcontinent and its dynastic, artistic, and literary traditions. This aura of the Ganges as an auspicious river, conceived as a goddess of bounty and good fortune, and as the ultimate prize of kings and emperors who dared to rule over her fertile valley, lingered long after the Gupta Dynasty came to an end.[3]

Cities, Guilds, Caravan Trails

The history of the last kings who succeeded the great Mauryan emperor Asoka still remains unclear. This is largely because of conflicting Buddhist and Jain accounts. This much is known, however: within fifty years of Asoka's celebrated reign, the great Maurya Empire had unraveled beyond recognition, overtaken by a new dynasty known as the Shungas. During these twilight years the empire shrank to the region from which the founder, Chandragupta Maurya, had launched his kingdom. The great poet Banabhatta's verses celebrating the reign of King Harshavardhana of Kanauj, written during the seventh century C.E., provide a brief historical sketch of the political coup that ended the writ of the last Mauryas. It seems that Pushyamitra Shunga, commander of the Mauryan army, murdered the Mauryan ruler Brihadratha as he was inspecting his troops. The Shungas came to power in 187 B.C.E., seizing the effects of what was left of the last Maurya line. They held the core territories of Pataliputra, Ayodhya, and Vidisha, and possibly parts of the Punjab. Pushyamitra claimed to have defeated the king of Vidharbha and fought with the Bactrian Greeks and other foreign powers of the frontier.

If the great philosopher of yoga Patanjali is to be believed, tribes from across the Himalayas had started encroaching into northwestern India around this time. In his celebrated commentary on Panini's grammar, the *Mahabhasya*, Patanjali illustrates the use of imperfect verbs with sentences that refer to barbarian Greeks (*yavana*s) and their cavalry regiments besieging Saketa (Ayodhya) and Madhyamika (near Chittor, Rajasthan).[4] It is impossible to figure out whether he was drawing these examples from actual, recorded history, but the manner in which he deplores grammatical errors and debased words spewing from barbarian tongues supports the idea that people living during the ascendancy of the Shungas were quite aware of the threat of the Greeks, Bactrians (*bālhīka*s), and Scythians (*śaka*s) across the Himalayan frontier.[5]

The Shungas, who had unseated the last Mauryas, held on to the heartland of the Ganges plains and Magadha for about a hundred years before being uprooted by the the Kanvas (73 C.E.) in a bloody palace coup. Toward the end of the millennium, the Bactrian Greeks extended their empire all the way to Mathura, but they failed to subdue Magadha and seize Pataliputra. This was because they remained divided into two major warring branches, soon to be edged out by the Shakas and Pahlavas (Parthians), who in turn were routed and chased out of northern India by the Central Asian Kushans (Yueh Chi, Chinese) during the first century C.E. The Kushans established a vast trans-Himalayan empire and commanded territories along the Ganges plains to the easternmost cities of Sarnath and Banaras and parts of Bihar, evident by the large deposits of their coins that have been found in Ghazipur near Patna.[6] A minor branch of the Kushans ruled parts of the valley until the early fourth century C.E., followed by Lichchhavi and Maukhari chieftains who also laid successive claims over the prized region of Magadha.[7]

The middle country thus survived the rise and fall of the political fortunes of many tribes and kingdoms during these long centuries of unrest. Historical sources are too scanty to reconstruct how smaller kingdoms and

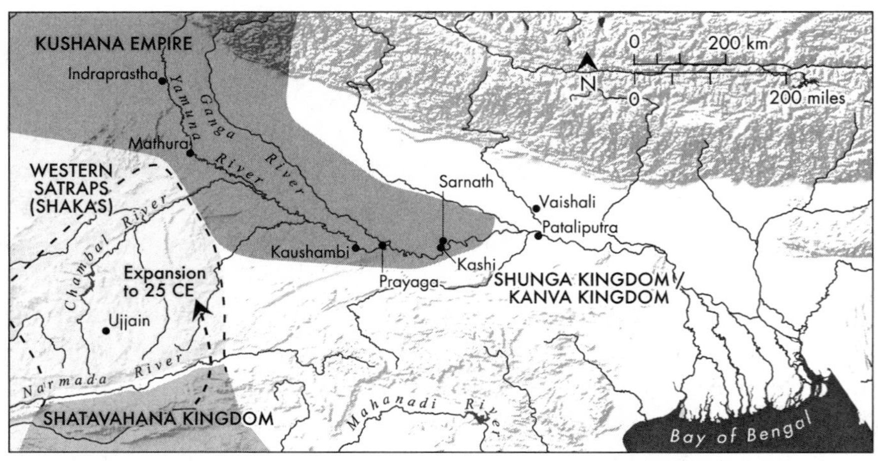

The Kushan Empire in northern India. Map by M. Roy Cartography.

chieftaincies fared in the region of Magadha. A motley collection of coins and artifacts, especially figurines and seals in terra-cotta, that have been unearthed at archaeological sites such as Ahichchhatra and Kaushambi provide some clues to these local struggles for power. Such a long period of political uncertainty disrupted but did not necessarily uproot age-old patterns of trade and exchange. The cities of Hastinapur, Mathura, Sankisa, Ahichchhatra, Saketa, Ayodhya, Kaushambi, and Bhita continued to rely on the time-tested routes of the old caravan trade that sustained the various guilds of perfumers, metalworkers, bankers, and transporters. An inscription in the Brahmi script dedicated to Manibhadra, a popular *yakṣa* deity worshipped near Masharfa in the present-day north Indian state of Uttar Pradesh, that can be roughly dated to the first century B.C.E. reveals that a *vedīkā*, or elevated enclosure, was erected to this cult figure by a householder (*gahapati*) known as Gotiputa, son of Kusipala. A devout Jain, Gotiputa was by profession a leader of caravans (*sārthavāha*) and a devotee of Manibhadra, who seems to have been the main guardian spirit of all caravans and caravan traders.[8] Archaeological evidence from the region of Mathura during this period, especially Naga serpent figures, indicates a concentration of wealthy patrons belonging to the Bhagavata and Naga sects, many of whom were also known to have close associations with large trading guilds.[9] The prominence of Jain merchants is evident from the numerous *āyāgapaṭa*s, the ornamental, votive stone slabs commonly used in forms of Jain worship.

Yaksa head from Bhita, Uttar Pradesh, ca. 119–199 C.E.
Photograph courtesy of the American Institute of Indian Studies.

These, along with sculptures of unadorned Jaina *tīrthaṅkaras* dating to the Kushan period, have been found in large numbers at digs for ancient Jain sites like the Kankali Tila mound at Mathura.[10]

The vibrancy of such urban centers was first revealed by the excavations that were carried out at the ancient fortified settlement of Bhita near the banks of the River Yamuna near Allahabad led by Alexander Cunningham in 1872.[11] *Bhita* denotes a mound or foundation, and the site of the citadel of Vicchigrama was a great find, because it revealed a largely unbroken record of urban settlements dating from the Maurya period to the Gupta period.[12] Bhita seems to have survived both depredations and rebellion, and there is evidence of parts of the city being destroyed and rebuilt. These reconstructions show a remarkable degree of continuity in architectural design, including the use of bricks and brick fragments from previous structures. Archaeologists have found a number of town houses and neatly planned, broad thoroughfares running parallel to each other.

Remains of the oldest residences from the Maurya period at Bhita suggest that extended families of traders and artisans lived together under the same roof in a series of adjoining rooms built around a central courtyard. Surveying their layout and plan, the stalwart British archaeologist John H. Marshall in the 1920s thought that they resembled the floor plans of a typical Buddhist monastery, which were originally "copied from this type of domestic house."[13] Many belonged to merchants and guild members, evident from seals inscribed with the words *nigama* ("guild") and *kulika* ("banker"). Entrances to a courtyard and a veranda have also been found with wheels of a child's toy cart nearby, along with steatite caskets decked with floral designs suggesting funerary rites.

During the early second century C.E. a new house was built nearby that belonged to a merchant named Navadeva, whose name appears on an ivory seal. Navadeva must have been a wealthy man because clay seals found in his house are also present in a row of shops (dating to the Shunga period) that stood on one of the main thoroughfares of the city. His house was separated by a narrow lane from the house of another banker, Shreshti Jayavasuda (fourth or fifth century C.E.), whose name appears on an ivory die. He must have been a devotee of Vishnu because his trading seal sports the tortoise (*kūrma*), one of the principal avatars of Vishnu.[14] The house of a merchant, Pushyavriddhi, stood across the shops. The foundations of this residential complex were laid during the Maurya period, and it was occupied continuously through the first century B.C.E. It was abandoned for a while during Kushan times but seems to have been restored during the early period of the Gupta Empire. One of these structures also contains a small room that housed images of Shiva, his consort Parvati, and Shiva's mascot, Nandi the bull, suggesting a separate space for the resident deity worshipped by the family. Parts of the excavated city also show signs of looting and the possible sack of entire neighborhoods. In such crises, some inhabitants seem to have left in a hurry, abandoning their ancestral idols and everyday items, including copper utensils left on the stoves.

The digs of Bhita offer a rare glimpse of city life in the post-Mauryan era. Archaeologists such as F. R. Allchin have described them as the prototypical "town houses of the Ganges plains" that also provided the blueprints for the rectilinear Buddhist monasteries of a later period.[15] Similar patterns of urban settlement have been seen at Kaushambi, Ahichchhatra, Rajghat, and Vaishali: a square plan, houses with multiple rooms occupying three sides of a large, open courtyard laid with bricks, covered by roofs made out of terra-cotta tiles, and equipped with drains, wells, and storage. Domestic and public ring wells dating back to Mauryan times may have been used for drawing water or drainage for latrines.[16] Most of these sites reveal layers that can be identified with the Shunga, Kushan, and Gupta eras, and the archaeological evidence shows the clout of artisans, traders, and merchant guilds of the Ganges plains that endured the rise and fall of multiple regimes over the course of several centuries. They also suggest an intimate association between forms of piety and wealth evident in seals and other imprimaturs of mercantile guilds and banking houses. Shunga seals are inscribed in Prakrit, whereas Gupta-era seals are typically in Sanskrit. Some belonged to worshippers of Vishnu, evident from motifs such as wheels and conch shells. Others carry the impression of bulls, Shiva's favored mascot. Some bear the insignia of Gupta officials in charge of religious foundations, keepers of peace, and representatives of specific administrative districts of the empire.

Ramparts of Brick and Wood

Excavations of cities that flourished in the floodplains of the Ganges during this period are far from complete, but they generally confirm the impression that people living in them had long been acquainted with long-distance trade and the diversity of artisanal production. At the same time, they were also prepared to defend their settlements. Cities were often known by the rivers along which they were laid, and those along the Ganges were especially well connected with trade, markets, and pilgrimage. In many places the Ganges also provided a natural defensive barrier and was also a source of fine clay for the erection of embankments.

The grammarian Patanjali mentions that Hastinapur was located along the Ganges, as was Varanasi, while Pataliputra was by the banks of the Son (*anu gaṅgaṃ hastināpuram, anu gaṅgaṃ vārāṇasī, anu śoṇaṃ pāṭaliputram*).[17] Mathura on the Yamuna was well known in antiquity for its kettledrums, textiles, and coins that Patanjali wrote admiringly about, and also for its extensive ramparts that were added during the Shunga period.[18] Similar attempts at defense can be seen in cities of the western Ganges plains, most visibly in Ahichchhatra, located in the ancient kingdom of Panchala, and its sister city Kampilya (Kampil), laid out on the banks of the Ganges between the present-day urban centers of Badaun and Farrukhabad and placed strategically on ancient caravan routes connecting Mathura to Shravasti. Cunningham, who wrote one of the earliest archaeological reports of Ahichchhatra, noticed a series of extensive mounds rising from the banks of the old abandoned course of the Ganges strewn with ancient bricks and brick fragments, which local people believed were remains of the great palace of Raja Drupada, a rival of the Pandava brothers in the *Mahabharata*. The name *Ahichchhatra* can be taken to mean "snake canopy." The city is supposed to have been protected by the cult and legend of a great Naga.[19] The ramparts have been measured at about three and half miles in circumference, with ruins and foundations of stupas and monasteries. Cunningham saw similar structures at the site of the ancient city of Sankisa, nestled in the fertile tracts between the Ganges and the Yamuna, with multiple heavily fortified gateways, and remains of monasteries and temples inside the citadel.[20] Subsequent digs have shown that successive phases of fortification stretched across the Kushan and Gupta periods.[21] In Kosam, which is the site of the ancient city of Kosambi (present-day Kaushambi), remains have been found of earthen ramparts spread across a four-mile circuit reaching thirty to thirty-five feet from ground level, reinforced with bastions rising from fifty to sixty feet for watchtowers, and parapets made of brick and stone. There is also archaeological evidence of deep trenches dug for moats around the fort. Dieter Schlingloff, in a recent study of ancient Indian cities reexamining the formulas of urban defense in Kautilya's *Arthasashtra* against the current archaeological evidence, points out how the architects of Kaushambi remained faithful to classical templates for the defense of their city, fashioning a moat that narrowed near the entrance of the city from a width of 473 feet to about 66 feet, protected by a drawbridge. One could enter by this moat through strategically placed causeways of tamped gravel, and it could also be flooded by dikes during times of siege.[22] The last ramparts of Kaushambi seem to have been added during the early sixth century C.E., which would coincide with the spate of Hun invasions.[23] The wealth and prosperity of the city were widely known. Among its major attractions was a great sandalwood statue of the Buddha described by the Chinese pilgrim Xuanzang, and one of the largest establishments of Buddhist monks flourished within its walls, towered over by the pillar erected by Emperor Asoka. It is no surprise that it withstood repeated sieges.

Despite such trouble, the agricultural surplus of the Ganges valley was still able to support these fortified settlements. Even during long periods of political unrest they enjoyed a relatively stable order of governance and taxation, evident in new terms for villages (*grāma*) and clusters of hamlets (*grāmahāra*).[24] Bhita had a population estimated at ten to twenty thousand, while Kaushambi at its height might have sustained ninety to a hundred thousand.[25] In the age of military raids, moats, ramparts, and walls had become a common feature of urban life, as granaries, guilds, deities, and sites of worship were vulnerable to plunder. We find towering walls in Varanasi during this period, as we do in Shravasti and Vaishali.[26] Defenses were often added in haste, with makeshift engineering rigs to ward off surprise attacks by elephants, battering rams, and projectiles. Cities close to the waterfront of the Ganges raised defensive walls as extensions of their embankments. Before the Gupta Empire gained full command, these cities continued to serve as an extended archipelago of refuge from roving bands.[27]

The kingdom of Magadha lay roughly in the eastern section of the lower Ganges plains, corresponding to the present-day Indian state of Bihar, and its capital Pataliputra was located at the confluence of the Ganges and the Son, which was at least 15 miles farther upstream during this time period. If Megasthenes's Greek text *Indika* is to be believed, during the reign of Chandragupta Pataliputra was 80 stadia (over 9 miles) long and 15 stadia (over 1.7 miles) wide.[28] Built in a rough rectilinear shape, it was protected by a wooden wall "pierced with loopholes for the discharge of arrows" and equipped with multiple gates and defensive towers. Remains of this palisade have been unearthed at Bulandbagh, near Patna, which reveal more than one line of stakes interspersed with impacted earth, the massive beams of teak being held together by iron dowels.[29] The palisade was fronted by a moat almost 600 feet wide and 45 feet deep that also received the city's sewage. During Maurya times Pataliputra was one of the largest cities of the world. We do not know much about how it fared between the Maurya and Gupta Empires, except that it suffered multiple sieges and a succession of palace coups.

Because of its historic and strategic importance, Magadha was still the prized territory for aspiring regimes, and it was especially vulnerable with the rise of the Satavahana power to the southwest. Further east, Kalinga, situated in the coastal belt between the Ganges and Godavari Rivers, famously seized by Emperor Asoka after his last and bloodiest campaigns, reasserted itself as an independent kingdom under King Kharavela of the Cheta Dynasty. Kharavela amassed a powerful army and inflicted a humiliating defeat on Rajagriha, Magadha's old capital, unleashing panic among its inhabitants by driving his royal elephants inside the walls of Pataliputra, forcing the Shunga ruler Bahasatimitra to bow at his feet.[30] With this act he made the hearts of all the rulers of the northern Indian plains (*uttarapatha*) tremble with fear.[31] Kharavela's war elephants were reputed to have been washed in the torrents of the mighty Ganges, his army overrunning the southern flank of the river.[32] In his Hatigumpha inscription, in Prakrit with Brahmi script, laid out on an overhanging rock inside one of the caves in Udayagiri in the present-day state of Odisha, Kharavela announced his daring feats—a testament to the fact that the well-watered plains of middle-Ganges were still subject to an ongoing struggle for political supremacy, with Magadha, its most emblematic kingdom, casting a long historical shadow.

The Resurgence of Magadha

The *Vayu Purana*, which mentions the imperial Guptas in its genealogical inventory, suggests that they were destined to rule over the major kingdoms of the middle Ganges plains:

> *anugaṅgā prayāgamca śāketam magadhāṃstathā*
> *etān janapadān sarvān bhokṣyante guptavaṃśajāḥ*[33]
>
> Prayag, Saketa, and Magadha along the Ganges
> Scions of the Gupta line will enjoy these kingdoms

While the origins of the dynasty remain obscure, it is clear that the Guptas consolidated their sovereignty over Magadha over at least three generations. The earliest mention of a king (Shri Gupta) is found in the account of the Chinese pilgrim Yijing, who traveled to India in the seventh century C.E.

to study at the monastery of Nalanda. Yijing says that the king very kindly donated a temple for Chinese pilgrims in Mrigashikhavana with a grant of twenty-four villages, and that this spot was more than 200 miles east of the monastery of Nalanda. The Guptas may have already established their hold over the eastern tracts of the Ganges from present-day Bihar Sharif to Murshidabad in Bengal.[34] The exact succession of this dynasty and the extent of its territory are not known, but it was likely poised for political expansion during the early fourth century C.E. before Chandragupta I in 320 C.E. took the title of "king of kings," *mahārājādhirāja*. While scholars disagree on whether he actually took Pataliputra by force and proclaimed a new empire based in Magadha, there is little doubt that the Guptas established their dominance in the region through a timely and strategic alliance of marriage with the powerful house of Lichchhavis, who commanded territories to their north. The title of *mahārāja*, by which Chandragupta's father King Ghatotkacha came to be known, was in fact a subordinate rank, and the Guptas most likely were Maurya feudatories.[35] The marriage of Chandragupta to the Lichchhavi princess Kumaradevi might have been a calculated political move to elevate the status of the Guptas. The Lichchhavis of Nepal, as described earlier in this book, were a political oligarchy of undisputed pedigree and power, and they commanded the foothills of the Himalayas and the northern stretch of the Ganges plains. The alliance gave the Gupta Dynasty new prestige, which is clear from the gold coins of Chandragupta bearing the name of the queen and her paternal clan, and medallions commissioned by Samudragupta in honor of his parents, featuring the caption *licchavayaḥ*, indicating that the prince was descended from the ruling confederacy of the Lichchhavis.[36] Securing such a powerful ally must have helped the Guptas dispel the threat of Shaka satraps and the last remaining Kushan vanguards of the northern Indian plains. This was the foundation that Samudragupta inherited, which helped him consolidate his kingdom and launch daring military expeditions.

Samudragupta's military raids detailed in the pillar inscription of Allahabad, composed by court poet Harisena, have been the subject of much discussion by epigraphic experts ever since it was brought to public attention in 1834 in the pages of the *Bengal Asiatic Society Journal* by early antiquarians of British India such as T. S. Burt, J. Troyer, and James Prinsep. The inscription appears on a rounded monolithic sandstone column dating back to the third century B.C.E. carrying a famous edict of Emperor Asoka addressed to the rulers of Kaushambi. The column was inscribed after Samudragupta's

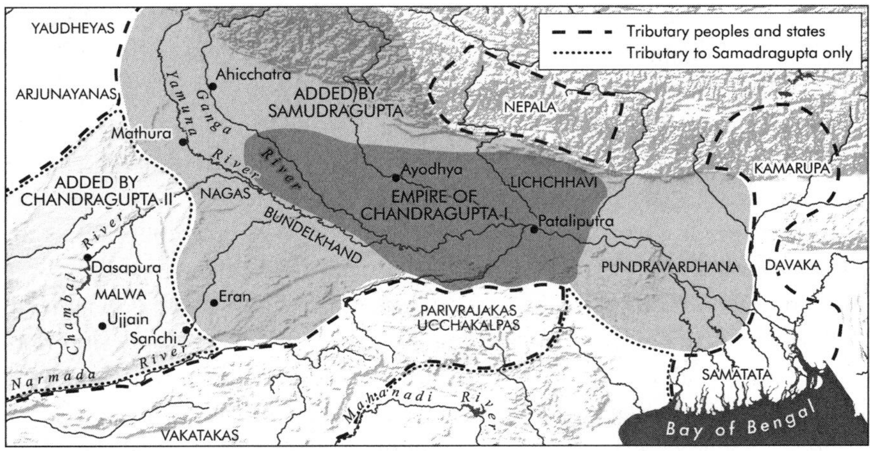

Expansion of the Gupta Empire. Map by M. Roy Cartography.

death, possibly during the accession of his successor Chandragupta II to the throne of Magadha. It contains a detailed account of the marches of Samudragupta's army and provides a remarkable inventory of the various regions and kingdoms of India during the mid-fourth century C.E. The fact that it was carved atop a column erected by Asoka, who reigned more than six hundred years earlier, is a testament to the Gupta Empire's own sense of legacy and grandeur.

Harisena lauds Samudragupta as a mighty warrior whose body was scarred from battles and wounds received from battle-axes, arrows, spears, pikes, swords, and many other weapons. He defeated, captured, and freed many powerful neighbors such as King Mahendra of Kosala and the Vyaghraraja of Mahakantara. He violently overthrew the nine great kings of the *āryavarta*. Samudragupta wanted to secure the north before venturing farther west and south. Bulandshahr is located west of the Gangetic plains, and Mahakantara, which literally means "great wilderness," suggests the extensive forests and scrubs of central India. Samudragupta also seemed to have persuaded "all the kings of the forested countries (*sarvvāṭavikarājasya*) to become his servants."[37] Samudragupta's empire was thus bounded on the east by the delta of the Ganges and the floodplains of the Brahmaputra in Assam, toward the west the powerful Vakataka kingdom of the western Deccan, and the Shaka and Kushan satraps of the northwest. It was girded in the north by the valleys and foothills of Nepal and Kangra, and in the south it extended along the green, hilly belt of the Coromandel Coast and the coastal road down which the Gupta army marched to defeat the kingdoms of Vengi and Kanchi, carefully avoiding conflict with the Vakatakas.[38]

It is not clear whether the Guptas were able to exact the lasting political allegiance of far-flung kingdoms visited by Samudragupta's nimble army, but their successors did inherit an impressive roster of tributary kingdoms. At its core the Gupta Empire occupied the entire stretch of the northern floodplains of the Ganges, which included the ancestral domains of Magadha. The Allahabad inscription acknowledges this imperial geography by fusing the mythical image of the celestial river with the figure of the emperor. The supreme qualities of the emperor that were manifest in every direction—generosity, military strength (*bhujavikrama*), steadfastness (*praśama*), and knowledge of the teaching of the scriptures—thus cleansed and sanctified the three worlds (*bhuvanatrayam*), just like the waters of the Ganges gushing forth in every direction after they were released from the confines of Shiva's great matted locks.[39]

Graven Images

Inscriptions, coins, sculptures, and dedications that have survived from this era indicate a widespread currency of images and motifs that suggest generic Hindu expressions of divinity. This elaborate alphabet of icons must have long preceded the years of Gupta reign, but they were certainly circulated and patronized, not only by the Guptas but also by subordinate regimes, and in turn by common subjects. In temple art and votive figures, as well as in the design of mundane and secular objects, we find the various avatars and emblems of Vishnu, the patron deity of the Guptas, such as the winged mascot Garuda, along with conch shells, lotuses, maces, and the discus (*cakra*). The resurgence of this iconography provides the context for the precipitation of an image of the River Ganges as a female deity and one of the enduring symbols of imperial might. However, before we explore this further, it is necessary to explain how such figurative and plastic art gained popularity and patronage in northern India, especially with the emergence of the Buddha image in a recognizably human form from the first century C.E. onward.

For more than a century, a lively debate has continued over the degree to which the use of the human image of the Buddha was forbidden for both monastic and lay communities.[40] We still do not know how pervasive these proscriptions were, or whether they were responsible for what scholars at the turn of the century such as Alfred Foucher considered the essentially aniconic character of Buddhist art in South Asia, where the presence of the Buddha was implicit in symbols and settings but never present in a fully articulated human form. These are still matters of contention among historians of Indian art and religion. Judging by the artistic styles of the great stupas of Sanchi, Bharhut, and early Amaravati farther south, it would be safe to assume that the Buddha was not yet a subject of direct emulation, although customary images of bodhisattvas may have begun to circulate among the laity.[41]

During the last phase of Maurya rule, the veneration of the traces and footprints of the Buddha, along with the relics of his principal disciples, became a common expression of popular piety in India. It is difficult to pinpoint exactly when, alongside freestanding figural sculptures of *nāga*s, *yakṣa*s, and *dvārapāla*s (doorkeepers), the first graven images of the Buddha entered the world of plastic arts. However, such assimilation may have taken place even before prescriptive teachings of Indian Mahayana Buddhism were firmly set in place.[42] Opinions diverge about to what extent the human likeness of the Buddha in Indian sculpture originated in the schools of Indo-Grecian art associated with the region of Gandhara in present-day Afghanistan, or whether a prototype of the Buddha image emerged from the sculptural traditions that culminated in the famous schools of Mathura and Sarnath.[43]

Ananda Coomaraswamy in his brilliant and contentious essays on the origins of the Buddha image in India argued that as a definitive figure of art, it came into circulation during the period of Gupta rule. In his view, while such iconography may have originated with the Kushans, by the time of Gupta ascendancy the Buddha figure had become familiar and widespread.[44] The circulation of the Buddha image inaugurated an era of anthropomorphic art "whose influence extended far beyond the Ganges valley."[45] It inspired similar figural representations of Hindu gods and goddesses. This was not so much the product of a Hindu "renaissance," thought Coomaraswamy, but a significant convergence of various artistic traditions and influences. Such art does not seem to have been exclusively Hindu or Buddhist in practice, although we do not know much about the production or patronage of these schools. Much more significant was a heightened aesthetic consciousness derived from Persian, Indo-Grecian, and north Indian influences, refined further during the height of the Shaka and Kushan advances in the northwest.

Coomaraswamy in his discussion of Gupta period sculpture spoke of the emergence of a "hieratic canon"—a normative order of artistic representation—captured by the sheer vitality of the various schools of sculpture that came into fruition at this moment in history. This was a result of centuries of artistic experiment, he suggested, evident in the manner and style by which the various *mudrā*s of the Buddha came to be represented. The *dharmacakra mudrā,* the gesture of the turning of the wheel, where the thumb and index fingers of both hands touch to make a circle, was a classic example of such iconographic convention. Coomaraswamy suggested that this was a natural progression from the early depictions of *nāga*s, *yakṣa*s, and the early prototypes of Shiva.[46] Whether we accept the view of such a long antecedence of iconic forms or attribute a new spurt of innovative genius to the master

Head of Buddha Sakyamuni, Gandhara region, Hadda, Afghanistan, fourth–fifth century C.E. Photograph © Museum Associates/Los Angeles County Museum of Art.

sculptors of Sarnath or Mathura, the weight of recent scholarship points to an overlap of Buddhist, Jain, and Hindu deistic images, seen early in the work of the Mathura school of the Kushan period.[47]

Art historians have long considered sculptures of the Buddha from Bodh Gaya ascribed to the early Mathura style as bearing the definitive signs of a new artistic standard.[48] These images carry the imprint of Kushan-era figures, evident most clearly from the fashioning of the garment.[49] They also depict the Buddha head haloed by the nimbus, a signature disc adorned with decorative edges.[50] Here a frontal human body is presented without the monastic robe, lustrous and ornamented. This changing embodiment of the Buddha form is particularly instructive in considering the anthropomorphic images that characterize Gupta-era sculpture. Details of hair, or the fluid lines with which clothes and drapery are represented, anticipate the explosion of sensual figures, not only of the Buddha but of gods, goddesses, and guardian figures that came to adorn the stupas and temples of Gupta and post-Gupta India.

Many of the small republics of northwestern India such as the Yaudheya and the Kuninda that threw off Kushan overlords by the second century C.E. retained the circulation of Kushan-style copper coins. They put the symbols of the goddess Lakshmi in their currency, along with images of Shiva and Kartikeya.[51] Coins are a reliable indication of the ubiquity of the Lakshmi image across northern India. Copper *kārṣāpaṇa* currency of the Shunga period bear images of Lakshmi standing on a lotus, holding a lotus, or in a typical posture flanked by elephants holding pots, sprinkling water over her.[52] These motifs, especially the images that feature elephants performing ritual ablution (*gajalakṣmī*), show a remarkable continuity of the Lakshmi motif found at the great stupas of Sanchi and Bharhut, in sitting or standing postures, or with elephants sprinkling consecratory water over her head.[53] Elephants are the companions of Indra, one of the ancient gods of rain. The lotus was an enduring aquatic emblem of immaculate birth and regeneration. The lotus on which the Buddha sits or stands is not simply a "botanical specimen," as Coomaraswamy pointed out.[54] His feet do not touch the earth, just as the lotus seems to spring from water itself. Over time the lotus became a highly refined cosmic and existential motif, and we find it in profusion accompanying the images of Lakshmi, the goddess of abundance and fortune, throughout the Gupta period. Gupta clay seals discovered in Vaishali that were used in official documents for mercantile and land transactions show the use of images of Lakshmi standing in a lotus pond attended by *yakṣas* with cornucopia or attended by elephants.[55] The iconography and stylistic features of these early images of the goddess were also influenced by the *yakṣī* figures from Buddhist stupas such as Sanchi.[56]

Similar figures of female *yakṣīs* and guardians (*dvārapālikās*) served as models for the earthy and sensuous figures of the Rivers Ganges and Yamuna that appear in the caves and temples of the Gupta period. One of the most memorable sites where such images appear is the imposing and magnificent rock relief of Vishnu in the form of the boar avatar (*varāha*) in

Vishnu as Varaha Avatar lifting Bhudevi in his tusks, Udayagiri Caves, Madhya Pradesh, ca. 402 C.E. Photograph by Frederick Asher.

the Udayagiri caves near the ancient city of Vidisha in present-day Madhya Pradesh.[57] Here, a colossal human form with the head of a boar lifts the earth—represented in the figure of a goddess (*bhudevī*)—up from the depths of the oceans in its tusks. A phalanx of gods and goddesses stand witness to this tumultuous event. A great *nāga* serpent of the netherworld waits as a supplicant at the feet of the central image, which seems to loom over the four oceans of the world. On an opposing facade the goddesses Ganga and Yamuna are carved in feminine form standing on their respective mascots

Figures of Ganga and Yamuna at the Udayagiri Caves, Madhya Pradesh, ca. 402 C.E. Photograph by Frederick Asher.

(*vāhanas*), the *makara* and the tortoise, against ripples suggesting that the two rivers are on the path to the confluence and destined for their onward journey toward the great ocean. The donor of this cave was a feudatory of the Gupta monarch Chandragupta Vikramaditya, and the fact that this territory had been wrested away recently from the western Shaka satraps gives us an idea of how such a powerful image of Vishnu, the resident deity of the Guptas, might have been associated with the power and aura of the dynasty.[58] The Ganges and Yamuna Rivers were the gateways and thoroughfares of the Gupta kingdom, and river passages were guarded vigilantly. It is no surprise that by the same token Ganga and Yamuna were also refashioned as figures that guarded the doors and passageways of Gupta-era temples.

Imperial Icons

The Kushan rulers of the northwest began issuing gold coins commemorating rulers and their respective regimes, and the practice spread to succeeding dynasties in India, especially the imperial Guptas.[59] Early Gupta coins fashioned from gold were known by the Kushan term *dīnāra*, which has been linked to contemporary coins of Rome (*denarii aurei*).[60] Much like during Kushan times, such currency bore the imprimaturs and epithets of each Gupta emperor. They also retained some reliefs and legends from the Kushan period, such as fire altars, columns, and nimbi around the heads of dynastic figures. They also added new elements such as columns of the bird-mount of Vishnu, Garuda. Such coins served as dual emblems of piety and conquest, pressed with a view to proclaim the territorial reach and prowess of the Gupta suzerains, anticipating no doubt the circulation of currency beyond the immediate frontiers of their empire. This is one reason why the imperial hunt was such a familiar feature in numismatic art. The tiger-slayer coins of Samudragupta—on which the emperor, brandishing bow and arrow, is shown trampling a tiger—can be seen as celebrating the victories of his army in the forest kingdoms of the Ganges valley, especially Samatata in deltaic Bengal, the natural habitat of the predator.[61] Similar claims seem implicit in the lion-slaying coins of Chandragupta Vikramaditya, a clear nod to his exploits against the Shaka satraps in western India, and those of Kumaragupta hunting rhinoceroses, which would have been abundant in those times throughout the forest belts of the Ganges plains.

The goddess of plenitude and fortune, Lakshmi, appears routinely on the coins of Samudragupta, typically *en face* seated on a throne; decked in a

Tiger-slayer type Gold Dinar of Samudragupta, with Ganga astride Makara on the obverse, ca. 335–375 C.E. Courtesy Pankaj Tandon, coinindia.com.

flowing robe; adorned with earrings, necklaces, and armlets; holding a lotus stem in her left arm; and accompanied by the legend *parākarama* (prowess).[62] A similar figure with a lotus in hand, feet resting on another in full bloom, has been found on the obverse of coins featuring the emperor wielding a battle-ax, with the epigram *kṛtānta-paraśu* (ax of death). On one of the tiger-slayer coins of Samudragupta, instead of Lakshmi on the obverse, we find a figure of Ganga astride her vehicle, the snouted crocodilian creature, at times resembling a dolphin, known as the *makara*. She holds a lotus with a long stem in her left hand. Her right hand is empty and outstretched, reaching out toward a standard topped with a crescent, decorated with long, flowing streamers.

Examined carefully, early coins bearing the impression of Lakshmi as goddess of prosperity (known as Shri as well) that also appear at the time of Samudragupta's reign reveal a strikingly similar design.[63] The figure of Ganga as a goddess here seems to mirror the figure of Lakhsmi. In the coins of Kumaragupta I of the horse-rider type, we find a seated Lakshmi feeding a peacock, while in coins of the tiger-slayer variety we see Ganga standing on the snouted vehicle *makara*, also feeding a peacock. On gold coins of the same period depicting an imperial rhinoceros hunt, again, we can see Ganga in a pose very similar to Lakshmi, holding a lotus in her right hand, standing on a *makara* with an attendant holding a parasol over her head. Similar figures of a female deity on a lion couchant indicate that the same prototype may have been used for the likeness of Durga or Parvati, the consort of Shiva and the goddess of victory. The debates surrounding these numismatic icons have not been entirely resolved, but evidence from these finely

Rhinoceros-slayer type Gold Dinar of Kumaragupta, with Ganga astride Makara with parasol on the obverse, ca. 415–455 C.E. British Museum. Photographs by Shailendra Bhandare.

crafted coins clearly suggests that along with Lakshmi, Ganga as a female deity had become an established figure of imperial sovereignty and fortune.

During the heyday of Kushan rule, caravan trains from all over India traveled along the ancient highway between Varanasi and Taxila, bringing ivory, elephants, spices, cloths, salt, musk, saffron, and indigo to Afghanistan and Central Asia, returning with lapis lazuli, turquoise, ceramics, wine, and an abundance of gold and silver coins.[64] Among many carved ivory artifacts most likely used as part of wooden furniture, now kept in the National Museum of Afghanistan, are three distinctive figures of a female deity astride the mythical aquatic Indian *makara*.[65] These have been cataloged rightfully as *yakṣī* figures, but they also provide an early indication that figures of such female guardians inspired Indian artists and artisans who carved similar images for Ganga, Yamuna, and Lakshmi. The city of Mathura was under a long period of Kushan occupation, and thus it is of little surprise that local ivory objects would end up in the Kushan capital of Begram (ancient Kapisa in Buddhist accounts) in Afghanistan.[66] Just as the Kushans put their deity Oaxso, the river god of Oxus, on the reverse of their coins, the Guptas did the same with Lakshmi and Ganga as aquatic icons embellished with lotuses and *makaras*.[67] The influence of such iconography is evident in the early second-century Kushan sculpture from the same region of Mathura, where images of mother-goddesses have been found with victuals in one hand and a water ewer in the other.[68] Two similar figures appear during the same period (mid-second century C.E.) from the Satavahana kingdom in the Deccan,

Makara, the vehicle of the goddess Ganga, Indian Museum, Kolkata. Photograph by author.

found on a railing among the Amaravati ruins. In this instance, they take the form of voluptuous river-goddesses, standing on the fabulous *makara* with a typically upturned snout.[69] They stand on either side of a *nāga* figure upholding brimming pots of water.

Early Gupta temples found in Eran, Sanchi, Tigawa, Nachna Kuthara, or Bhumara, flat-roofed structures without a spire, with prolonged lintels extended beyond doorjambs, are adorned with statues of the river goddesses Ganga and Yamuna guarding their entrances.[70] Inside cave nineteen in Udayagiri we find early examples of these river maidens on door frames and doorjambs along with male guardians bearing pitchers, decked in garlands, and flanked by aquatic mascots. In these renditions, as scholars like Odette Viennot have suggested, they have not yet been fully established as river deities and retain many of the traditional characteristics of tree spirits (*vṛkṣadevatā*).[71]

Gupta temple sculpture from the fifth century onward routinely features the motif of the two rivers. Deftly fashioned terra-cotta figures of Ganga and Yamuna recovered from the site of a Gupta-era brick temple in Ahichchhatra (late fifth century C.E.), now preserved in the collection of the Indian National Museum in New Delhi, are perfect examples of their fully realized forms. The figures are shown wearing a gossamer upper garment and carrying life-sustaining vessels of water. Yamuna stands on a lifelike tortoise, and Ganga rides a *makara* rendered in fine detail with distinctively crocodilian scales.[72] We can compare these figures to the exquisitely crafted image of Ganga found in Besnagar, which is now at the Boston Museum, standing underneath the dense foliage of a mango tree. Here she holds in her right hand a full vessel (*pūrṇakumbha*), and she is riding a long-snouted *makara* leaping on stylized waves.

The Parvati temple of Nachna Kuthara, near the world heritage site Khajuraho in the state of Madhya Pradesh and dated to the second half of the

Ganga and Yamuna as guardians of the Nachna Kuthara Parvati Temple, Madhya Pradesh, ca. fifth century C.E. Photograph by Frederick Asher.

Ganga on Makara overlooking the entrance of the Vishnu temple, Deogarh, Madhya Pradesh, ca. fifth–sixth century C.E. Photograph by Frederick Asher.

fifth century C.E., a two-story structure surrounded by a covered passage, sports ornate *yakṣa*s on inner doorjambs with profuse foliage sprouting from their navels, and a second frame decorated with two river goddesses and their attendants. Ganga and Yamuna are featured in similar fashion on the doorways of the Dashavatara Vishnu temple in Deogarh, Uttar Pradesh, dated to the early sixth century C.E.[73] We find this theme repeated in almost all later Gupta temples, and the motif seems to have traveled as far east as

Assam, as evident in the doorway that still stands among the ruins of the Dah Parvatiya temple in the present-day city of Tezpur. Here the figures are shown with distinct nimbi, holding long flower garlands in their hands, with beautifully executed wild geese in full flight approaching them.[74] The recurrence of these images flanking the doorways of so many Gupta-era temples, along with *makara* and tortoise designs lavished on doors, brackets, and lintels, shows clearly how attentive Gupta-era sculptors were to this recurrent motif, much like contemporary masters of the mint who set similar figures on the obverse of the most important gold coins.

The Vakatakas of the western Deccan, close contemporaries and allies of the Guptas, embraced the same iconic river goddesses and their vehicles in the sumptuous rock-cut cave sculptures of Ajanta and as consorts of Shiva in the colossal sculptures in the island cave of Elephanta.[75] Water from these sacred rivers was seen not just as the source of salvation but as the essence of natural potency. By placing these sculptures on doors and entryways, sculptors sought to capture the grace and power of these rivers for the greater glory of royal patrons. As Stella Kramrisch pointed out, visitors and devotees entering through these portals had to pass their threshold, almost mimicking a ritual of initiation for the privilege of viewing the deity within, which would be typically placed deep inside the nave.[76] Rivers were placed at the gates of temples to guard against contaminating elements from the profane, external world outside.

Symbols of Consecration

By the second century C.E., the waters of the Ganges became an indispensable ingredient in the sumptuary rituals of kingship in northern India. Early evidence, though scattered, comes from the little-known line of Naga kings of central India, whose ancestral seats were the settlements of Vidisha, Kantipuri, Mathura, and Padmavati. Most of these domains were subjugated by the Gupta emperor Samudragupta after he defeated Ganapatinaga, one of the powerful kings of that lineage. The Puranas suggest that they may have been worshippers of Shiva belonging to the Bharasiva clan. One such king, Bhavanaga, married into the formidable Vakataka ruling dynasty of central India, contemporaries of the Guptas. The martial legacy of the Nagas is commemorated in more than one of the twelve copperplate inscriptions of the Vakataka king Pravarasena II. Two of these plates, one found near Berar in the Deccan and the other in Siwani (Seoni) in central India (first deciphered by Cunningham in 1836), contain a striking reference to the consecration of great Naga kings of the past. They suggest that during coronation the kings-to-be carried a replica of the Shiva lingam on their shoulders and sprinkled their foreheads with the untainted waters of the Bhagirathi (Ganges) obtained by dint of their valor.[77] These inscriptions also suggest that the performance of Vedic horse sacrifices, followed by ritual ablution, were also a part of the coronation.

The fifth-century Sri Lankan Pali chronicle *Mahavamsa*, compiled by Buddhist monks of the Anuradhapura Vihara, details an exchange of gifts between Mauryan emperor Asoka and the Sinhalese king Devanampiya Tissa. The text suggests that Asoka's envoy and offerings were also a gesture of recognition of the accession of the Sri Lankan king, and they give us an idea of the items directly associated with the ritual consecration of a monarch, if not exactly during Mauryan times, then surely by the time the text was put together. Among these were a ceremonial sword, a parasol, golden slippers, unguents, red ocher, water from the mythical lake of Anotatta—source of all the great rivers of the world in Buddhist cosmology—and a conch shell with a right-hand swirl containing water from the Ganges to be poured over the head of the king. These items suggest an elaborate set of rites involving the preparation of the body of the king, including anointment, ablution, and the sprinkling of water from a sacred source. While some of these gestures can be traced back to the coronation rites discussed in the later Vedic texts, especially the Rajasuya sacrifice that required the ministration of water from many different sources—rivers, ponds, whirlpools, and the sea—the requirement of waters from the Ganges suggests a new recognition of their potency.[78]

The Sanskrit word for coronation (*abhiṣeka*) is derived from a root verb associated with the action of pouring out, scattering in droplets, sprinkling, emitting, or impregnating. This etymology could be extended to include anointment, inauguration, and consecration. The same word was used to describe the sanctification of Hindu deities, especially the ritual bathing of divinities and their images as part of the accoutrements of worship. The solemnity of the consecration ritual struck the imagination of the great Sanskrit playwright Bhasa, who is reputed to have lived during the second or third century C.E. Bhasa's play *Abhisheka*, which draws on the *Ramayana* for its plot and characters, explores this theme of coronation for the adulation of its main protagonist, Rama, god incarnate.[79] Rama was denied his rightful coronation by the machinations of his stepmother, who persuaded King

Dasaratha to banish him and his wife, Sita, to the wilderness. Bhasa's play focuses on the fight between Vali and Sugriva of the monkey clan over the crown of Kishkindha, in which Rama intervened. After Vali, the tyrannical elder brother, is defeated and killed, Rama anoints Sugriva as the king of the forest kingdom. At the end of the play, after the demon Ravana has been slain and Sita rescued from his clutches, Rama prepares to reclaim his ancestral kingdom. But his wife, Sita, is suspected of infidelity and compelled to take the test of fire to prove her chastity. Before she can be immolated, the fire god Agni intercedes. He identifies Rama as the incarnation of Vishnu in the flesh (*Nārāyaṇa*), and the play ends with the veneration of the prince of Ayodhya, the god Agni himself performing the *abhiṣeka* of Rama as the new king.[80] The celebrated poet Kalidasa's epic verse *Raghuvamsa*, a lengthy tribute to the exploits of the long lines of kings of the solar progeny (including Rama), ends with a poignant description of the death of the king and the pregnant and grieving queen assuming the mantle of the kingdom—being inundated by the consecratory waters being poured on her from golden jars in keeping with the long-established family custom (*vaṃśābhiṣekavidhi*).[81] Kalidasa speaks of true kings as *atharvavidā*—crowned according to the rites prescribed in the *Atharva Veda*. Such rites were believed to enhance the longevity of the new king, protect him from enemies, and invite the most powerful gods to render him invincible in battle.[82] He also describes veneration of the king on his inauguration day, with aged relatives waving sheaves of barley and barks of the banyan tree and sacred waters being poured on him as he ascends the ivory throne. Prisoners and caged birds are set free, men condemned to death are spared for a day, full-grown bulls are unyoked, and cows are unmilked. Kalidasa describes the waters of *abhiṣeka* splashing on the head of the monarch shining like the river Ganges thundering on the matted locks of Shiva.[83]

These dramaturgical and literary references to ceremonial lustration are significant, although it is difficult to pinpoint exactly when coronations began to be viewed as the conferral of divine qualities. Such rituals had been devised long before the advent of the imperial Guptas. Inscriptions, coins, architecture, and sculpture from the Gupta period suggest, however, that a sea change had taken place in what sociologist Jan Heesterman called the "diffusion of power and authority" in the conception of kingship and its relationship to ancient Vedic sacrificial rituals.[84] Military power and the capacity to overwhelm adversaries through the force of arms or the ability to dictate the fate of tributary kingdoms and feudatories, while they catapulted rulers like Samudragupta to imperial stature, may not have been seen as sufficient means to create a lasting legacy for successive generations and regimes. Such power and fortune had to be acknowledged and ritualized through a strategic resumption of *śrauta* consecratory rituals such as the *aśvamedha* horse sacrifice, which were also featured prominently on coins and medallions. These rituals were not just reserved for the accession of a new ruler to the throne but were replicated at various points during the entire course of the reign. In thinking about the foundations of kingly authority in the European tradition from the point of view of both temporal and immortal power, we are used to thinking about divine right and the king's two bodies: one human and perishable and the other immaculate and everlasting. The aura of imperial kingship in India, by contrast, seems to have been shared across many other entities, including gods of war, goddesses of fortune, and elemental water and earth bodies representing a cosmos of fecundity and plenitude—in effect, a great chain of earthly and celestial beings.[85] In this exercise the Rivers Ganges and Yamuna as bearers of water, fertility, and blessings, much like archaic tree and water spirits, became the lasting aesthetic and iconic symbols of the Gupta and post-Gupta imperial orders.

A Contested Terrain

The great Chinese traveler and Buddhist scholar Faxian, from the Shanxi province of Jin China, visited India between 399 and 414 C.E., following the trail and reliquaries of the Buddha and his disciples. He had survived an arduous trek across Central Asia, through the Swat Valley and Gandhara into the Punjab, and he traveled through many of the important cities and pilgrim centers of the Gupta Empire during the reign of Chandragupta II. In his travelogue Faxian uses the contemporary description of the upper and middle Ganges valley as the middle country (*madhyadeśa*) discussed earlier in this book. He describes the passage on foot from Mathura via Sankisa to Kanauj on the Ganges, and thence to Ayodhya and Vaishali onward to the capital Pataliputra, then back westward again to Varanasi. He also describes Bodh Gaya (place of Buddha's enlightenment) and Tamralipti in deltaic Bengal, from where he set sail for the monasteries of Sri Lanka.

Faxian's account reads like a pastiche of history, myth, and folklore, animated by a geographical landscape of piety and wonder, and he writes about the Ganges valley strewn with richly endowed *vihāra*s and legendary monasteries. His observations suggest that despite their dynastic veneration of Vishnu, the Guptas ruled over what was still a largely Buddhist society.[86] Faxian's account describes the spot where one of Buddha's closest disciples, Ananda, traveling from Magadha to Vaishali, attained nirvana. His body was cleft in two parts, one appearing on each bank of the Ganges, ensuring a fair distribution of relics. He also writes about the ruins of Asoka's great palace in Pataliputra, its walls, doorways, and sculptures still standing by the side of the flowing Ganges. The monastery set up by Asoka was still flourishing there with a settlement of six or seven hundred monks.[87] He also gives a vivid description of a festival held during the second month of each year, near Vaishali near the confluence of the Gandaki and Ganges Rivers, roughly eighteen miles from the boundaries of Magadha. Faxian writes that at the meeting point of five rivers he witnessed a procession of images carried by a four-wheeled moving tower, 22 feet tall, with five separate platforms secured with bamboos tied together.[88] He counted at least twenty such carriages, each decked in brightly colored linen, with statuettes of deities fashioned out of gold, silver, and glass placed inside. Under embroidered canopies of vibrant and flowing silk, images of the Buddha were placed in each of the four corners. Much like in the parade of a Hindu god or goddess today, he saw crowds throng to light lamps at night, with music playing and offerings being made to the Buddha and many other deities. The ceremony was attended by visitors from faraway countries, nobles, and wealthy householders who were the patrons of various charities.[89]

Faxian's descriptions give us a view of the riverside as a bustling concourse of pilgrims, travelers, and traders. Feudatories and officials of the Gupta Empire commanded the fords, crossings, and roads across the major channels of the Ganges and its tributaries, and the legacy of such territorial preeminence continued long afterward. Provinces along the main course of the river were some of the most important administrative divisions of the empire (*bhukti*s): Pundravardhana (northern Bengal), Vardhamana (western Bengal), Tira (northern Bihar), Nagara (southern Bihar), Shravasti (Awadh), and Ahichchhatra (Rohilkhand).[90] The prosperous region of the Ganga-Yamuna Doab, known at the time as Antarvedi, was placed during Skandagupta's time under the formidable officer (*viśayapati*) Sarvanaga, who was likely a descendant of the great Naga line of kings.[91] During the gradual decline of the empire, however, the valley began to lose its autonomy. The last descendants of the Guptas (known as the Later Guptas) who held on to Magadha lost many of these key provinces to new and upstart powers such as the Maukharis, whose stone inscriptions suggest that they had wrested control away from the Guptas in key areas near Lucknow, Jaunpur, and the Gaya districts in the present-day states of Uttar Pradesh and Bihar. In 554 C.E. Isanavarman, a former tributary who had made a name fighting the Huns, raised his standard against the Guptas and took the Gupta-style title of "king of kings" (*mahārājādhirāja*). Epigraphic evidence suggests that in one of these encounters Kumaragupta of the Later Gupta lineage was defeated and, overcome by his loss, took himself to the confluence of Prayag. There, decorated in flowers as if he were plunging into the Ganges, he plunged instead into a fire kindled by dried cow dung.[92] His son Damodaragupta, leading a direct attack against a line of Maukhari elephants, was also killed.[93]

Toward the end of the sixth century C.E. the Maukharis emerged as undisputed rulers of the upper-Ganges tracts, ultimately driving the remaining Guptas out of Magadha. The long and unquestioned writ of the Guptas over the Ganges valley had thus come to an end. They became a secondary power holding on to bits of territory in Odisha and Bengal, with a new outpost in the region of Gauda in the north.[94] There were many contenders for supremacy in the valley of the Ganges after the eclipse of the Gupta Empire. The Maukharis of Kanauj held sway over the old provinces of the Gupta Empire in the middle country, and they commanded territories as far east as the Brahmaputra in Kamarupa (Assam) and as far south as the Vindhya Hills. They would face a fresh challenge in the west from a new dynasty known by the name of Pushyabhuti (*puṣpabhūti*), which had raised its standard in the region known as Thaneswar (Sthanvisvara) under King Prabhakaravardhana. The Maukharis and the Pushyabhutis fought over the supremacy of the valley for almost three generations. By 605 C.E., the Pushyabhutis defeated the Huns; took hold of Sind, Rajputana, Malwa, and Gujarat, and secured an advantageous alliance with their Maukhari rivals, with the marriage of Princess Rajyashri to the young ruler of Kanauj, Grahavarman.

A series of calamities befell the Pushyabhuti dynasty at this time. The king died suddenly, leaving the prince regent, Rajyavardhana, distraught; the Guptas of Malwa attacked the region of Kanauj, killing Grahavarman and taking the queen, Rajyashri, hostage. Rajyavardhana, at the command of a large army, set out to punish this act of aggression against his brother-in-law, but after routing the enemy he was lured into false confidence by Shashanka, the ruler of Gauda in southeast Bengal, and murdered in cold blood. The widowed Rajyashri, thinking of ending her life, had cloistered herself in a Buddhist monastery at the eastern outcrop of the Vindhya Hills. Prince Harsha and his troops encamped by the Ganges caught up with her and persuaded her to return to her kingdom. She asked Harsha to take charge of affairs in Kanauj, and he acceded to the throne guardedly, giving himself a new title of Shiladitya.[95] Harsha resolved to avenge the death of his brother by making Gauda "resound with the fetters on the feet of all the kings incited to insolence."[96] The story of Harsha's rise to preeminence is well known, corroborated by many sources, including the official eulogy of the poet Banabhatta, known as the *Harshacharita*, and the account of Chinese traveler Xuanzang, and only a brief sketch can be offered here.[97]

Harsha, merely sixteen years old, with his maternal cousin and general Bhandi, embarked on a military campaign that went in every direction (*digvijaya*). For six years his "elephants were not unharnessed nor the soldiers unbelted" and vultures with their heads bloodied in human remains circled the capitals of rival kings, until all his rivals finally accepted his lordship in the north (*sakalottarapatheśvara*).[98] Neither Harisena nor Xuanzang says much about his vendetta against the king of Gauda, Shashanka, who seemed to have ruled untroubled until 619 C.E. in contemporary Bengal. We also know that Harsha's troops were no match for the great Chalukya king Satyasraya Pulakeshin II, whose army defeated him on the banks of the Narmada River.[99] Harsha succeeded in establishing a measure of dominance over most kings and chieftains of the valleys of the Ganges and Indus that had not been witnessed since the time of the Guptas. He also extended his sway in eastern India by securing a valuable alliance with the ruler of Kamarupa (Assam), Bhaskaravarman. While he defeated many kingdoms of the north, he did not seize their land, ruling instead as a titular overlord. Constantly on the move, with awnings, screens, tents, and pegs for makeshift camps, he kept a personal vigil on the distant parts of his sprawling empire.[100] Harsha was keenly aware of the outward display of power and piety, and toward this end the waterways of the Ganges provided a perfect setting for a grand staging of his elaborate rituals of tribute and charity.

Empyrean Display

Harsha arranged for a royal march along banks of the Ganges from the Rajmahal Hills in the east to Kanauj in the west as a preamble to the unveiling of a new religious summit held in the capital. According to Xuanzang, he mobilized a large section of his army for the occasion, including hundreds of war elephants and a large flotilla of boats.[101] He placed his friend and vassal the king of Assam at the head of the cavalcade, followed by a long train of religious men: devout Jains, Brahmin priests, various sects of Shiva worshippers, eminent Buddhist *arhat*s, all marching in unison toward the imperial capital of Kanauj. At regular stretches during this march Harsha dismounted and walked on foot, golden drums sounding as the distance of each *krośa* (about 2,000 yards) was covered, an honor reserved strictly for the emperor.[102]

Thanks to the detailed observations of Xuanzang, who also participated in this ceremony as the king's guest, we can reconstruct parts of this remarkably ostentatious spectacle. The royal procession skirted the southern bank of the Ganges, headed by the two kings side by side. Soldiers followed on boats and elephants, accompanied by music from flutes and harps and the beats of drums, with crowds swelling to several hundred thousand. The march took place over the course of three long months, while invitees gathered in Kanauj. Among them were eighteen kings, thousands of monks, eminent Buddhist scholars from the leading centers of learning such as Nalanda, and notable Brahmins. Seating was arranged for two thousand people in two covered halls along the river, with a full statue of the Buddha wrought in gold and an improvised palace from where the emperor would emerge, dressed as the deity Sakra, with the prince regent decked out as the lord Brahma. In their flamboyant costumes, Harsha and his son led the procession carrying a statue of the Buddha, holding a canopy and umbrella over it. At the end of the march the image was placed on a throne, with offerings made by the emperor and his retinue, and accompanied by Xuanzang, who seems to have presided over the ceremony himself. Copious amounts of silver, gold, jewels, and fine cotton were distributed to mark the occasion.

The ceremony was not without incidents. A dispute erupted between the Mahayana patrons and the Theravada adherents. A fire broke out in the main dedicatory tower where the Buddha image had been placed, and a visibly distraught emperor was seen rushing toward it. The blaze, which turned out to be an act of attempted sabotage, subsided miraculously, but as Harsha was stepping down from the tower an assassin came at him with a knife. Harsha, unshaken, quickly disarmed him. It was soon found out that he had been dispatched by a group of disgruntled Brahmins—Xuanzang calls them heretics—who felt that they were no longer feted or revered in the new regime. After the facts had been divulged, Harsha punished the leader of the plot and banished the other five hundred Brahmins to the frontier provinces. The Brahmins might have had genuine grounds for complaint if Xuanzang's estimate of the amount of charity lavished on the Buddhist establishment is to be believed.

Harsha chose the confluence of Prayag for grand ceremonies of public charity appropriately titled *mokṣa* (salvation). A large stretch of fine, sandy bank—an area measuring about five square miles on the west side of the confluence, which today is the site of the great pilgrimage of Kumbh—came to be known as "the arena of charitable offerings" during Harsha's reign.[103] Invitations were sent far and wide to Buddhist monks, Brahmins, ascetics, orphans, the needy, and the recently bereaved. Here again, enclosures and temporary structures were set up with chests of jewels, gold and silver coins, and a multitude of precious garments for distribution to the masses. This grand potlatch also served as a makeshift stage for some of the most important vassals of the empire to exhibit their allegiance to Harsha. Harsha's tent was pitched along the north bank of the Ganges, the tent of the great king of Vallabhi lay to the west of the confluence, and the king of Kamarupa camped on the south side of the Yamuna.[104] This was also a perfect opportunity to exhibit the imperial army, caparisoned war elephants, and boats.

On day one, an image of the Buddha was installed in a thatched building, with a distribution of valuable garments. On days two and three, images of the sun god Aditya and Shiva were taken out, and Brahmins were presented with gifts. Day four was reserved for a select group of ten thousand Buddhist monks, who were each presented with gold, a pearl, and a cloth and plied with drinks, food, flowers, and perfumes. For the next twenty days other religious orders, along with the poor, orphans, and refugees from faraway countries, were honored. The emperor made it a condition that everything must be given away except for the horses, elephants, and military equipment, including personal jewelry, clothes, and necklaces from the royal collection.

What is remarkable in this pageantry is the elevation of the emperor to an iconic, quasi-divine status.[105] Harsha's court poet Banabhatta (known also as Bana) describes the young emperor lounging on his magnificent throne, attended by subjects, vassals, and attendants, his winsome body shining like a resplendent canopy. Lit by the luster of his jewels, like a lotus pond "embraced by the royal goddess of prosperity," it was as if the goddess Shri had taken him in her own arms, blessed him with all the marks of kingship, and, against all odds, placed him on the throne of Kanauj.[106] Bana also calls him an avatar of all the gods mingled in one, his crowned head lustrated by fortune, the crown jewels adorning his topknot shining as if he were being consecrated by the combined waters of the Ganges and the Yamuna flowing on their own from the confluence at Prayag.[107] This recurrent theme of divine interdiction should not come as a surprise given the nature of Bana's unabashed flattery, but the symbols of kingship he touches on playfully blur the distinction between the supernal and mortal attributes of the king. When Bana intones that the emperor's sovereignty resounded across the four oceans of the world, and that his most auspicious marks were those of the lotus, conch shell, fish, and *makara*, we know that he is alluding to some of the classic attributes of Vishnu.[108] Late in his account it is revealed that Harsha's ancestors of the Pushyabhuti lineage, owing to a set of rites performed by the famous Tantric guru Bhairavacharya and their steadfast devotion to Shiva, were preordained to produce a scion with such divine qualities.[109] It was by the strength of this favor that King Prabhakaravardhana on his deathbed named Harsha as the one "accepted by Goddess Shri of her own accord" (*svayameva śriyā gṛhīta*) rather than his valiant and unlucky brother.

In retrospect, it seems natural that the ruling house of Thaneswar, poised to gain the upper hand in the valley of kings, would entertain such grand and boastful epithets. More than a hundred years before them, the mighty Gupta emperors had often compared themselves to gods. Both Chandragupta Vikramaditya and his grandson Skandagupta took the lofty title of Devaraja (king of the gods). Skandagupta's Junagarh inscription boasts that the goddess Lakshmi herself chose the emperor as her husband, rejecting all other kings. However, Harsha seems to have proclaimed a more studied image, even dressing up as a god while offering homage to the Buddha, playing the role of a royal patron and a humble bodhisattva at once. Not just the unmistakable imprint of *harṣadeva* on his coins but especially the king dressed up as the god Sakra during the procession of the giant statues of the Buddha point to a concerted display of public adoration that had not been seen before on the imperial stage of Indian politics. Harsha's exhibition of exalted piety seems to have been further encouraged by the surge of image worship that was redefining popular culture during the heyday of Mahayana Buddhism in northern India.

Elusive Fortunes

During the early decades of the seventh century, many kingdoms of north India with imperial ambition—the Maukharis of Kanauj, the Maitrakas of

Valabhi, the Later Guptas of Malwa, the Pshyabhutis of Thaneswar, and the Gauda in Bengal—tried to claim the fertile plains of the Ganges. This was still the coveted Middle Country, the heart of Buddhist pilgrimage, blessed with some of the most fertile soil in India, endowed with the largest cities and markets, and knit together by the sinews of trade and transport. The rise of Kanauj to imperial glory was based on the fact that it commanded such a large swath of the valley. It was, however, a hard-fought privilege. While the later years of Harsha's reign is seen as a time of peace, tolerance, and public charity, it should not blind us to the political rivalries, intrigues, and violence that accompanied Harsha's rise to power.

A major rival was Shashanka, the king of Gauda, Bengal, who had plotted the murder of Harsha's elder brother. Shashanka rose from the position of a tributary chief to gain a stronghold in the region of Magadha, and he won a number of swift campaigns in Varanasi, Kushinagar, Gaya, and Pataliptura, ending with the annexation of Gauda in Bengal, where he built his capital city, Karnasuvarna.[110] Poet Banabhatta and Xuanzang the Chinese monk, while reviling Shashanka in their accounts (Bana calls him the "vile Gauda serpent"),[111] suggest at the same time that he was a formidable adversary who had also earned the reputation of a devout worshipper of Shiva and an arch-enemy of the Buddhists. Unlike Harsha, Shashanka tried to drive fear into the kings of the valley by desecrating some of the major Buddhist sites. Xuanzang, discussing the distribution of Buddha's relics after his nirvana, states that the envious Shashanka "destroyed the religion of the Buddha," broke up the convents, and dispersed the establishments of monks.[112] He cut down the great Bodhi tree under which the Buddha had found enlightenment, digging up and setting fire to its roots, but he could not destroy it. After Purnavarma came to the throne of Magadha, he bathed the roots with milk and miraculously brought it back to life, and he built a protective wall around the tree after it grew back.[113] In Magadha, Shashanka attempted to destroy a prized stone relic with Buddha's footprint that had been enshrined by Emperor Asoka. He tried to smash it to pieces and throw them into the Ganges, but again by the grace of the Buddha, it reappeared unharmed in the same place.[114] It is hard to overlook the Chinese pilgrim's animus, but these atrocious acts of defilement he describes suggest that Shashanka must have tried to gain an upper hand over the powerful Buddhist establishments of the lower Ganges valley. We do not know how Buddhists fared in his capital Karnasuvarna in north Bengal, which was renowned for the Raktamrittika Vihara monastery, a great center of learning. We do know from his coins that he sought to project his legitimacy as a ruler blessed by the goddesses of fortune and prosperity, much like the legendary Gupta overlords. Gold coins attributed to Shashanka follow closely the numismatic tradition of the Later Guptas and other rulers of Magadha that preceded him. One of these is the archer-type coin, with the king standing with a bow and arrow, with Shiva's bull, Nandi, and on the obverse, a goddess with nimbus, seated on a lotus, a lotus-stalk in her left hand and a fillet in her right.[115] The second is a coin featuring Shiva with a nimbus, reclining on his bull, with his left arm raised and the moon above on his left. On the obverse of this coin, again, is the goddess of fortune—Lakshmi or Shri—astride a lotus, holding a lotus-stalk in her left hand, which rests on her knee, with two elephants on either side sprinkling consecratory waters over her crowned head. Shashanka could not hold on to Magadha, but Gauda would not yield to Harsha until his sudden illness and death, which Xunzang describes as a curse that befell the unbelieving Bengal monarch after his numerous misdeeds. After Shashanka's demise, circa 640 C.E., Harsha was finally able to annex the kingdom of his bitter rival.[116]

Harsha seems to have been keenly aware of the bounty of the Ganges and the mythic potency of its waters. During his lavish *abhiṣeka* ceremonies the wives of his most important vassals poured consecratory waters from the river on his crown from special golden pitchers.[117] Like Emperor Asoka, he built thousands of stupas and resting houses all along its banks.[118] His

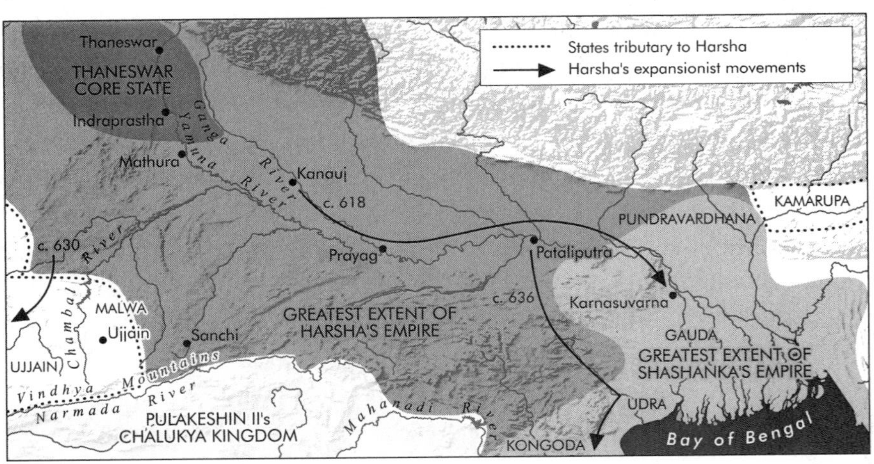

The empire of Harshavardhana of Kanauj. Map by M. Roy Cartography.

sumptuous palace in Kanauj on the west bank of the river was decorated by sculptures of crocodiles, fishes, and tortoises, with festive *makara* spouts gushing with scented water cascading into pools.[119] Bana, in his account of Harsha's life, quips that the goddess of fortune is fickle and does not stay for long in the same place.[120] He seems to have been unusually prescient. Harsha's empire fell apart within just a few years of his death.

CHAPTER 7
CRUCIBLE OF EMPIRES

The Ganges near Kanauj that the Chinese traveler and scholar Xuanzang saw was pure and blue like the ocean, its banks full of fine-grained sand. He describes it as the legendary "river of religious merit" that had the manifest power to wash away countless sins.[1] He also saw large crowds at the major pilgrimages along the river, such as the Ganga-Yamuna confluence at Prayag, full of merit seekers gathered to assuage the wrongs of their lifetimes. Some had come for penance and mortification, some to fast unto death in the hopes of getting to heaven quickly. Bathing in the Ganges was synonymous with the acquisition of merit as a kind of spiritual collateral for divine judgment after death, and the regimes that succeeded the Maukharis and Pushyabhutis of Kanauj fought over the distinction of protecting the many pilgrimages and sacred cities that dotted the Ganges valley. The Gupta Empire had left behind a rich and variegated iconography of the Ganges. Some of these representations, especially of the river as a female guardian or as the heavenly companion of Shiva, became standard figures installed in temples throughout the Indian subcontinent.

These images also began to appear routinely in temples of the far south. The Pallavas of Kanchi, who ruled from the valley of the Kaveri River and the emerald strip of the Coromandel Coast, paid rich tribute to the Ganges in their rock-cut structures. Their rivals, the Chalukyas of Badami, who held the territory between the Krishna and Narmada Rivers in western and central India, did the same. The Rashtrakutas, who displaced and succeeded the Chalukyas, also designed temples to create replicas of the mythical mountain Kailasa, Ganga's celestial abode. For an entire century they fought with the Gurjara-Pratiharas of western India and the Pala Dynasty of Bengal over control of Kanauj and the western Ganges valley. A ruler of the Chola Dynasty whose empire stretched across most of the southern peninsula and who sent overseas military expeditions to Sri Lanka and Indonesia took the title "the conqueror of Ganges." Having vanquished the major powers of the north, he brought back great quantities of sacred water as a prize to his newly constructed capital, to be stored in a massive temple complex that still dominates the surrounding countryside.

This chapter describes how the image of the river, its valley, cities, and kingdoms, became an inextricable part of the political imagination for a succession of imperial regimes. It was not just the sacredness of the river or the antiquity of its Hindu pilgrimages and bathing steps that inspired such a vision of territorial power. It was also the rich trove of agrarian surplus from its fertile basin that helped sustain the armies and courts of the Gurjara-Pratiharas and their vassal kingdoms, the Gahadavalas of Kanauj, the Turkish sultans of Delhi from the twelfth to fourteenth centuries C.E., and indeed the great Mughal Empire. An essential thread in the flow of riparian history narrated in this chapter is the rise, expansion, and political struggles waged over one of the most densely populated river valleys in the world.

Imperial Rivalry

Within a hundred years of the end of Harsha's empire, the arena of imperial rivalry had shifted to the south. In the political drama that unfolded over the seventh and eighth centuries C.E. in the Deccan, the Western Gangas, the Pallavas of Kanchi, the Chalukyas of Badami, and Pandya kings farther down the peninsula were embroiled in a struggle for political supremacy that brought about an extensive period of intrigue and military raids. The Western Gangas, based in present-day Karnataka, were once vassals of the Pallavas but asserted their independence with the rise of the Chalukyas in the western Deccan and eventually allied with the Chalukyas against the Pallavas. Around 617 C.E. the Chalukya forces under Pulakeshin II had forced their way into the outskirts of Kanchi, the capital of Pallava king Mahendravarman I. While they could not take the capital, they inflicted heavy losses on the Pallava forces at a place called Pullalur. To avenge this debacle the Pallava king Narasimhavarman I attacked the Chalukya capital Vatapi (Badami) in 642–643 C.E. and burned it to the ground.[2] After the sack of the city, he took the title of Vatapi-Konda, "the conqueror of Vatapi." The two regimes were henceforward locked in a series of deadly raids and counter-raids that lasted for generations.[3] Under Vikramaditya I (great-grandson of Pulakeshin II), the Chalukyas would finally overcome their rivals and expel them from their homeland.[4]

During these fierce contests, the temple served as a direct symbol of political power. The Pallava king Narasimhavarman commissioned the Mallikarjuna temple in the heart of the Chalukya capital, with an inscription proclaiming his military triumph over his Chalukya rivals. In a fitting riposte, Queen Lokamahadevi, wife of Vikramaditya I, had the Virupaksha temple built in 745 C.E. to commemorate the victory of her husband, "captor of Kanchi," over the Pallavas.[5] In both these monuments, the River Ganges appears as a deliberately chosen motif. In the Mallikarjuna temple, Ganga and Yamuna are installed as guardian deities.[6] In the Virupaksha temple, Ganga is represented as flowing in all three directions—heaven, earth, and the netherworld—along with a recapitulation of the story of its descent.[7] In fluid dynamic sequences carved meticulously in sandstone, the old myths are retold. The Vasus come to life in the act of being liberated from the curse of sage Vasishtha. There is a penitent Bhagiratha waiting to consecrate the remains of his ancestors, the sons of Sagara lying in a heap of ashes beside an enraged sage Kapila, and Ganga flowing into Shiva's matted locks before splashing onto earth. Paintings of Ganga as Shiva's consort once adorned the pillared courtyard (*maṇḍapa*) of this temple, which can still be recognized in the fragments that have survived on the underside of the eaves. On the facade of the nave (*garbhagṛha*) of the Papanatha temple in the Chalukya capital of Patadakkal, adorned with finely carved pilasters on both sides, Ganga and Yamuna appear astride their respective vehicles, the crocodile and the tortoise.

The recapitulation of stories from old Indian mythology on the walls of rival structures gives us some idea of how the Ganges was viewed, not only from the perspective of traditional cosmology but also as a manifest emblem of royal power and protection. As Cathleen Cummings has recently pointed out in the case of the Virupaksha temple, the site of its construction was also the place where the king was crowned—with the grand entrance to the temple facing the banks of the Malaprabha River.[8] Indeed, there are panels dedicated to Shiva and Parvati playing a game of dice, scenes from a chariot drive, and images of Vishnu as the dwarf avatar (Trivikrama) who once covered the three worlds in one stride—each composition dedicated to a particular ritual associated with the coronation of Chalukya kings.[9] The Pallava rock-cut sculpture of the descent of the Ganges at the shore-temple in Mahabalipuram, discussed earlier in this book, can be seen as a similar attempt to exalt the status of the dynasty. As art historian Padma Kaimal has suggested, this rendition of the myth of the Ganges—which can also be read as homage to Arjuna, the warrior-hero of the *Mahabharata*—gives us an insight into how the Pallavas sought to incorporate the tale of Ganga's descent into their new lexicon of sovereignty.[10] In this sense, the architect's homage to the river may also be seen as a prayer to the actual rivers of the region for the well-being and prosperity of the kingdom. The Pallava kings wanted to extend their domains beyond the fertile basin of the Kaveri River, into the valleys of the Tungabhadra and the Krishna toward the west and north. An inscription of Simhavishnu, one of the early rulers of the dynasty, states clearly that he seized the territory of the Cholas of the north country "embellished by the daughter of Kavira," that is, the river Kaveri, "whose ornaments are the forests of paddy" and where one can find the "brilliant groves of areca palms."[11]

A copper plate inscription of the Pallava king Nandivarman dating to the eighth century C.E. clearly shows an attempt by his scribes to relate the story of the descent of the Ganges to the descent of royal lineage. It states that the Pallavas were a powerful and untainted race of warriors, a part of Vishnu incarnate (*viṣnoraṃśavatāra*), who had demonstrated unrelenting courage in their conquest all parts of the terrestrial sphere, enforcing rules of the caste order (*varṇa*), and they had descended on this earth just like the Ganges (*gaṅgāvatāra*) to purify the world.[12] In the light of such a claim, the proliferation of Ganges images in Pallava temples, either as freestanding guardians or displayed along with Shiva, makes perfect sense. In the cloisters and panels of the freestanding masonry temples built by the later Pallava kings we can see a further elaboration of this theme, with the rendition of Shiva as Gangadhara, catching the celestial river in his matted hair to break the impact of its fall on creatures of the earth.[13] An early execution of this form can be found at the Tiruchirapalli rock-cave temple, where Gangadhara stands with his foot on the dwarf Apasmara, a symbol of folly and ignorance. At

Image of Shiva Gangadhara at the Tiruchirapalli Temple, Tamil Nadu, Pallava Dynasty, early seventh century C.E. Photograph by Padma Kaimal.

the Kailasanatha temple in the Pallava capital, Kanchi, Shiva holds out a single lock of his hair to receive the river hurtling down from heaven. Such images were meant to augment the reputation of the ruling dynasty, which is one reason why many of these temples were attacked and desecrated during times of war.

The appropriation of the Ganges motif during the long and unfinished struggle for dominance between rival powers of the Deccan—from the Narmada River in the north to the Kaveri in the far south—points to a different kind of war waged over icons and meanings. The patronage and construction of royal temples had become essential elements of statecraft. The evolution of the Pallava temple, for example, from smaller rock-cut caverns and enclosures to massive freestanding structures on stone platforms endowed with lofty towers (*vimāna*) and stepped cornices rising skyward toward elaborate golden capstones show this most clearly. The careful placement of guardians (*dvārapālas*) and guardian deities at the entrances of Pallava temples was an attempt to fashion the gates of the temple as entrances to the kingdom itself.[14] The Pallava ruler Mahendravarman I, who initiated many of the dynastic feuds with the Pandya and Chalukya kingdoms, was an indefatigable patron of such structures.[15] Many of them were named after his honorific titles (*biruda*s) in Sanskrit, Tamil, and Telugu, such as the upper-cave temple at Tiruchirapalli, famous for the image of Shiva as Gangadhara, bearer of the Ganges.[16]

These were massive undertakings that involved breaking, cutting, carving, and polishing giant slabs of stone or the refashioning of freestanding monolithic structures as in the chariot-shaped Pancha Ratha temples of Mahabalipuram. Many of these projects outlived the rulers who commissioned them, and some lie unfinished to this day. The ubiquity of the Ganga image in Chalukya and Pallava temples as door-guardian, goddess of bounty, or loving companion of Shiva entangled in the coils of his matted hair, not to mention the frequent invocation of the river in royal inscriptions, reminds us of the suffusion of myths, icons, and texts in the fabrication of political legitimacy. Placed within the sanctum of the king's temple, the Ganges—the river that is supposed to emanate from the Milky Way galaxy—implied a communion between the ruler and the realm of Shiva in the northern mountains of Kailasa. The temple was a visible embodiment of the heights of royal power, reinforced by endowments, gifts, and charity. It was precisely because the temple and the deity enthroned therein were extensions of the king's consecrated body that they were objects of great political value and vulnerable to attacks during times of war. For the regimes that succeeded the Pallavas and Chalukyas in the south, pillaging and desecration of idols and artifacts became increasingly common, sometimes involving distant regimes of the north. The construction, endowment, and defense of large temples, in this regard, were practices dedicated largely to the extension of the political aura and ritual preeminence of the ruler.

The Curse of Kanauj

Kanauj in the ancient times was known as Kusumapura, or "city of flowers." Legend has it that its ruler Brahmadatta had a hundred daughters. A holy man from the banks of the nearby Ganges, known for his austerities and spiritual power, smitten by their beauty from afar, came to the king and asked him for their hand. The youngest daughter agreed to the proposal, but the rest, frightened by his wild looks, turned him down. Enraged by their refusal, the sage cast a spell that instantly made them ugly and disfigured, and that is how Kusumapura became Kanyakubja (Kanauj) or the "city of hump-backed women."[17] Kanauj came into prominence as the ancestral stronghold of the Maukhari rulers, and Harsha made it the center of his sprawling empire. Although Harsha's dynasty came to an unexpected end after his death, the image of the imperial city on the Ganges had left a lasting impression on the Indian political imagination. Xuanzang provides some memorable descriptions of the city on the west bank of the Ganges, with its battlements, towers, and moats, along with lakes and lotus ponds shining like mirrors, abloom with flowers and lined with orchards of abundant fruit.[18] Standing high on a cliff, overlooking the passage of boats, pilgrims, and traders, and commanding the fertile plains of the upper Ganga-Yamuna Doab, it enjoyed a strategic military advantage over the rest of the valley.[19] At the same time, Kanauj became one of the most frequently plundered cities of the northern plains. This was not only because of its historical value as the old capital of the Maukharis and Pushyabhutis but also because of its preeminent position on the banks of the Ganges. The river, emanating from the great mountain Kailasa, as Ronald Inden has insightfully suggested, inspired not only contemporary notions of Indian cosmology but also the conceptions of both royal and territorial power.[20] All aspirant kingdoms and empires, especially from the Deccan and farther south, tried to emulate this idea of a sacred, riparian capital. Kanauj in this regard was now the most legendary city on the Ganges and therefore a much sought-after political prize.

The fate of Kanauj hung in the balance after the dissolution of Harsha's kingdom, and during the period of provincial rule that followed, it was plunged into a seemingly endless spiral of violence.[21] Rulers and chieftains of Assam, Magadha, Gurjara, and Kashmir were involved in raids, counterraids, and intrigue during a long period about which, except for some Jain sources, the historical record is largely silent.[22] During the first half of the eighth century, the upstart Yasovarman occupied Magadha, where he founded a new town, after which he defeated the king of Vanga (Bengal), seizing a huge cache of war elephants.[23] He was in turn challenged by the formidable king of Kashmir, Lalitaditya. The great Kashmiri historian Kalhana in the *Rajatarangini* records that Kanauj and parts of the upper Ganges valley came under the sway of Kashmir for some time. When the Rashtrakutas of the Deccan, a newly risen power who had driven out the Chalukyas, laid siege to the city, it was being held by the Ayudha Dynasty—a minor power who were no match for the ruthless invaders from across the Vindhya range.

In the period roughly between 750 and 900 C.E., Kanauj became a trophy fought over by three major competing empires. In the west, power lay in the hands of warrior clans who had banded around a ruling dynasty known as the Gurjara-Pratiharas, *pratihāra* denoting "sentinel." It has been suggested that the Gurajara-Pratiharas were of the same bloodline as some of the later Rajput clans including the Chahamanas, Paramaras, and Solanki (eastern Chalukyas), who trace their origins to the fire clans (*agnikula*) that hailed from the well of fire (*agnikuṇḍa*) at Mount Abu. Historians have speculated that they might have had ties to the Huns and the pastoral nomads of western Gurjara stock. The Gurjara-Pratiharas came into power during the second quarter of the eighth century C.E., and if Arab sources such as those of Sulaiman and Al-Biladuri are to be believed, they checked the power of the Arab settlers of the Indus Valley on more than one occasion. Under Nagabhatta I, they extended their rule over Malwa, Rajputana, and Gujarat.

Their rivals to the east, the Palas of Bengal, had been thrust into power after a prolonged period of political unrest following the death of the great Shashanka. Contemporary accounts describe this time period as the advent of the "rule of fishes" (*matsyanyāya*).[24] Invaded repeatedly by Kanauj and Kashmir, and even an expeditionary force from China, the Buddhist sources suggest that the people, anxious about the fate of their country, invested power in the hands of the able local chief Gopala. His successor, Dharmapala, sought to extend the authority of the Pala Dynasty across the Ganges valley. Claiming to be the overlord of Kanauj, he quickly set up a string of vassals, including chieftains from lands as far north as Nepal. The Palas were the last known Buddhist rulers of eastern India, known for their patronage of the great Somapuri Vihara in Varendra (present-day Paharpur, Rajshahi) and the Odantapuri Vihara in Bihar. They proclaimed that they had seized *āryāvarta* by force. Their expansive title "Lord of the Five Gaudas" (*pañcagauḍeśvara*), which belied contemporary political realities, implied that they also commanded the realms of Punjab, Kanauj, Gauda, Mithila, and Utkala.[25] They also claimed that their dynasty had been blessed by the Ganges. The poet Sandhyakara Nandi, writing in the early eleventh century C.E. in praise of Ramapala, described the line of Pala kings as immaculate (*dhavala*, white), just like the river of the gods (*sūrasindhu*).[26]

The Rashtrakutas (*rāṣṭra*, "kingdom"; *kūṭa*, "peak, summit") were feudatories of the Chalukyas, who had migrated from the Latur region of Ellichpur, which is the source of the Tapi River in western India, around the early seventh century C.E.[27] One of the early monarchs of the dynasty was Dantidurga ("one whose elephant is his fort"), a maverick, battle-tested leader and architect of memorable victories in territories as far away as Kalinga on

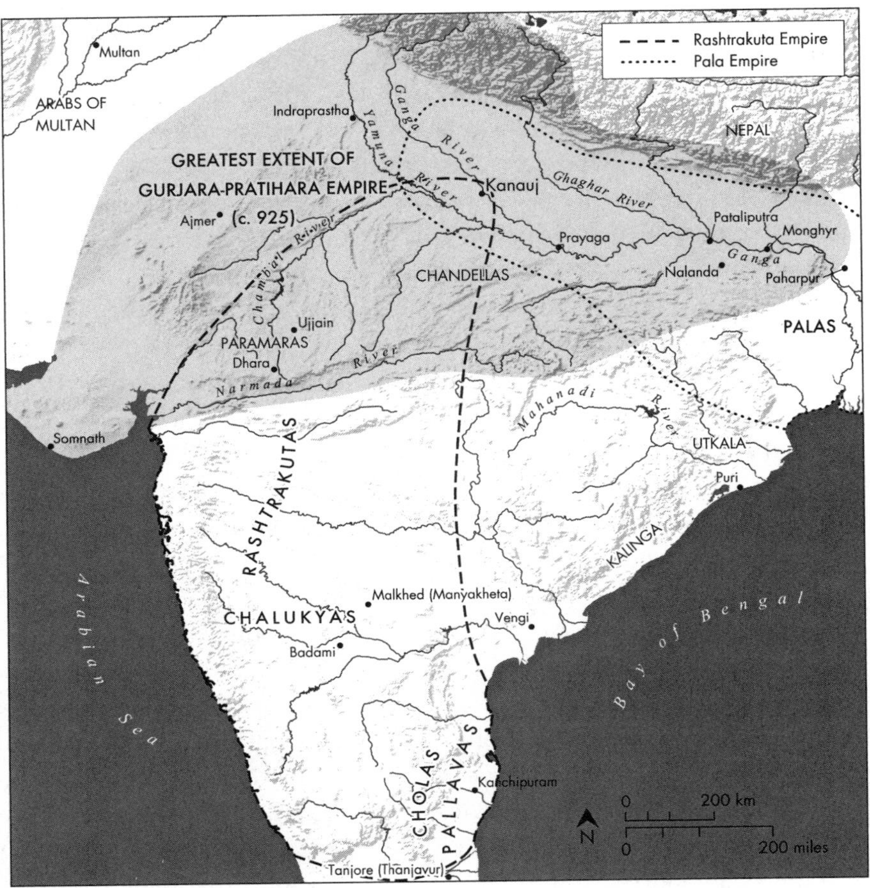

The Gurjara-Pratiharas, Rashtrakutas, and Palas and the struggle for Kanauj. Map by M. Roy Cartography.

the eastern seaboard and Kosala on the northern plains. By 735 C.E. he had defeated Kirtivarman II and taken over most of the Chalukya Empire, after which he took the customary titles of "king of kings" (*mahārājādhirāja*) and "supreme-lord" (*parameśvara*).[28] Under Dantidurga and his son Krishna I, the Rashtrakutas commanded much of the present-day Indian states of Gujarat and Maharashtra, having routed and displaced the kingdom of Vengi ruled by the Eastern Chalukyas, and were poised to intervene in the major theaters of conflict around the Ganges valley.[29]

In many respects, the epic fight over Kanauj that occupied these three major powers—the Gurajara-Partiharas, the Palas, and the Rashtrakutas—

was not just a test of military strength, but a long-drawn struggle for pre-eminence over the valleys of the Ganges and the Yamuna.[30] It was an all-out contest that involved not only the clash of arms but the burning and looting of cities and temples, pillaging of objects and artifacts including images of gods and goddesses, the commemoration of successful campaigns with the dedication and building of temples, and grand public consecration of rulers and military heroes. Detailed historical evidence for such activities among these major kingdoms is limited, and we do not know how everyday life in Kanauj and its environs was affected in these raids. During the period of the "Kanauj triangle," every political regime in India nursing imperial ambitions wanted to march into the imperial city and make a political statement. As a result, Kanauj became one of the most ravaged and despoiled cities in India, which might be one of the reasons why ruins of the old city are sparse. A list of the notable invaders of Kanauj and their origins over the centuries shows a remarkable succession of sieges:

Ambassador Wang Xuangze, Tang China, 648 C.E.

Lalitaditya Muktapida (724–760 C.E.), Karakota (Kashmir), 741 C.E.

Vatsaraja (780–800 C.E.), Gurjara-Pratihara, late eighth century C.E.

Dhruva Dharavarsha (779–793 C.E.), Rashtrakuta, circa 780–784 C.E.

Jayapida Vinayaditya (779–810 C.E.), Kashmir, circa 780 C.E.

Dharmapala (ca. 780–815 C.E.), Pala, late eighth century C.E.

Govinda III (793–814 C.E.), Rashtrakuta, circa 800 C.E.

Nagabhatta II (805–833 C.E.), Gurjara-Pratihara, 814 C.E.

Indra III (914–929 C.E.), Rashtrakuta, 916 C.E.

Dhanga (950–999 C.E.), Chandella, mid-tenth century C.E.

Mahmud of Ghazni (998–1002 C.E.), Ghaznavid, early 11th century C.E.

Chandradeva (1090–1103 C.E.), Gahadavala, late eleventh century C.E.

Qutbuddin Aibak (1206–1210 C.E.), Ghurid, early thirteenth century C.E.

The repeated sacks of Kanauj raise a number of questions about the conventions governing pillage and distribution of the spoils of war that cannot be fully answered here. The history of spoliation and seizure of objects is still a relatively unexplored subject in Indian history, and we do not know how far back in time we can trace the desecration of temples and looting of idols and artifacts as acts of war. We know from the Allahabad pillar inscription of Emperor Samudragupta that after he deprived his enemies of their sovereignty, they were expected to "offer themselves" (*ātmanivedana*) to the emperor, which can be taken to imply that they were also required to present themselves to the Gupta imperial court.[31] They were also expected to offer their daughters or maidens to the king along with other gifts. We also know that his successor, Chandragupta II, boasted of his "standards of victory and fame" (*yaśavijaya patākā*) won in numerous battles, which possibly included the fallen standards of his enemies.[32] In Bana's account of Harshavardhana we find a detailed description of the spoils of Rajyavardhana's last great victory in Malwa before his murder being brought in by their faithful commander and cousin Bhandi for inspection at the court of his younger brother Harsha.[33] The booty included part of the defeated king's army and equipage, numerous war elephants "great as moving boulders," prized war horses caparisoned and decked with ornaments, pearl necklaces snatched from the bosoms of "love-intoxicated Malwa women," royal white fly-whisks made out of yak tails, the vanquished king's white umbrella with a golden stock, female captives beautiful as celestial nymphs, lion-crested thrones of gold, and couches and stools. The victorious army had also seized the treasure chests of Malwa laden with "wreaths of ornaments" along with detailed inventories of their contents. These spoils of war were accompanied by a long line of retainers of the Malwa ruler captured alive, their feet bound together in iron chains.

His characteristic poetic flamboyance notwithstanding, Bana provides a vivid and realistic account of the rich harvest of the Malwa campaign circa the seventh century C.E. We can draw a parallel from the near-contemporary Roman Empire, which had established a long and rich tradition of the display of objects seized in war as a vital aspect of its exercise of power. Seizure of such objects as talismans from the defeated was also a symbolic appropriation of the strength of one's enemy, which is why weapons of fallen combatants were so prized by Roman patricians and plebians alike.[34] The sack of temples and capture of idols is particularly noteworthy in the fierce Roman campaigns waged against the Greek islands of the Aegean, intended to strip the enemy of its divine protection. Graven images of conquered regimes ended up in the imperial treasury as well as in private collections as cumulative inventories of past laurels. At times, such *spolia* also served to reaffirm the supreme virtues (*pietas*) of Roman gods and Roman ways of worship. The display and public affirmation of trophies in the Temple of Jupiter, and the adornment of patrician villas with shields, enemy standards, or the broken-off beaks of vessels destroyed in naval combat, suggests a penchant for the relics of worthy foes, especially those collected from distant and exotic frontiers of the empire.[35] The Roman case is instructive in thinking about societies that were oriented toward routine warfare. Imagine the plunder pouring into the capital during the Greek campaigns—arms and armor from enemy chieftains slain in distant battlefields, idols seized from temples, agoras and columns to be reinstated in Roman shrines, and paintings, sculptures, and other objects of art. As Margaret Miles points out, the most sought-after prizes of war were objects from personal collections, referred to as *spolia in se*: statues, paintings, vessels of precious metal, textiles, and carved objects of wood or ivory, dedicated to sanctuaries and buildings.[36] The capture of such objects by Romans in Sicily pricked the conscience of Cicero, recoiling from the wanton plunder of preciosities as thievery punishable under Roman law. Similar questions about the justification of warfare and looting, we shall see, is key to an appreciation of what transpired in the kingdoms and cities of the Ganges basin during the long centuries that were engulfed by the desperate struggle over the control of Kanauj and the valleys of the Ganges and Yamuna Rivers.

Raiders of the Northern Realms

It is said that when Dantidurga, celebrating his victories over the kingdoms of Kalinga in Odisha and Kosala in the north, performed the great "golden egg" (*hiraṇyagarbha*) sacrifice in the capital city of Ujjain consecrating his rebirth as a true Kshatriya warrior, he compelled the subjugated Gurjara-Pratihara kings and other vassals to participate in the ritual as humble doorkeepers.[37] Warring Indian polities of this period were preoccupied with investitures of supremacy and subjection. Military conquests were tied to symbols of divine, cosmic, and temporal power, along with symbolic representations of conquered territory. Marching into Kanauj at the head of a victorious army was perhaps the most coveted feat of all, attempted repeatedly by the three major contending regimes discussed earlier.

The Rashtrakuta king Dhruva Dharavarsha, who had acquired the unusual eponym (*biruda*) of *Kalivallabha*—meaning both "favorite of the age of the Kali (the age of apocalypse)" and "lover of destruction"—marched into Kanauj around ca. 780 C.E.; the city was held at the time by a vassal of the Gurjara-Pratihara king Vatsaraja.[38] Vatsaraja had apparently become intoxicated with power after defeating the Palas of Gauda in Bengal and was caught unawares.[39] In the engagement that followed, Dhruva not only defeated Vatsaraja but seized the two moon-white canopies that the Gurjara-Pratiharas had taken from the Pala king of Gauda, Dharmapala.[40] Decorative canopies and umbrellas displayed over the head of a traveling monarch were the most common insignias of sovereignty in contemporary India. By taking these items by force, Dhruva sealed his writ over the northern realms. A later inscription makes it clear that the Rashtrakutas were especially proud to announce their triumph over the land of the Ganges and Yamuna, having captured the canopies of their rivals, each like a lotus, bearing the will of Lakshmi, the supreme goddess of fortune.[41]

The Chalukyas too were known for their war standards (*dhvaja*) marked with the crest of the boar—Vishnu's fearsome avatar. On the battlefield they brandished the *pālidhvaja* ("flags in rows"), an assemblage of multiple banners captured from defeated rivals in a single file.[42] When the Chalukya forces under Vijayaditya crossed the Vindhya Hills and pushed north, routing the kings of the Ganges valley, they added to their standard flags with signs of the Rivers Ganges and Yamuna. It is striking that two great rivers—most likely drawn with the symbols of the *makara* and the tortoise—were claimed as royal insignias in the battlefield, along with white umbrellas, conch shells, double drums, peacock tails, spears, and thrones. The Rashtrakutas seem to have been quick to seize on the idea of this standard adorned with the crests of enemies felled in battle. It is said that when the when the forces of Krishna I overcame the Chalukyas, they carried off the family fortunes "adorned with a garland of waving Palidhvaja flags."[43] The Rashtrakuta imperial standard wielded an image of the eagle-winged Garuda, Vishnu's triumphal mascot rising from the lotus. Branching out from this central staff were the pennants and emblems of the conquered dynasties. The Nesarika grant of Govinda III lists the insignias appropriated by the Rashtrakutas from thirteen defeated kings.[44] Among these were fish of the Pandyas of Tamil country, bulls of the Pallavas, tigers of the Cholas, elephants of the Gangas, bows of the Keralas (Malabar), and different kinds of boars of the Andhras, Chalukyas (Vengi), and Mauryas.[45] Included in this inventory was also a doorkeeper carved on an oblong tablet used as a standard by the Gurjara-Pratihara Dynasty ruling in Kanauj. There were figures taken from Sinhala (Sri Lanka) as well, and an image of the goddess Tara (*tārābhagavatī*) seized from King Dharmapala of Bengal, which might also have been used as a standard by the Palas on the battlefield.[46]

Chalukya and Rashtrakuta battle standards give us an insight into the signs and representations of conquered territory in ninth-century India. Seizure of such emblems was part of the extended politics of spoilage, which also included the vanquished rulers' coins being restruck by the usurping regime.[47] We can see this as a form of stylized violence that had become the norm in a society accustomed to warfare as display. The capture of standards bearing the emblems of the Ganges and Yamuna and their addition to the Rashtrakuta staves can be seen as an attempt to establish audacious dynastic claims over the Ganges plains.[48] Such territorial ambition can be seen clearly at play in an epigraphic record recalling Dhruva Dharavarsha's military success against the Western Ganga kings of Gangavadi (in Talakad, present-day state of Karnataka) invoking the name of the enemy and the enemy country (Ganga, Gangavadi), both of which derived from the name of the river. The phrase in question employs the pun deftly, stating that the Rashtrakuta emperor impeded the Ganga kings and their successor, which also summons the image of Shiva stopping the flow of the River Ganges to save the earth. Indeed, Dhruva took the title of supreme lord (*parameśvara*) and stopped the march of Ganga king Sivamara II's army, taking him prisoner.[49] A later inscription applauding his victories over the kings of the north states that he snatched from his enemies the Rivers Ganges and Yamuna "with their charming waves" and adopted the insignia of the two rivers in his standard of war.[50]

During the thirty-second year of his long rule in Bengal, King Dharmapala seems to have intervened repeatedly in the politics of Kanauj, ultimately installing his vassal Chakrayudha as its new ruler. He persuaded a long list of kings—including the rulers of Avanti and distant Gandhara—to attend the coronation of his protégé. Pala-era sculptures of Ganga as a *makara*-riding female deity that have survived, as in the life-size, full-bodied image found in the Rajshahi district, most likely a part of the entrance to a votive structure, temple, or stupa, suggest that they too were keen to lay claim to the Ganges valley.[51] However, they failed to hold on to their status as the kingmakers of Kanauj. Around 815 to 816 C.E., Nagabhatta II defeated the Pala ruler somewhere in eastern Bihar near Munger, successfully marched into Kanauj, and ousted the Pala vassal—a signal act by which the Gurjara-Pratiharas claimed supremacy over the north and the Ganges valley, assuming lofty titles such as *paramabhaṭṭāraka* (supreme overlord).[52] Their hold over Kanauj, once again, did not last for long. In ca. 806–807 C.E., Govinda III, another Rashtrakuta ruler, struck deep into the northern territories with his formidable army. This daring exploit is recorded in the Sanjan Copper Plates of Amoghavarsha I. It says that Govinda's horses lapped the water of the springs gushing from the Himalayas, its caverns resounding with the music played during his consecratory ablutions. As he performed the Vedic rituals worthy of a true king, he justly resembled the mighty Himalayas in renown.[53] An inscription of Emperor Govinda V dated to 933–934 C.E., which claims somewhat routinely that the Rivers Ganges and Yamuna flowed through his palace, also pays homage to this grand feat performed by his illustrious ancestor Govinda III, who wrested the two celestial rivers from his enemies and added them to his standard.[54]

Such back-and-forth over outlying areas of warring kingdoms reveal the essential geographical conceit of the Rashtrakutas—their attempt to transport the entire northern realm, the *āryāvarta,* along with the valleys of the Ganges and Yamuna southward, to consecrate their new capital in Manyakheta (Malkhed), rivaling Kanauj between the floodplains of the Krishna and Godavari Rivers. The most visible part of this effort lies at a site 18 miles northwest of the present-day city of Aurangabad in the famous cave-temple complex known as Ellora, known in ancient times as Elapura. Sheltered in the ragged and rocky formation of the Western Ghats known as the Deccan Trap, dug out by hammer and chisel from sheer basalt, are a series of grottoes, caves, and temples. Before it was taken over by the Rashtrakutas, this had been an important Buddhist and Jain pilgrimage, patronized by the Chalukyas. Many of the caves are Buddhist *caitya*s with pillared halls, *vihāra*s, and monasteries dating to the early seventh century C.E.[55] Rashtrakutas, who started out as feudatories of the Chalukyas, seized on this old site for a new spurt of temple building, placing their favored deities and icons at its center. Early Rashtrakuta pioneers such as Dantidurga would have been conscious of this historical reclamation, and the fact that these rocky outcrops were located near the pilgrimage of Grishneswara, especially coveted by worshippers of Shiva of the "radiant lingam" (*jyotirliṅgam*) whose pilgrim trails included a number of scenic waterfalls, streams, and rivulets.

Work at Cave 15, dedicated to Shiva, began under Dantidurga; an inscription (741–742 C.E.) tells us that the Rashtrakuta warrior had defeated the Chalukya king Kirtivarman II and conquered major kingdoms such as Kanchi, Kalinga, and Kosala, and that the temple was donated after a consecratory bath at the auspicious caves of Guheswara (*guheśvara-tīrtha*)—which was, most likely, the site of the old temple of Grishneshwara dedicated to Shiva.[56] Dantidurga's uncle and successor, Krishnaraja (Krishna I, 757–772 C.E.) built the great temple of Kailasa, now Cave 16 of the Ellora complex, which by his own admission was the rare treasure of the Elapura Hills (*elāpurācala*). His architects and stoneworkers had created an abode for

Shiva (*śivadhāma*) of unparalleled beauty, the likes of which had not been seen in art before.[57] The image was bedecked with gold and jewels, adorned with the moon and the River Ganges in his hair, and showed the cosmic poison that turned his throat blue.[58] The scale and sculptural feats executed by workmen in the quarrying and finishing of monolithic structures and images at the Kailasa temple are indeed capstones of Indian art and architecture. What is particularly noteworthy is the evidence of a number of cisterns and aqueducts that were fashioned to capture and redirect the flow

Ganga standing on Makara at the Kailasanatha Temple, Ellora, Maharashtra, Rashtrakuta Dynasty, eighth century C.E. Photograph by Anannya Deb.

of rainwater for ritual ablutions. The refrain of running water would have been a part of the intended effect of the overall architectural design, meant to remind pilgrims and subjects that the great Rashtrakuta warrior-kings had brought Mount Kailas and the heavenly streams Ganges and Yamuna into his empire—a theme recapitulated throughout the entire temple complex.[59]

Cave 14, known also as Ravana Ka Khai, features the goddess Lakshmi flanked by elephants (*gajalakṣmī*) seated on lotuses with attendants holding pitchers of water, as elephants shower waters of lustration (*abhiṣeka*) over her head. It also has an image of Vishnu in the company of the twin goddesses of earth and fortune. The shrine of Shiva and Parvati has a sculpture of Mount Kailasa being rattled by Shiva's notorious and fearsome devotee, the mighty ten-headed demon Ravana. Cave 15, the Dasavatara Cave, named after Vishnu's avatars, where a visit of the Rashtrakuta king Dantidurga is mentioned in an inscription dating it to the eighth century C.E., opens into a large hall raised on a plinth in the central courtyard, with life-size figures of Ganga and Yamuna as doorkeepers placed on either side of the entrance. The temple at Cave 16 is a freestanding shrine excavated out of sheer basalt, with huge carved panels on either side of the entrance to the structure. Once again, Lakshmi is presented being bathed by elephants in a large lake covered in delicately carved lotus leaves that appear to float at her feet. Here both the conch shell and the lotus are given added prominence as emblems of prosperity for the king and his realm. Three separate panels frame images of the rivers Ganges, Yamuna, and Saraswati, suggesting a meeting point of the three at the great confluence of Prayag. Towering over the entire scene is a triumphal pillar, essentially a rock-hewn standard or flagstaff (*dhvajastambha*). The Rashtrakuta architectural projects of Ellora are a reminder of the grand political ambition of a regional power envisioning a pan-Indian empire, proclaiming the conquest of all four quadrants of the world (*digvijaya*) and eminent clusters (*maṇḍala*) of kings, including those of the Ganges basin. As Inden suggests, through these daring feats the Rashtrakutas had in fact bettered their predecessors, the Chalukyas, in establishing a rival *mappa mundi* and claim of universal sovereignty.[60]

Spoils of War

The roar and tumult of the battlefield might not have always reached the contemplative quiet of temple interiors, but many temples were built directly from the spoils of war and dedicated as memorials to successful campaigns. This convergence of godly and temporal aspirations was brilliantly exploited by the Cholas of Tanjore in present-day Tamil Nadu. The Cholas had been a tributary power allied to the Pallavas of Kanchipuram, and they wrested a measure of autonomy at the expense of their patrons, also defeating a rival power to the far south, the Pandyas of Madurai. The Pallavas were finally overcome at the end of the ninth century C.E. The Cholas demonstrated the mettle of their military force when the powerful Rashtrakutas, flexing their muscle under Krishna II, invaded their land in 916 C.E. During this war, which proved indecisive, Parantaka Chola inflicted heavy losses on the Rashtrakuta army. By the mid-tenth century C.E. the Rashtrakutas had been eclipsed by a later line of Chalukya rulers, now split into two rival regional branches—one in present-day Karnataka with the capital Kalyani, the other centered in Vengi, Andhra Pradesh. The Cholas exploited this rivalry and pressed their advantage. During the late tenth and early eleventh centuries the Cholas and Chalukyas were embroiled in a series of battles that signaled the rise of the Cholas as a significant power in southern India. In one of these encounters the Western Chalukya ruler Satyasraya, losing heart at the advance of the Chola king Rajaraja I's swarming forces, fled the battlefield in ignominy. A contemporary inscription attests that rampaging Chola elephants caused panic in the ranks of the Chalukya army on the banks of the Tungabhadra River. Rajaraja, seated on his warhorse, singlehandedly checked the onslaught of the onrushing Chalukya forces, like Shiva Gangadhara restraining with his matted locks the force of the Ganges.[61] The Chalukya general Kesava was taken prisoner. Rajaraja's son Rajendra Chola would later take a vow to capture the capital, Manyakheta, that the Western Chalukyas had snatched from the Rashtrakutas.

The city of Manyakheta, the fallen Rashtrakuta capital, suffered the fate of Kanauj. It had been already attacked and looted once before by the Paramara ruler Harsha Siyaka (Siyaka II) of Malwa in about 972–973 C.E. It was subsequently invaded and annexed by Tailapa II, the Western Chalukya ruler.[62] The conflict between the Western Chalukyas and the Cholas did not bode well for the city. In the Karandai Cankam plates we find verses that describe the city of Manyakheta consumed in flames, burned down by Rajendra Chola's vengeful army. It speaks of women darting about in panic as high buildings went up in smoke, like lightning flashing in dark clouds.[63]

The Cholas looked upon warfare and plunder as routine exercises. The destruction and ravaging of the Chalukya capital was no exception. When Rajaraja invaded Sri Lanka with a navy in 933 C.E., setting up Chola provinces in the northern part of the island, the Chola army laid siege to the Buddhist monastery and *vihāra* of Anuradhapura. The Sri Lankan chronicle *Mahavamsa* records that in the wake of the Chola attack on the island, the ruler fled to the jungle in terror and was captured alive by the Chola army. It states in unflattering detail that the invaders from the north "took away all valuables in the treasure house of the King and plundered what was there to plunder in vihara and town."[64] Rajaraja's men captured the king and his queens, along with the royal crowns. They also seized the lucent Indra garland that had been part of the Pandyan regalia left behind by King Rajasimha. They took the Buddhist relic of the torn strip of cloth, the relic chamber, and many golden statues. The army pillaged monasteries with abandon. The capital after the campaign looked "as if it had been plundered by demons [*yakkha*s]."[65]

There is no doubt that triumphal display of *spolia* was an integral part of warfare and dynastic rivalry. Richard Davis has drawn attention to the looting of cultural artifacts, especially temple idols, as a habitual feature of warfare in southern India. Royal inscriptions, Davis points out, routinely proclaim the forcible seizure of such objects from rulers defeated in battle. Inscriptions from the south wall of the Kalyana Pasupatisvara temple of Karur (in present-day Kongu Nadu), identifies Kudalasangamam, meeting point of the Tungabhadra and Krishna Rivers, as the site of Chola emperor Virarajendra's victory over the Chalukyas. In this battle the entire Chalukya camp fell directly into the hands of the Chola forces.[66] The inscription provides a graphic inventory of the booty. Virarajendra seems to have taken from the fallen king his principal wives, his family heirlooms, insignia of royalty such as conches, parasols, trumpets, drums, canopies, ceremonial white yak-tail whisks, and the vaunted boar pennant.[67] They took over an entire herd of war elephants, including the prize she-elephant called Pushpaka, along with a stable of warhorses. Most remarkably, they had carried off the ornamental *makara*-crested gateway of their shrine. The victorious king, cheered on by his retainers, marking this moment, put on a champion's crown inlaid with precious rubies and other diadems. The inscription goes on to suggest that under the banner of the ferocious Chola tiger, Virarajendra's rampaging army had cut off the garlanded heads of many enemy combatants. This boast gives us the ultimate motif of decapitation as the victor's prerogative. We know from the title of *cōḻataḷaikoṇḍa*, or "seizer of Chola heads" taken by rival Pandyan kings, that the image of the severed head of the king, or its equivalent, the fallen crown, is more than just a poetic image.[68] It is said that after Parantaka's son Aditya II killed the Pandyan king in battle, he brought back his severed head and set it up as a pillar of victory in his capital.[69] The bowed head of a vassal was also a key sign of sovereignty. Kings defeated in battle were made to acknowledge their status by placing their heads at the feet of the emperor in the Chola court, and the ruler often mounted the state elephant by planting his foot on the head of a chosen vassal.

The scale and magnificence of the Chola temples, distinguished by their shining cupolas of gold, were in many respects a direct allusion to the king's crowned head. As A. K. Ramanujan pointed out, devotional poetry of the era looked askance at the accumulated riches and hubris of kingly power. Medieval south Indian temples, Ramanujan writes, "looked remarkably like palaces with battlements . . . richly endowed and patronized by the wealthy and the powerful, without whom the massive structures housing the bejeweled gods and sculptured pillars would not have been possible."[70] Such temples were built to mark coronations, particularly auspicious regnal years, and memorable victories in battle. They were partly underwritten by spoils and revenue from newly acquired territories, which made plunder necessary for the ulterior ends of power, wealth, and virtue. The genre of classical Tamil poetry under the category *puṟam*—devoted typically to exterior, public matters, including those of kings and warfare—gives us great insights into the view of war as an inevitable aspect of human affairs.[71] Poems describe military sieges in graphic detail, with chariots, elephants, horses, battlefields, banners, and camps. A poem titled "Harvest of War," translated by Ramanujan in his anthology of poems from the Second Cankam, talks about the "unfailing harvest of victorious wars."[72] It gives us the fleeting tableau of an entire army on the move: the long parade of bull elephants, rows of shields and white flags, with birds of carrion circling overhead.[73] Such poetry also posed at times poignant questions over the moral consequence of such acts of aggression, as in the poem addressed to King Peruvaluti by Nettimaiyar:

> Is this right
> O lord rich in victories
> This ruthless taking
> of other men's lands . . .[74]

Basavanna and other Virasaiva saint-composers decried the arrogance, profligacy, and lavish displays of piety that were the hallmarks of the king's earthly power and prestige.

The Chola March to the North

In his twenty-ninth regnal year the Chola emperor Rajaraja I held a consecration at the newly built Rajarajeshvara temple in Tanjore.[75] He also unveiled a gold-plated finial for the central (*vimāna*) edifice, towering over two hundred feet. David Shulman suggests that there was always a measure of uncertainty about the reach of the king's absolutism, which is why temples were a "rhapsody to size."[76] To this we must add that the theater of wars, triumphal marches, and coronations never stopped at the Chola capital. Rajaraja's son and successor, Rajendra, had to match the feats of his father, and by most accounts he exceeded them in his lifetime. One of his most celebrated exploits was the military expedition into the heart of the northern country, beyond the Godavari River, through the hostile kingdoms of Kalinga and Odda. In anticipation of this march, work had already started on the construction of a new capital city. The mission seems to have been carefully planned by his military generals with a troop of outriders at the head of the force, followed by a backup reserve, and the strategy involved a military alliance with Jayasimha II of the Western Chalukya kingdom.

In a campaign that lasted for less than two years, the Chola army moved swiftly across a vast stretch of country.[77] A later inscription embroiders the account of this march with the observation that the Chola army was as massive as the slopes of the Himalayan range, their horses galloping forth like the waves of the Ganges, every corner of the earth resounding with the din of their advance. They marched through Vengi and the forests of Bastar, crossed the Indravati River by making a bridge out of elephants, and reached the country of Odda-Visaya (Odisha). They overran Kosala, exacting homage from ancient Brahmin establishments, and went eastward through Dandabhukti, the corridor connecting Odisha to Bengal. They subdued Dharmapala, a minor ruler in Bengal, and King Ranasura of southern Radha, arriving finally in Bengal during the height of monsoon rains. They easily defeated the Pala ruler Mahipala I, whose feudatory Govindachandra jumped off his elephant and fled in the thick of battle. The Chola forces captured bull elephants of rare strength, ladies of the court, and vast treasures. The Ganges was just beyond the northern edge of the Radha country, and the army finally reached the banks of the great river, most likely in the vicinity of Mithila and Varendra, "whose waters bearing fragrant flowers dashed against the bathing places."[78]

The Cholas were after a different kind of prize. Chola kings had always wanted to bless their kingdom with the sacred waters of the Ganges. The

Tiruvalangadu Plates attributed to Rajendra state that the emperor about to accomplish this extraordinary feat seemed to be "laughing at Bhagiratha who had brought down the Ganga by the power of his austerities."[79] The commander of the Chola army had been directed to collect the waters of the Ganges, and the epigraphic record suggests that defeated chieftains of kingdoms bordering on the holy river were made to carry the waters all the way south to the banks of the river Godavari, where the emperor was encamped.[80] The noted historian of the Chola Dynasty Nilakantha Sastri casts some doubt on these particular details, suspicious of the laudatory excess typical of royal scribes.[81] Notwithstanding the exact nature of the ignominy

Shiva Nataraja at the Brihadeshvara Temple, Chola Dynasty, early eleventh century C.E., Gangaikondacholapuram, Tamil Nadu. Photograph by Padma Kaimal.

heaped on the kings of the Ganges valley, this forcible appropriation of the Ganges waters by the Cholas was by all accounts an extraordinary act. The water was brought for the ceremonial tank of the king's new Brihadishvara temple in the new capital named after the northern expedition *gaṅgaikoṇḍacōḻapuram*, the city of the "captor of the Ganges" in latter-day Tiruchirapalli. This was perhaps the crowning title added to a long list of accolades, including the Mudikondai or "crown-seizing" Chola, whose predecessors had captured the crowns of the Chera and Sinhala kings.[82] This was more than just a homage seeking the blessings of the Ganges. It was a dramatic display of military power, capped by a wondrous "pillar" of victory made out of stored Ganges water (*gaṅgājalamayam jayastambham*), connected to a temple tank and a man-made lake fed by the Rivers Vennar and Kollidam proclaimed as the new "Ganges of the Cholas" (*cōḻagaṅgam*).[83] The act was further sanctified by the migration of devout Shaiva Brahmins from cities along the Ganges resettled by the emperor in the ancient Tamil city of Kanchi.

Brihadeshvara Temple at Gangaikondacholapuram, Tamil Nadu, commissioned by Rajendra Chola, ca. 1035 C.E. Photograph by Padma Kaimal.

Not only was Rajendra's new temple able to secure jars of real Ganges water, as Richard Davis points out, it also served as a gallery of objects looted during the king's various conquests: a sun pedestal; images of Durga and Ganesh from the Chalukyas; Shiva's bull, Nandi, from the Chalukyas of Vengi; stone images of Bhairava and consort from the country of Kalinga; a bronze image of dancing Shiva from the Palas of Bengal; and many more items stolen or plundered by generations of Chola kings.[84] Set in this context, the river water appropriated by Rajendra Chola, along with the gods, goddesses, and other dynastic emblems of vanquished kingdoms, were nothing less than divine *spolia*. This is also why the Cholas brought back for display at their capitals gateways, arches, and guardian images from the various kingdoms they invaded, including the majestically carved *toraṇa*s of overseas kingdoms like Kadaram in present-day Kedah on the Malaysian littoral. The audacity of such exhibition, as Daud Ali has suggested, lies precisely in the refashioning of political sovereignty through a reassignment of both geography and cosmology. Like the Rashtrakutas before them, the Cholas attempted to reorient their kingdom as the new center of the known world with its hierarchy of kingdoms that comprised Bharatavarsha, along with the Himalayas and the Ganges, displacing the hierarchy and geography of the various kingdoms.[85] Crossing the rivers of various rival kingdoms—Godavari, Krishna, Tungabhadra—over the backs of war elephants, ritual bathing in rivers and streams during military campaigns, and the impulse to coerce kings and chiefs from the shores of the mighty Ganges to carry its water can thus be seen as part of the spectacle of military conquest. This was also seen as the consummation of a Chola ancestral myth, which extolled the story of King Chitradhanvan. Just as King Bhagiratha in penance brought down Ganges upon earth, the great Chitradhanvan, who wanted everlasting fame, brought the celestial river to his dominions in the form of the maiden Kaveri. The inscription celebrating Rajendra's military feats speaks of the virtual transubstantiation of Kaveri waters into the waters of the Ganges. In bringing back actual Ganges water and letting it mingle with the tributary waters of the Kaveri—at the exact site of the royal temple—a dynastic pledge had come true, securing the exploits of Rajendra for all posterity.[86]

The versatility of Chola symbols of sovereignty makes it difficult to single out the River Ganges as the object of dynastic glory, but they clearly chose the river as one of their key icons of territorial prowess, signifying a realm of unalloyed purity and power. Here the Chola reappropriation of the emblem of the Shiva Nataraja, representing what Padma Kaimal translates aptly as Shiva's dance of "furious bliss" (*ānanda tāṇḍava*), is immensely instructive.[87] The dancing Shiva motif evolved with the expansion of Chola power through the tenth and early eleventh centuries across the Kaveri delta. The finished bronze sculpture, displayed at various places in the Chidambaram, Rajarajeshvara, and Brihadishvara temples, features a dancing figure with long matted tresses and garlands flying outwards, and on Shiva's right entangled in his hair is a little personified figure of Ganga. Her body is human above the waist but piscine below. Her palms are folded together in respectful prayer. Shiva's crown is adorned with *datura* flowers and *kondrai* leaves that grow nearby.[88] Also suggestive are the various guardian sculptures holding poses similar to that of the lord of the dance. Shiva here is the

Shiva as the Lord of Dance, with Ganga caught in his flying locks, copper alloy sculpture, Tamil Nadu, ca. 950–1000 C.E. Photograph © Museum Associates/Los Angeles County Museum of Art.

ultimate icon of potency, and his dance the epitome of the destructive and generative power of the kingdom itself. By bringing his Himalayan haunts of Kailasa and the heavenly river locked in his hair, the Chola kings had succeeded in harnessing the purity, fertility, and awesome potential of the natural and cosmic orders for the grandeur of their kingdom and the well-being of their subjects.

Advent of the *Turushkas*

When Mahmud of Ghazni had secured his hold on the Indus Valley as the new eastern flank of his sprawling mountainous empire—before he began to besiege and plunder the richest cities and temples of western India—the valley of the Ganges was ruled by the Gahadavala Dynasty from the cities of Varanasi and Kanauj. Their rivals in the north were the last of the Gurjara-Pratiharas and a subsidiary line of Rashtrakutas. To the south of the River Yamuna lay the lands of the Chandellas and the Kalachuris. The Gahadavalas came to power in 1090 C.E. during a time of great political upheaval. Kanuaj passed through many hands; it was first held by the Chandellas and then the Kalachuris. The Gahadavalas provided a degree of political stability in the region and became known as the protectors and patrons of Kanauj and Varanasi for over a hundred years.

Royal charters and plates give us a glimpse of these two preeminent cities and their pilgrims—acolytes of the various Hindu sects, Buddhists, and Jains. During the Gupta period the imperial family sponsored the worship of Vishnu and Shiva's scion Kartikeya, which is evident from coins found in the Rajghat area in present-day Banaras.[89] As far back as Harshavarshana's reign in Kanauj and Varanasi we see a mélange of Buddhist and Shaiva practices, and by the eighth century C.E. various forms of Vajrayana Buddhism seem to have been firmly established in these cities. Mahipala I, the king of Gauda in Bengal, along with his brothers Sthirapala and Vasantapala, had donated handsomely for the construction and repair of various Buddhist monuments. During the early eleventh century C.E., Varanasi and nearby Sarnath were bustling centers of commerce, travel, and pilgrimage, with a profusion of *vihara*s and stupas, which indicates the continuous influence of Mahayana monks in the religious life of the city known for Shiva worship.[90] One of the most imposing structures visible on the banks of the Ganges in Varanasi was the Kalachuri ruler Lakshmikarna's temple known as the Karnameru ("Mount Meru of Karna").

Mahmud knit together a new Islamic kingdom across the foothills faults of the Hindu Kush, bridging the mountain passes between the valleys of the Kabul and Indus Rivers, through carefully planned military raids. Parts of India beyond the Indus were known to the wider Islamic world through the work of Arab mapmakers, but only in outlines. The interior parts of *al-Hind* was still enveloped in tales and exotic hearsay. The famed Arab geographer and historian Al Masudi visited the valley of the Indus in 915–916 C.E., which was roughly the beginning of the Gurjara-Pratihara ruler Mahipala's reign. His great work *Muruj-ul-Zahab* [Meadows of Gold] speaks of the al-Balhara (Rashtrakutas) of the southern realms and Bavura (Gurjara-Pratihara), the ruler of the distant and fabulous city of Kanauj, surrounded by warlike kings.[91] Perhaps it was this reputation of wealth that enticed Mahmud and his band of military adventurers in Ghazni. A description of Mahmud's siege of Kanauj can be found in the Ghaznavid court historian Al-Utbi's *Kitab-i-Yamini*. Utbi notes that Kanauj was terra incognita, located deep in the heartland of India, almost a six-month journey from Ghazni in Afghanistan, and it was only the zeal and loyalty of Mahmud's commanders that made the expedition to that city possible.[92] In order to reach the high ground on which the imposing ramparts of the city stood, the Ghaznavid troops and mercenaries had to ford the Ganges. They knew that the river they were crossing was held in high esteem by the people of Hind and was treated with great dignity by all the chiefs and kings of India.[93] To the Hindus, Utbi notes, it was like a wellspring of eternity. When they washed their dead in these waters, their souls were cleansed of all past accounts of accumulated sins and evil deeds. Devout Brahmins came from far-off places to bathe in the Ganges for salvation. Seven battlements of Kanauj could be seen rising from the banks of the river. Utbi recounts that there were a thousand temples of idol worship, with great hoards of money and jewels at the temples, which the local Hindus boasted were more than a thousand years old. The presence of Mahmud's troops had struck terror, and knowing that a siege was imminent most people took their money and belongings and fled. Utbi does not say much about the siege itself, except that it was quick and lasted for a day. Terrifying as the attack was, the inhabitants of Kanauj had witnessed many such plunders before. As in most of his campaigns, Mahmud turned the booty of Kanauj into money to be used for the construction of his fabled capital in Ghazni, with its grand palace, its fabulous Friday mosque, and streets and markets bustling with merchandise, mercenaries, and slaves.[94]

Mahmud's army did not venture as far as Varanasi, which boasted the most richly endowed temples in all of northern India. However, it was attacked by a detachment of Turkish cavalry under the command of Ahmad Niyal Tigin, the governor of Lahore, in 1033 C.E., who broke the resistance of local Thakur chiefs, seized a large amount of treasure from them, crossed the Ganges, and found himself along with this crack troops at the gates of the holy city. The invaders were able to lay siege only for half a day, and fearing a reprisal, they quickly looted the shops of the jewelers, perfumers, and drapers and broke camp.[95] The city seems to been unscathed by this event. Its continued prosperity during the Gahadavala ruler Govindachandra's reign is evident from the long inventories of taxes taken out on trade and pilgrimages that appear in the contemporary records. During these years of the early twelfth century C.E., the legendary temple of Kashi Vishwanath was endowed.[96] The other famous temple from this period was the Adi Keshava, which stood next to the bathing steps of Adi Keshava Ghat.[97] Some of these steps on the river are still there, but many have disappeared or have been overlaid. The Adi Keshava Ghat at the meeting of the Varuna River; Vedeshwar and Kapalamochana, near present-day Rajghat; and Kotitirtha, Trilochana, and Swapneshwar, near present-day Kedar Ghat, are gone now, but they appear time and again in the epigraphic record, especially in the context of the donation of land and entire districts dedicated to their support. Patrons donating toward religious establishments followed prescribed ritual ablutions at designated *ghāṭ*s during auspicious solar and lunar eclipses. Some of the massive steps and parapets of Varanasi descending to the waters of the Ganges were designed specifically for such ritual display of exemplary piety. King Chandradeva in 1100 C.E. was weighed at one of the *ghāṭ*s in one thousand pieces of gold and silver, and donated thirty-two villages to five hundred Brahmins. Similar grants were made to the Gahadavala family priest Jagu Sharman and his son Praharaj. Prince Jayachandra in 1168 C.E. was directed by his father to bathe in the Ganges by the steps of the Adi Keshava Ghat and take the vow of serving Lord Krishna, securing taxes for the upkeep of the Aghoreshwar, Panchokar, Laudeshvar, and Indramadhav temples entrusted to Praharaj and his family.[98]

After the sack of Kanauj, a number of prominent Brahmin families sought refuge in Varanasi. Contemporary accounts speak of Varanasi as a city unsullied by the ravages of time, where people lived in a state of perpetual blessedness.[99] It was also the new Gahadavala capital. King Govindachandra earned the distinction of having protected the city from an invasion of the Turks. The poet Bhatta Lakshmidhara calls him the valiant vanquisher of the *hammīra*—the Sanskrit adaptation of the Turkish *āmir* (nobleman or warrior).[100] He fought with Mahmud's nephew Masud Salar Ghazi, who was later reputed to have attained martyrdom, adored by his Sufi followers as Ghazi Mian. Govindachandra was also known to have settled a number of Turkish subjects in his kingdom, subjecting them to a "Turk-capitation" (*turuṣkadaṇḍa*) in lieu of their alien religious practices.[101] The Gahadavalas had a formidable military force, having crossed swords with the Kalachuris, the Chahamanas, and the Paramaras of Dhara. Jayachandra earned the nickname of Pangula, or "the lame one," because his army was too large and unwieldy, and when he advanced in battle he had to stick close to the Rivers Ganges and Yamuna so that his soldiers would not suffer from want of water.[102] The ultimate mantle of the protection of the Ganges valley thus fell on the Gahadavala Dynasty, a distinction that so many kingdoms had sought in the past. The Ganges and its rich floodplains continued to inspire the panegyrists of various Indian dynasties, large and small. Vijaysena, architect of the Sena Dynasty in Bengal, is described by the poet Umapatidhara as the king whose crown is sprayed by the waters of the Ganges (*gaṅgāśīkara*), whose fleet of boats sailed west to conquer kingdoms along the course of that river.[103] A praise of the Paramara kings of Malwa describes the horses of King Vakyapati as having drunk the waters of the Ganges and all the oceans.[104] An inscription of the Tomara Rajas of Delhi boasts that they kept the Turks (*hamvīra*) away from the banks of the Ganges.[105]

Mahmud of Ghazni did not wish to expand his empire much farther beyond the Punjab. After his dynasty fell, a new line sprang from among their military commanders based in the region of Ghur, in the mountain city of Firozkuh. The Ghurid Empire was roughly a contemporary of the Seljuk regimes in Iran, Iraq, Syria, and Anatolia. The Ghurids captured Lahore in 1186 and engaged in a long and protracted struggle for dominance with the various Rajput groups, especially the Chahamanas. They also attempted to push southward but were defeated by the Chalukyas in Gujrat. The Ghurids sought to establish an Islamic regime that included the building of congregational mosques where the name of the ruler could be recited as the defender of faith during the Friday congregational prayer (*khutbā*)—which placed architecture at the center of manifest authority. At least five major Ghurid mosque complexes, made out of brick, mosaic, and terra-cotta, have been located across northern India, along with fragments of Hindu and Jain temples dating to the eighth or ninth century C.E.[106]

Shihabuddin Ghuri famously defeated the Chahamana ruler Prithviraj in a pitched encounter at the second battle of Tarai in 1193 C.E., ending an unusually concerted resistance from the various warring Rajput clans. Chand Bardai's famous epic *Prithviraj Raso* and surviving Persian sources do not agree about whether he was killed in battle or blinded and taken as captive—to either Ajmer or Ghazni.[107] The end of Prithviraj has been emblazoned in the annals of Indian history partly because of the eloquent tribute paid to Rajput resistance by the herald of Rajput lore Colonel James Tod in his *Annals and Antiquities of Rajasthan*. After the fall of Delhi, Shihabuddin advanced on Kanauj, and Jayachandra (or Jaichand) was slain at Chandawar near Etawah during this encounter—his body identified by a bit of gold thread by which his teeth were fixed. His head was supposedly hoisted on a petard and brought to the victorious sultan.[108] Legend—at least in the Jain sources—has it otherwise, suggesting that he was thrust with a dagger and tossed into the torrents of the River Yamuna from the back of an elephant. His oldest son died in battle and his older wife committed suicide. His younger wife, with child, submitted to Shihabuddin, and the Turkish army finally entered Varanasi.[109] Tod, typically, does not divulge his sources, but writes that the king drowned in the Ganges, a "traitor to his nation" paying the price for having plotted against Prithviraj and his clan, nursing an old family vendetta. The price of defeat was unimaginable. After Varanasi fell to the Turks, there were "scenes of devastation, plunder, and massacre that lasted through the ages," laments Tod, and "all that was sacred in religion and celebrated in art was destroyed by these ruthless and barbarous invaders."[110]

A Valley of Idols

British colonial officials such as Tod, intent on documenting the ravages of Islam in India, were quick to level the charge of intolerance and ruthless iconoclasm at the new regime in power in the valley of the two sacred rivers. The evidence seems ready at hand, especially in latter-era Persian accounts that dutifully record the work of the destruction of temples and idols. Hasan Nizami, in his history of the Ghurid general Qutbuddin Aibak, states that during the siege of Varanasi nearly a thousand temples were razed to the ground (*zamīn-dūz*) and mosques raised on their sites, and new dispensations of Islamic faith and Islamic law were established.[111] Ibn Asir (*Kamil-ut-Tawarikh*) notes that the booty that fell to the conquering army was laden on the backs of 1,400 camels.[112] This number is reported as "four thousand" by Muhammad Qasim Ferishta.[113] Similar lists of conquest and pillage at Gwalior, Malwa, Kalinjar, Multan, Sialkot, Ajmer, Sarasti, Meerut, Delhi, Bihar, Lakhnauti, Banaras (Varanasi), and Badaun adorn the pages of the standard historical compendium of the era, the *Tabaqat-i Nasiri*.[114]

These rounded figures are wildly erratic, and they may well have been conventional—perhaps even prescribed—sets of numbers. Nizami is very clear on the point that each and every campaign against infidels must be recorded by historians, and the story of the sultan's deeds circulated far and wide. There was a clear precedent in the use of set figures, emulated from standard classics such as Al-Tabari's account of the major battles during the Islamic conquest of Persia.[115] Such made-up numbers should not surprise us. Customary figures were also standard fare in European histories as late as the eighteenth century, and as F. P. Lock has suggested, Edward Gibbon showed considerable mastery in such rhetorical use of numerals.[116] Suffice it to say that while the court panegyrists of the likes of Harsha's Bana or Pulakeshin's Ravikirti employed elaborate devices of literary allusion and ornamentation in exaggerating the deeds and feats of kings, the great chroniclers writing in Persian employed their own unique flourishes to elevate their patrons and the significance of their acts. It is difficult to believe, therefore, that the Turkish advance ended overnight the patronage of Hindu temples or the extant forms of pilgrimage and worship.

Conspicuous desecration of temples and idols were aspects of prescribed Islamic orthodoxy, as well as routine acts of plunder. Only recently have historians of architecture begun to salvage the complex history of *spolia* from the dusty annals of colonial and early nationalist history writing. James Fergusson, the doyen of architectural historians in nineteenth-century colonial India, insisted on certain rules of thumb separating elements of the Hindu from what he called the Saracenic style. In the early mosques, he notes, the Muslim rulers "appropriated the remains of Jaina architecture to save themselves the trouble of erecting the whole building from original materials."[117] Vexed by the resulting jumble of what appeared as historically and artistically incompatible, Fergusson searched for the exact seams where the mutually incompatible forms of Muslim and Hindu architecture had been joined. Looking at the Jami Masjid at Kanauj, converted from a Hindu temple by Ibrahim Shah of Jaunpur in the early fifteenth century C.E., Fergusson wrote that this was "undoubtedly a Jaina temple, rearranged on plan similar to a mosque in Cairo."[118] The roof and dome were pure examples of Jain architecture, so that "no trace of the Moorish style is seen internally," but the exterior was "purely of Mahomedan architecture." Similarly at the great Qutb Minar, near the roof and in less-visible areas he noted fragments of cross-legged figures belonging to the Jain saints, which he called the "Hindu remains." Hybrid architecture thus became the indisputable, outward proof of Islamic conquest and Hindu adversity.[119]

A history of the assimilation of Islam into the folds of local and regional communities in India, especially across the extended plains of the Ganges, begs a serious reexamination of the subcontinent's embattled past, which has been the subject of such acrimony and dispute in the fields of history and contemporary Indian politics. Military conquest, iconoclasm, and selective appropriation of architectural space are all parts of a long-unfolding cultural and social evolution. Contentious as it may be, the selective reuse of buildings and building materials from Hindu, Jain, and Buddhist antiquity is an essential part of this history, just as the appropriation of Rome and Roman antiquities was instrumental to the making of Carolingian monuments in Europe. They bear witness to the role of architecture in absorbing bits and pieces of the material past, which, as Maria Hansen has pointed out, is an essential element of all forms of architectural *spolia*.[120] In this regard, structures like the Arch of Constantine in Rome and its layers of recycled fragments of previous reliefs pose a fundamental challenge to the authenticity of historical objects. They remind us that fragments put together in different epochs have their own unique aura and significance. The plinths, pillars, and ornaments reused in the "conquest mosques" of the Ghurid and Sultanate periods are both prizes and relics, but they are also acknowledgment of previous regimes, bridging the recent past of Islam and the receding past of previous idolatrous regimes. As art historian Barry Flood suggests, iconoclasm, which includes the work of destruction, desecration, defacement, recycling or redefinition of images, has always been a difficult, daunting, and unending task, fraught with ambiguity.[121] While the tearing down of religious monuments was a common feature of warfare in contemporary India as well as in the wider Islamic world, it is important to emphasize the specific historical context in which such acts took place, along with the afterlives of the vandalized structures and looted objects. Richard Eaton, who has written extensively on the subject, points out that the desecration of temples should also be considered political acts calculated to supersede the authority of former rulers.[122] Qutbuddin Aibak demolished temples in Kanauj and Varanasi, and his successor, Iltutmish, destroyed a three-hundred-year-old temple in Malwa, and most memorably, the grand temple of Mahakala at Ujjain, transporting the main deity and a statue of the legendary King Vikramaditya (Bikramjit), the founder of the Hindu solar calendar, to the gates of his congregational mosque in Delhi to be trampled underfoot by visitors.[123] According to Ferishta, these images were first demolished at the gate of the mosque.[124]

The inscription found on the inner lintel of the eastern gateway of the Qutb Mosque in Delhi notes that material from twenty-seven temples (*bist va haft ālāt-i butkhānah*) was used for the structure that was endowed by Qutbuddin Aibak.[125] The Qutb complex and the Jami Masjid in Delhi were completed in two phases, first in 1191–1212 C.E. by Qutbuddin Aibak, followed by his slave and successor Iltutmish in 1229–1230 C.E., with stones, pillars, and carvings from plundered temples. The imposing structure, with its lofty 240-foot-high tower made out of reddish sandstone and capped by white marble, has been long seen as synonymous with the absolute power of the Delhi sultans and their efforts to erect a memorial to the triumph of Islam. This framework of interpretation has been challenged by historians such as Sunil Kumar, who points out that we know very little about the history of such constructions, but the creative freedom with which native Hindu artisans incorporated the fragments of old temples into the new structure, and the fact that along with the Turkish chieftains, subordinate Hindu *rāut*s and *ṭhākur*s took part in such expeditions, should caution us against viewing such rebuilding activity simply as acts inspired by aversion to the idol worship of Hindu subjects. The new Islamic regime, despite its vaunt of a pious and unyielding authority, was, as Kumar puts it, part of the "complex, fragmented political and religious world" of twelfth-century India.[126] Pillars and defaced sculptures reused in the Qutb mosque not only reproduce *kīrtimukha* faces from which flowers and creepers emanate, there are also crocodile-crested gateways (*makara-toraṇa*) accompanied by carved bits of stone that once belonged to sculptures of the Rivers Ganges and Yamuna.

Notwithstanding the initial burst of image-breaking to announce the supremacy of the new creed, the dismantling of temples, as we have suggested, was a time-tested weapon of political intimidation with which the Ghurids, followed occasionally by the Khalji and Tughluq sultans of Delhi, served notice to local Hindu kings and chiefs, trying to keep their autonomy in check. Mathura, Kanauj, and Varanasi had been secured, and also the stronghold of Kalpi along the rich Ganga-Yamuna Doab. The chieftains of Badaun, who foreswore not to let the Turkish *hammīra* cross the banks of the most sacred of Indian rivers, had been proven wrong.[127] The most important cities and forts of northern India came into the possession of the Delhi sultans and their dependents. In Bengal an expeditionary force under Bakhtiyar Khalji put the ineffective ruler of Lakhnauti (old Gauda) of the Sena Dynasty to flight, sacking Nadia, the prosperous city and center of Brahminical learning on the banks of the Ganges.[128] A portion of Bengal came under the control of Bakhtiyar, who established the garrison towns of Lakhnaur and Devkot near latter-day Dinjapur. Muizzi and Shamsi slaves, many taken from the campaigns of the army in Afghanistan, Punjab, and northern India, were elevated to the position of military elites, who were now in charge of some of the most lucrative fiefs along prize tracts watered by the Ganges.

CHAPTER 8

THE MAKING OF THE AGRARIAN HEARTLAND

The *Matsya Purana* says that near the confluence of Prayag in present-day Allahabad there was a shrine and pilgrimage site known as the Hamsaprapatana or the "Descent of the Swans." During the onset of winter, multitudes of migratory swans alighted there on the banks and shallows. On the eastern flank of the river was another sacred spot called Urvashiramana, whose ridge was festooned with white swans that nested there all year long.[1] The common word for swan in Sanskrit, *haṃsa,* may have included gray-white bar-headed geese (*Anser indicus*) and greylags (*Anser anser*) that have been migrating for millennia between the valleys and plains of India across the Himalayan range and summer breeding grounds of Central Asia, along with whooping (*Cygnus cygnus*) and mute (*Cygnus olor*) swans.[2] The *Matsya Purana* is a repository of myths and tales compiled over many centuries; its origins are dated uncertainly between the eighth and thirteenth centuries C.E.[3] The lakes, marshes, grasslands, and forest canopies of northern India mentioned in this text have all but disappeared today.

This chapter takes up the historical significance of such epochal changes that have taken place in the landscape of the great northern Indian plains

Greylag geese (*Anser anser*) taking flight. Photograph by Debal Sen.

with the expansion of empires and the intensification of cultivation. We often forget how common it was for travelers, pilgrims, and traders to traverse forests and sparsely inhabited wilderness to move between the cities, villages, and markets of northern India. As late as the mid-nineteenth century the hunting-gathering tribes and nomads of forest India were seen as a threat to settled agrarian society. We consider here the gradual, irreversible transformation of the agrarian ecology of the Ganges plains, yielding to a society of landholders, peasants, traders, and military elites and a gradual upheaval of the balance between forest, scrub, and tillage; wildlife and domesticated animals; and nomads and settlers during colonial times.

Meticulous descriptions of forests, trees, animals, and birds can be readily found in the compositions credited to the Kanauj court poet Banabhatta, or Bana, one of the towering figures of Sanskrit poetry and drama. Bana is famous not only for his study of Harsha and his exploits, but also for his romantic drama, *Kadambari*, dated to the early seventh century C.E., which was likely left unfinished at the time of his death and completed by his son. *Kadambari* is set in the idyllic principality of Vidisha, with a complicated plot involving an eponymous princess and her courtly companions, woven around the themes of love, infidelity, and intrigue. Although Vidisha is rightfully located in central India, Bana's descriptions of the countryside are suffused with references to the lush valleys and forests around the Ganges. He describes rice fields covered with red lotus and patchworks of entire forests of *karañja* trees (*Millettia pinnata*) abloom with white and mauve flowers, *kuśa* grass tall in the autumn, and raging forest fires of the dry seasons.[4] Villages in the *Kadambari* seem to glisten in the sun in a well-watered valley, with plentiful rice and beds of cumin irrigated by water wheels and, farther upland, laced with fields of wheat and beans. Green plantains shade rural settlements, alongside herbs, vines, pomegranate, saffron, and citron, as do coconut and date-palm groves, offering cool coconut water and sweet dates for exhausted woodcutters.[5]

There are also resplendent vignettes of wildlife. Flamingoes descend on the waters of the Ganges in such numbers that its entire surface turns white, floating like the lotus and mocking the sun with their moonlike glow. Caravans of wild geese return to the plains in autumn, when the Ganges shrinks from the flush of the monsoon, with the mud receding, exposing sand banks covered with flowers, inviting the spotted deer to come out and play.[6] In the *Harshacharita*, Bana discusses the capture and training of wild elephants for war and transport.[7] He also mentions the hunting of rhinos and tigers with bows and arrows in thickets of tall-stemmed reeds.[8] Farther south of the valley, in the rise of the Vindhyas, people set up tiger traps near pastures to keep newborn calves safe.[9] Bana's passages remind us how close the edge of the forest once was to the banks of the Ganges. He describes the royal forest guards of seventh-century India, who watch over crops and villages for foraging deer and rampaging elephants. He also gives us clear details of the hard life of small farmers who live at the peripheries of the forest, especially where the hard soil is difficult to break up by spade or plow, where it takes time and labor to clear scrub and fields of prairie grass (*kāśa*) and to loosen the soil with steers and make it cropworthy with manure, and then protect the young seedlings from the scourge of rabbits and antelopes.[10] Bana also describes tribes of hunters and trappers who weave intricate, camouflaged screens behind which to hide and shoot game with bow and arrow. They also design nettles, nooses, and traps made out of gut and sinew, and they gather young bamboo for the best arrow shafts. Bana mentions birders on the lookout for partridges, falcons, and *kapiñjala* cuckoos, with birdcages hanging from the shoulders of their young boys, and clever dogs in tow to track down fallen birds. He also speaks of woodcutters and collectors of fruit, flax, hemp, honey, feathers, wax, and bark.

These images dating back to late Indian antiquity give us an idea of an ecological setting where villages and cultivated fields were enveloped by forest and scrub. How this balance began to tilt in favor of peasant society is a story that can only be told over an extended environmental time scale. For a long stretch of time the expansion of land under the plow was an uphill battle with the forest. As late as the sixteenth century the Chagatai conqueror of Hindustan, Babur, observed the abundance of flora and fauna of northern India. He took careful notes on species that were unfamiliar and exotic, especially birds such as adjutant storks, *sāras* cranes, black ibis, spot-billed ducks, black-hawk eagles, wild fowls, peacocks, and parakeets.[11] He also saw herds of wild elephants near Kalpi on the Ganges. In the districts of Kara and Manikpur in present-day Allahabad, he saw thirty to forty villages earning their living by capturing and taming elephants. He also saw islands on the Ganges with dense vegetation, and from his campsite at night he heard lions and rhinos. He noted blue bull antelopes, hog-deer, monkeys, and squirrels.[12] In Jaunpur and Banaras he saw the Ganges full of crocodiles that made fish burst out and leap onto boats right out of the water and saw alligators carry off soldiers. He also saw blind Gangetic dolphins, their dark heads bobbing in and out of the waters, their snouts mimicking that of a crocodile, with rows of tiny white teeth.[13]

The Persistence of War

The story of intensive settlement of land across the valley goes back to the beginnings of the Delhi Sultanate. There is much that we do not know about the political upheavals in the Ganges basin during the period of Turkish conquests. After the initial military forays, what sort of political authority was Delhi able to exercise over its outlying provinces? How difficult was it for the new military elites, Muizzi and Shamsi *malik*s and *amir*s, to gain control of the villages and districts and establish a regular share of revenues from Hindu chiefs and village headmen? The significance of the emergence of the Delhi Sultanate lies not just in its encounters with hostile groups such as the stubborn forest-dwelling Mewatis or unruly Rajputs—but also in its prolonged quest for political and material stability in the northern Indian heartland. A major transformation in the cultivation, distribution, and exploitation of resources was initiated along the 300-mile stretch of territory between the Ganga-Yamuna Doab and the head of the Bengal delta, effects of which lasted through the Mughal and British periods of rule.

The Ghurids established what historians once described as a "slave dynasty," because power was held initially by Turkish slaves (*ghilmān*) as military commanders, some of whom rose to high positions. The Muizzi slaves—named after Muizuddin Ghuri, joined later by Qutbi and Shamsi slaves from the retinue of Qutbuddin Aibak and his successor Shamsuddin Iltutmish—were all part of a tightly knit military elite, many having been captured and trained during early and extensive campaigns in Afghanistan and northern India. Despite their lack of social and legal status, many such slaves climbed up the ranks of the army because they could be trusted much more than family or clan members during times of political strife or bloody conflict over territory and resources.[14] In the early days of the Delhi Sultanate, rewards for service and loyalty were handed out in the form of military and administrative assignments known as *iqtāʿs*.[15] The transition from raid, plunder, and the redistribution of spoils (*ghanīmah*) to regular taxation of agrarian produce, in which the Ganga-Yamuna Doab and stretches of cultivable tracts farther downstream along the valley would play a significant role, proved decisive not just for the early sultans but for other Turkish and Afghan groups to follow. Exclusive command over large reserves of cultivable land (*khāliṣah*), subjugation and eventual absorption of petty kingdoms and chieftaincies, and the settlement of new villages and towns were key to this transition. While periodic plunder of outlying kingdoms took place throughout the thirteenth century, it was land revenue that supported the muster, equipment, and training of troops, which in turn extended the frontiers of cultivable land.[16]

Historians have debated the origins and workings of such military-revenue assignments and the degree of autonomy in fiscal or military affairs that they allowed for the new class of elites in the countryside. There has also been lively disagreement about the extent to which Turkish rule in Delhi was able to establish a measure of autocracy beyond the centralization of military command (*ariz*). Recent studies have questioned its purported absolutism, suggesting instead a fragile polity rife with internal dissension among the elite and beset by frequent coups.[17] Enslaved soldiers without prior social or legal status, holding office at the pleasure of their masters, may have been the most viable instruments of the newly defined power structure. They also served as a bulwark against the highly dissentious contest over privilege and resources among the freeborn military elite who dared to challenge the ruler's authority from time to time.

The Delhi Sultanate was a patchwork of military garrisons and urban centers, and its capital, Delhi, as Peter Jackson has shown, was a sprawling city built around the core of a vast military encampment—the army stationed there was estimated at half a million during the reign of Alauddin Khalji at the end of the thirteenth century C.E.[18] Military readiness was indispensable because of the looming Mongol threat on the western frontier that appeared during the reign of Iltutmish and continued for many decades, forcing an exodus of slaves, soldiers, notables, and literati fleeing across Himalayan frontiers into the Indian heartland. Raids of outlying Mongol bands based in Afghanistan had made the Punjab and the western flank of Delhi vulnerable during the reign of Ghiyasuddin Balban (1266–1286 C.E.). Toward the end of the century, the attacks were coordinated directly by the Central Asian khanate, and Mongol expeditionary forces appeared periodically in the Ganga-Yamuna Doab and even threatened territories lying farther east of the main course of the River Ganges.[19] This military exigency put a great strain on taxation and agriculture, and the sultanate attempted desperately to keep insubordinate chiefs of the far-flung provinces in line.[20]

Tribute and Tillage

W. H. Moreland in his classic treatise on Indian land revenue studied in detail the expansion of agricultural produce and taxation following the basic geographical features of the Ganges basin. During the period of the rise of the Delhi Sultanate, core areas of revenue collection began with the region around Delhi (*havālā-i Dehlī*) between the wooded foothills to the north and the haunts of the turbulent Mewatis scattered across the forested and rugged terrain toward the south. To the east was what Moreland called the "river country," that is, the upper part of the Ganga-Yamuna Doab comprising the provinces of Meerut, Baran (latter-day Bulandshahr) and Kol (present-day Aligarh)—an area that was brought directly under the control of Delhi in the late thirteenth century C.E. under Alauddin Khalji.[21] South of these lay the old cities of Kanuaj and Kara (near Allahabad). Farther east lay the fertile provinces of Awadh, Jaunpur, northern Bihar, Tirhut, and the kingdom of Lakhnauti in Bengal.

Money and produce poured into the treasury of Delhi in the form of *kharāj*, the main tax on agriculture, and *jiẓiya*, capitation levied on non-Muslims adjusted according to levels of wealth and income. Generally speaking, during the early period of Turkish rule in northern India, such taxes were largely skimmed off the top without much intrusion into matters of everyday administration—a situation that would change during the period of Khalji rule. This is one of the reasons that names of village headmen (*khuṭ*s and *muqaddam*s) begin to appear in the official records.[22] The proceeds

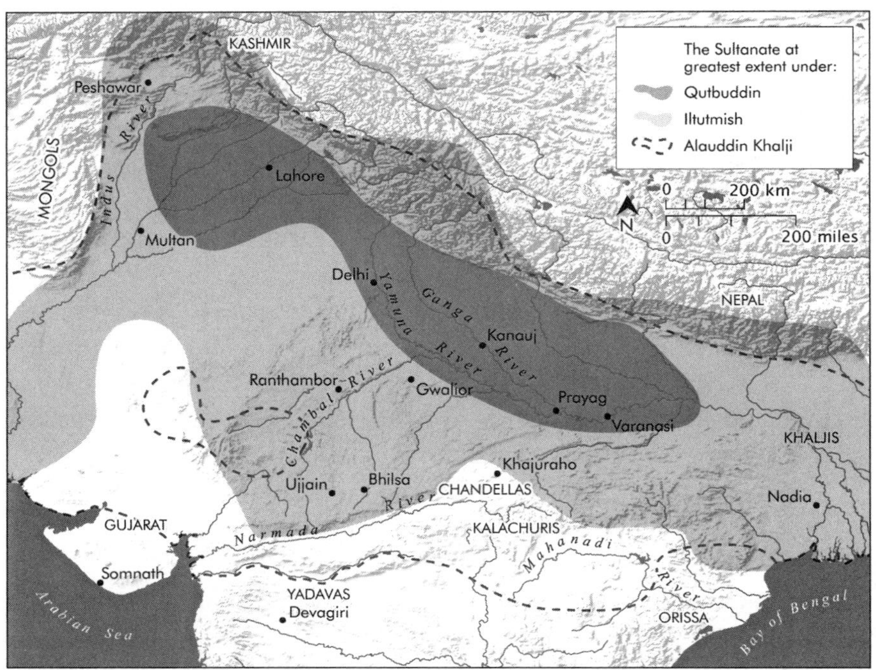

Northern India under the Delhi Sultanate. Map by M. Roy Cartography.

of each basic administrative division—loosely corresponding to the main provinces—were distributed as revenue assignments or *iqṭā's* to various military commanders. The leading military elite of this time were also often the largest *iqṭā'dār*s or *muqṭī's* of the realm. This has often been unsatisfactorily translated as a military-revenue "fief," which does not quite do justice to the fact that much latitude was given to chiefs of the military and their retinue to extract revenue from the provinces to which they were assigned.[23]

The freedom to squeeze revenue from peasants was eventually curtailed, although not easily, because many of the *āmir*s and *mālik*s (especially the Muizzi and Qutbi *jāndār*, or faithful) of the early sultanate periodically flouted the authority of Delhi and carved out independent territories or preyed on neighboring *iqṭā's*. This agenda of reining in freebooters and tightening control over the area under taxation began under Ghiyasuddin Balban, the last slave ruler, and was extended further under Alauddin Khalji and the Tughluqs who followed.[24] It was a difficult task. During the time of

Ghurid rule, Shamsi *iqṭa'dār*s commanded prize villages in the Ganga-Yamuna Doab, and Ghiyasuddin Balban settled his crack Afghan troops with prosperous villages in the vicinity of Bhojpur in the lower Ganges valley.[25] By the period of Khalji rule these early estates had become seedbeds of dissension. Alauddin Khalji took great pains to yank some of these tax-rich holdings, directing vizier Nusrat Khan to turn them into crown land or issue grants held at the mercy of the sultan that could not be resumed or passed on without notice. There was also an effort to bring unruly Hindu landlords and chieftains into compliance. Alauddin complained to the Qazi of Bayana that it was hard to break the power of the many *khuṭ*s, *muqaddam*s, and *chaudhurī*s, who had the freedom to ride fine horses, wear fine clothes, and shoot with the best Persian bows and arrows, without paying a single farthing in tribute or tax to Delhi.[26] Clearly, the sultanate, even at the height of its military power, did not command the resources to convert every species of tribute into regular taxation, and it is difficult to gauge the discrepancy between taxes demanded and taxes actually collected. Regional chieftains who fought against the Delhi sultans are termed as *rāi*s, *rānā*s, or *rāwat*s (leaders of cavalry) in contemporary Persian accounts.[27] They have also been described as "vassal chiefs" or members of an older "feudal hierarchy."[28] The merit of such a categorical view is subject to debate, but there is enough evidence to suggest that many of these petty rulers settled down in small market towns (*qasbā*), which were the seats of rural administration across northern India.[29] These were the likely predecessors of the vast and variegated group of large and small *zamīndār*s who became an indispensable class of landed intermediaries in Mughal India.

A Valley Besieged

Formidable as the sultanate may have been as a military formation, with the ruler endowed with undisputed authority, it probably never achieved a measure of absolute power. It certainly did not succeed in imposing Sunni orthodoxy on the majority of Hindu subjects under its rule. One of the earliest tracts on governance written during the time of Iltutmish, by Fakhr-i-Mudabbir of Lahore, noted that Turks were simply one tribe among many, which is why the burden of restraint in affairs of the state fell on the ruler himself. The wealthy Hindu Rais, he claimed, despite their vast body of soldiers and elephants, had been defeated and were now obedient to the sultan's desires.[30] Some had been converted, and Islam had been instated in the place of idols, and there was peace in the kingdom.[31] The incomplete surviving manuscript of the *Fatawa-i-Jahandari* [Edicts on Government] by the illustrious fourteenth-century historian Ziya-ud-din Barani (1285–1357), which is essentially an idealized set of precepts on governance passed on to his successors by Sultan Mahmud of Ghazni, gives us a fair idea of the unprecedented nature of the Delhi Sultanate as an Islamic state, where the duties of the ruler as a defender of the faith in a land of infidels are not clearly defined. Written after Barani had left the court of Delhi in disgrace, this is a rather murky view of the office and conduct of the sultan, which he considers unbecoming of the Qur'an's injunctions for the supreme leader of the faithful (*valī al-amr*), the vice regent of God, the emissary of the caliph, or the very axis of the earth (*quṭb-i 'ālam*).[32] The preeminence of Brahmins, Barani complained, had not been dealt with, and the sultan was simply accepting tribute and capitation from the infidels who worshipped idols and cow dung.[33] Barani also begrudged the fact that infidels openly worshipped stone images in Delhi and all other major settlements, and they sang and danced at public festivals without fear or hindrance. Many Hindus were showered with honor and distinction and elevated to high office. The base, the low-born, and the godless, unfit for any service, religious or mundane, had been given a "free hand."[34]

Barani's views can be supported by the fact that despite the pillage and desecration of temples during the early campaigns, various forms of Hindu public piety, temple building, and pilgrimage seem to have continued regardless during the first two centuries of Turkish rule. There is enough evidence to suggest that the busiest Hindu pilgrimages such as Varanasi, which was now a part of the larger city of Banaras (*banāras* in Persian accounts) enjoyed great prosperity in this period. The temple of Avimukteshwar may have been destroyed during the early Turkish sieges, but the imposing Vishweshwar temple was built during the reign of Iltutmish, and it is recorded that a wealthy merchant by the name of Seth Vastupal forwarded a prayer offering (*pūjā*) of 100,000 coins in its support.[35] During the early years of Alauddin Khalji's reign, the temple of Padmeshwar was erected on the banks of the Ganges. It also seems that temple building continued along the Ganges during the period of Tughluq rule, notably the temple of Manikarnikeshwar, built along one of the most famous steps of Varanasi in 1359 and endowed by a grant from a wealthy patron known as Vireshwar. Jinaprabha Suri's well-known fourteenth-century survey of the main Jain pilgrimages, *Vividha Tirthakalpa*, pays an eloquent tribute to Varanasi as the birthplace of the twenty-third Jain emissary Parshvanatha.[36] Suri describes the city as divided into four quarters, the oldest described as the "Varanasi of the gods" (*devavārāṇasī*); the palace of the reigning deity, Vishwanatha (*viśvanāthaprāsāda*); and the administrative quarters as the residence of *yavanas* (Turks).[37] Jain pilgrimage and learning thrived in the city through this period.[38]

The Turkish rulers of Delhi were not able to exercise unbroken dominance over the Ganges basin. After Illtutmish, the newly ascendant Ganga Dynasty of Odisha pressed their advantage over the eastern part of the valley. Bengal under the Senas had fallen to the cavalry of the renegade Bakhtiyar Khalji, who subsequently accepted the nominal authority of Delhi, but the political fate of Lakhnauti and Gauda still hung in the balance during the coup of Ghiyasuddin Balban. The Delhi Sultanate held on to the region of Lakhnaur, Lakhnauti, and Devkot, but upstarts like Samanta Rai, sensing the weakness of a Delhi preoccupied with the Mongols, wrested territories in the Rarha and Varendra regions of deltaic Bengal.[39] The city of Nabadwip on the Ganges, a noted seat of classical Sanskrit learning, likely remained under the writ of Delhi and paid taxes, but farther east smaller regimes were busily raising their own standards. Notably, the successors of ousted monarch of Bengal Lakshmana Sena restored a corner of their kingdom under Viswarupa and Kesava Sena, claiming to have beaten back the Turks (*gargayavanānvaya*) in the area of Dhaka-Vikrampur in present-day Bangladesh.[40] Turks held on to Magadha in southern Bihar by a narrow strip along the Ganges known as the country of the Uddandapur Vihar connecting the cities of Banaras, Shahabad, Patna, Munger and Bhagalpur, flanked in the south by warlike Gahdavala chieftains. Unlike the petty rulers of Bodh Gaya, resigned to the vassalage of Delhi, the sprightly Rajputs of Bundelkhand and Rewa overran the tracts south of the River Jumna, and venturesome Baghela chiefs held stretches of forested territory immediately to the south of the valley from Kalpi on the right bank of the Jumna all the way to the fort city of Chunar on the bend of the Ganges facing the Kaimur Hills.[41]

Taming the Wilds

Expeditionary forces from Delhi, time and again, had to travel by roads overrun by undergrowth and through dense jungle that provided cover to rebels, hostile local chieftains, and heads of militia. Such terrain compounded the distance between Delhi and far-flung independent chieftaincies, especially along the Bihar corridor that connected the middle Ganges valley to Lakhnauti in Bengal.[42] Barani in his *Tarikh* complains that the major roads of Hindustan had fallen into disrepair and resembled walking trails (*maslūk gushta-ast*).[43] Sultan Balban, facing war and scarcity, resolved to make the hinterland of Delhi accessible.[44] Jungles in the vicinity of Delhi and the Jumna valley, inhabited by the Mewatis, were flushed out. Balban also gifted towns and villages in the Ganga-Yamuna Doab to loyal chiefs, who in return were asked to lay waste to the villages of brigands and outlaws (*mufṣādān*) who routinely plundered (*tārāj*) and killed travelers and took women and children as slaves.[45] They were also directed to clear brush and settle the land with obedient peasants. A great numbers of slaves, cattle, and horses were captured in these raids. Once the roads were cleared, caravans and merchants could pass freely once again. Barani further notes that the route from Kampil to Bhojpur was also infested with robbers, and to counter them the sultan planted new forts in which Afghan garrisons were placed, supported by gifts of cultivable land. Similar lawlessness was dealt with in Katihar on the Ganges, and also in Badaun, and Amroha, where an advance guard of five thousand archers were sent to scour forests on both banks of the Ganges. In Badaun, woodcutters were sent to chop down trees in the forest and make way for the passage of the army. A large number of rebels were captured and killed during this drive. Heaps of the slain were displayed in the villages, and it is said that the stench of the dead spread all the way to the edges of the River Ganges (*kinārhā-i lab-i Gang*).[46] Alauddin Khalji led a similar expedition through Badaun, which began as a regular hunt but led to a punitive military exercise against what Abdul Malik Isami, the author of *Futuh us-Salatin*, calls unruly and hostile (*bad-mihrdār*) local Hindu chiefs (*sarāfarāz kāfūr*).[47] During Firuz Shah Tughluq's campaigns, along the Rajmahal Hills located at the point of entry of the Ganges into Bengal, roughly 25 miles downstream to Lakhnauti, a way was cut through the forest. Wherever soldiers found forest dwellers suspected of robbery, they strung them up high on trees.

These examples provide a glimpse of the limits of the agrarian economy of the thirteenth century C.E. Settlement and expansion of land under cultivation (*ābādī*) was key to the rise of rural and urban settlements (as in the suffix *ābād*).The sultanate initiated what Mohammed Habib once described as a long-term "rural revolution"—by which he meant the assimilation of the rural intermediaries, the regimentation of taxation on agrarian produce,

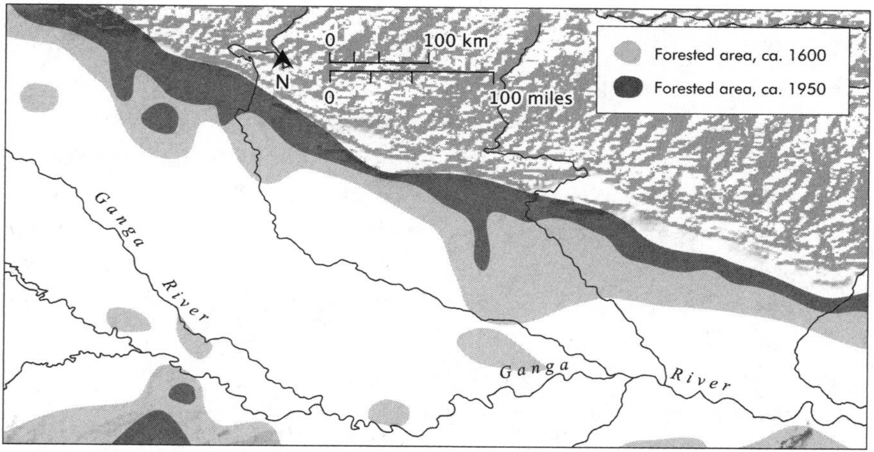

The forest cover of north India, ca. 1600–1950 C.E.
Map by M. Roy Cartography.

and the reclamation of wilderness.[48] These vast stretches of tropical and semitropical deciduous forests of the middle Ganges basin, from a period when northern India enjoyed a higher average annual rainfall and a higher water table, have all but vanished over the last two hundred years.

Even a cursory look at Captain James Rennell's maps of 1780s commissioned by the East Indian Company's government in Bengal shows that the routes between the Ganges and its major tributary, the Son, in the vicinity of Rohtas and Chunar were covered in stretches of impenetrable jungle. Captain Luis Felix de Gloss's survey team in the 1760s found it virtually impossible to find a route without cutting through stubborn and thorny thickets. Even near the old fort of Rohtasgarh, the jungle teemed with tigers, deer, peacocks, rhinoceroses, and bears. There were also forest tribes that greeted visitors with bows, arrows, and cutlasses. Even at villages nearer the banks of the Ganges, tigers menaced pack bullocks, and bears routinely mauled travelers.[49] The confluence of the Gandaki and the Ganges in Bihar was surrounded by fir and other conifers, and between the cities of Munger and Pipra was a profusion of reed jungles, infested with tigers, wild buffalo, and rhinos.[50] Samuel Showers, who charted the first military routes along the River Ganges between Allahabad and Banaras, and farther eastward from the mouth of the Gomti to the Karmanasa in Bengal, found "immense fields of thick grass" where an entire day's journey could be spent cutting a way through "one continuous jungle" marked with the footprints of tigers.[51] Some of these forests were still around during the late nineteenth century C.E. Edward Lockwood of the Bengal civil service, a hunter and wildlife enthusiast about a hundred years after the period of the first company surveys, described the teeming forests on the southern banks of the Ganges in the region of Munger, one of the most highly cultivated stretches of the northern Indian plains in modern times, and the profusion of ebony, sal, and *mahua* trees. He also wrote about the abundant birdlife supported by the deserted channels of the river and low-lying marshes and swamps where 100,000 ducks could be seen at once, "so close together that they almost hide the water; and as flock after flock pass overhead on being disturbed, the sound of their wings resembles waves breaking on a troubled shore."[52]

These forests and grasslands along channels of the Ganges were very old but not necessarily virgin. Forests were very much a part of the agrarian landscape of the northern plains. They had been cut and tended by local inhabitants over centuries and periodically torched by shifting cultivators and hunters. British foresters and conservators of the late nineteenth and early twentieth centuries, while condemning the crude and wasteful practice of shifting cultivation and the periodic burning of undergrowth, remarked on how rapidly the vegetation sprang back to life. Edward Stebbing in *Forests of India* disparaged the "pernicious system of shifting cultivation . . . in force for centuries" that claimed thousands of square miles of densely forested country; he observed not only that two rapid cycles of crops were raised from the ashen, nutrient-rich soil, but also how quickly the scrub regenerated after forest fires.[53] Dietrich Brandis, the great pioneer of Indian forestry, wrote about the abundance of the hardy sal tree (*Shorea robusta*) universally used for its timber in India. He had seen vast expanses of sal forests in their "natural state" along the sub-Himalayan foothills. In the right climate and soil, he found that sal "reigns supreme," especially along the banks of rivers where the soil was alternative beds of shingle and sand.[54] The seedlings were designed to survive and even thrive after the forest fires of the summer, remaining alive under the cover of tall grass.

This was an active, shifting ecology. Just as under favorable conditions of taxation and settlement, the margins of the forest would recede to make way for new agriculture, settlements suffering from famine, floods, or drought were quickly abandoned and reclaimed in turn by the forest. To this we must add the changing course of rivers flowing across the Ganges basin.[55] All the major tributaries of the Ganges, most notably the Sarayu,

Ghaghara, and Koshi, have shifted their courses over the centuries. Fourteenth-century accounts from the Tughluq period indicate large, dried-up channels of the Jumna where new forest had encroached. The changing hydrographic contour of rivers of the northern Indian plains suggests continuous human tending of water bodies and sandbanks: meanders, oxbow lakes, swamps, flood channels, and the use of irrigation channels fed by the early prototypes of the Persian water wheel (*noriā*).

Many of the major trading routes of India were also unflattering tracks of dirt and mud that passed through inhospitable terrain.[56] They were particularly treacherous during the height of the monsoons, when the entire drainage channel of the Ganges, including all the major tributaries, ran high or breached their levees. Barani in his *Tarikh*, describing one of the early marches of Alauddin Khalji from Kara to Delhi, mentions how difficult it was to cross the Ganges and Jumna overflowing their banks or to lead horses and elephants through roads clogged with mud and debris.[57] Bullock carts and pack animals were the tried and tested means of carriage and transportation, especially on unpaved roads that passed through rugged terrain and wilderness. The most prominent among such transporters were the Banjaras, who carried daily essentials such as salt, grain, and ghee, bringing them across long distances to the distant markets and villages of northern, western, and Central India—their *tāṇḍā*s spotted at the same time in Bengal, Agra, and Surat. Banjaras traveled in tightly arranged caravans at the head of ten to twenty thousand pack animals.[58] We do not know how this itinerant set of tribes of carriers of provisions evolved in India, but long before they were noticed by travelers from Europe, they had become an indispensable part of the landscape of inland trade in the company of their carts, oxen, wives, children, and watchdogs. The term "*banjārā*," it is said, originally derived from the Sanskrit *vānijyakāra*—traders. Some were hired by grain merchants, some were merchants in their own right. Thomas Roe, visiting Mughal India in the early seventeenth century, saw over ten thousand Banjara bullocks laden with grain assembled in one place.[59] Jean Tavernier, who visited India during the period of Emperor Shah Jahan, remarked that it was "an astonishing sight to behold caravans numbering 10,000 or 20,000 oxen together, for the transport of rice, corn and salt."[60] The presence of the Banjaras alone shows the proximity of the harvest, marketing, and distribution of agrarian produce to the margins of the forest. The movement of such great herds of cattle also indicates that agricultural fields were aligned closely to pastures, where the passage of cattle also meant easy availability of cow dung, an everyday source of fuel for the hearth as well as a prime nutrient for cultivable soil.

Salt of the Earth

The Ganges basin and its extensive alluvial floodplains saw the rise of one of the most vibrant and dense peasant societies in the world, comparable in dimension to the great expansion of peasant society along the lower Yangtze in China. Viewed over the course of a thousand-year period, an entire social strata of primary tillers of the land seems to have been established in the Ganges plains, emerging from a breakdown and reconstitution of smaller hunter-gatherer, tribal, and pastoral societies of antiquity—what Irfan Habib calls the great "caste peasantry" of northern India. While their caste positions and material conditions converged, they did not always shed their traditional identity or lineage, which explains the historical persistence of groups such as the Gond, Gujar, Jat, and Meo. These were the primary inhabitants who extended the frontiers of village India, cleared the lands, dug canals, tilled the soil, and brought their harvests to the weekly markets of rural India, ultimately tipping the ecological balance between land and human labor before the advent of fossil fuels and industrial production.

Population figures in India before the late-nineteenth-century censuses of British India are notoriously difficult to ascertain. Evidence from crop yields and land revenue figures of the Mughal Empire suggests a steady demographic growth, coinciding with the intensification of agriculture in the countryside, especially after the extensive land revenue reforms of Emperor Akbar. It is likely that the population of India rose roughly from 150 million in 1600 to 200 million in 1800; some scholars suggest a gain of around 30 million through the course of the seventeenth century.[61] Comparing these figures with estimates of the rising urban population of Mughal India, where the largest cities such as Agra and Delhi might have supported half a million inhabitants each, gives us an indication of the scale and long-term impact of the expansion of land under the plow at the expense of the forest.[62] The plains of northern India and the Ganges valley carried a disproportionate burden of this pressure of population on the land, which, despite famines and epidemics, increased steadily through the period of British rule.

One of the most densely populated regions of the northern plains was the Ganga-Yamuna Doab, where forests had been cleared and new villages settled since the period of the Khaljis and the Tughluqs. The population and prosperity of this fertile, revenue-rich area, however, was badly affected several times by war, shortages, and famines. It is difficult to assess the long-term demographic impact of pestilence and hunger, but one must assume that mortality rates, especially among children, the elderly, and the infirm, were high during these periods of crisis. Thousands perished during the major famine that took place in the environs of Delhi during the reign of Iltutmish, and again during the period of Khalji rule in 1291, which inspired some of Alauddin Khalji's drastic market reforms and interventions in the distribution of food grains. The most devastating famine took place during the reign of Muhammad ibn-Tughluq, which hit the region of the Doab very badly, raging over at least seven years, and a general scarcity lasted in the region from 1327 to 1340 C.E.[63]

To understand the significance of such transformation we have to consider changes taking place at the two opposing ends of rural society in northern India—the ancient communities that tilled the land and their old nemesis, the surplus-skimming landed elite. We have seen how large and petty indigenous chieftains, joined by military and slave elites, had evolved into a class of rural intermediaries, supported by the labor of poor peasants. The villagers of the northern Indian plains were no strangers to droughts, floods, and periodic famines. While the sultan's authority went largely unquestioned, common people did blame their ruler for scarcity and starvation. Barani speaks of the responsibilities of the ruler as the universal protector (*jahān-parvar*) in the first chapter of his book on princely conduct (*adāb-i salāṭīn*).[64] It was said that if just one old woman went to bed hungry in his realm, the sultan was expected to answer for it on his day of judgment.[65] According to Isami, during a period of great scarcity, when the news came to Alauddin Khalji's court that three old men had been crushed in a stampede that had broken out at a grain market (*manḍī*) during the time of famine (*qaḥṭ*) and drought (*khushkī*), he was so distraught that he gave up drinking wine.[66] Alauddin's forcible attempt to intervene in the demand and supply of grain across the markets of northern India was a response to this mounting threat of scarcity. While some of his measures were deemed harsh, including the planting of spies in bazaars and weekly markets, levying of fines, and physical punishment, Alauddin was able to bring the price of food grains and necessities down, curbing excessive hoarding and price gouging among traders and shopkeepers. One set of regulations was aimed at fixing the price of grains such as wheat, barley, and rice, which seems to have been largely maintained through the end of his reign.[67] A second regulation, carried out by trusted deputy Ulugh Khan, was the surveillance of markets by officers on horse and foot. The third regulation was to ensure a full stock of grain in the imperial granary in the event of floods, famines, and external invasions. A heavy burden fell on villages located on crown lands (*khālṣa*) in the Doab, who were asked to provide added taxes in kind. A set of subsidiary stores was set up in the villages in the region of Jhain near present-day Awadh, and sent to Delhi by caravan. A fourth regulation was passed to sequester the heads of Banjara pack-bullock caravans (*kārvāniyān*), who were dragged in chains in front of the market superintendents and detained until they agreed to a set of contractual obligations that required them to bring their wives, children, and cattle and settle along the banks of the Jumna River. A fifth rule made it illegal to hold on to grain and sell it above the fixed market price. The revenue officials were directed to ensure that peasants sold entire harvests of grain to the Banjaras at a fixed price, with no further stocking or reselling. The sultan demanded detailed and regular reports from the countryside on market rates and transactions. The list of prices for articles sold in the markets of northern India during this period included clothes, textiles, sugar, fruits and vegetables, oil for lamps, horses, cattle, slaves, and concubines.[68] Petty traders and shopkeepers were threatened with harsh physical chastisement if they were found cheating customers with false weights and measures.[69]

The severity with which Alauddin's men went after miscreants in the markets and villages, especially in the hinterlands of Lahore and Delhi, was prompted by the pressing need to provision garrisons that could not be maintained simply on salaries from the imperial treasury.[70] Attempting to bring down the cost of horses and equipage for soldiers, Alauddin tried to curb inflated prices by stripping middlemen who dealt with horse traders from Afghanistan and Multan. Isami suggests that Alauddin's regulation of the market price of food grain (*ghallā*) at prices current before the onset of dearness and famine (*nirkh-i qadīm*) succeeded in securing the availability of grain in the markets of greater Delhi throughout the years of drought (*khushk-sāl*).[71] These measures targeted the wealthiest traders and money changers of the Hindu mercantile castes, and intermediaries who stood between primary producers and the imperial treasury, including regular examination of the ledgers of village recordkeepers.

Alauddin's measures give us an early view of the conditions that dictated the subsistence of peasants. Delhi's appetite for taxes had grown, with new dues levied on dwellings in the village (*gharī*) and on cattle pastures (*charāi*). In retrospect, Delhi seems to have waged an extended war against both entrenched peasants and landlords. In the valley of the Punjab, between Lahore and Dipalpur, and along the most productive alluvial flats of the Ganga-Yamuna Doab, tax was taken in grain, sent to state granaries, and released in rural markets at low prices. The provincial heads and revenue assignees (*valīs* and *muqṭi's*) were allowed to collect their dues in money from the sale of grain at reduced prices at the weekly markets, and neither peasants nor petty traders were able to profiteer during times of drought and scarcity.[72] Such high levels of exaction, however, could not be enforced or sustained for very long.

Abdullah Mahru, a governor of Sind during the Tughluq period, author of the definitive treatise on epistolary style, *Insha-i Mahru*, which also contains many official letters written to chieftains, landlords, and Muslim clerics, comments in one of his letters on the futility of the legal status of the freeborn peasant (*ḥurr aṣl*), who barely scratched out a living and was all but chained to his plot of land.[73] The letter also suggests that when he fled his plot, revenues of the entire village, and ultimately the ruler himself, suffered greatly. Peasants struggling to pay taxes often abandoned their villages and fields and settled elsewhere. This was possible because of the low density of the rural population in relation to land. Peasants of the upper Ganges valley and the Ganga-Yamuna Doab exhibited the same behavior when another great famine broke out in northern India soon after Muhammad ibn-Tughluq's accession.

Barani, choosing his words with care, states that the Tughluq ruler's own ill-advised decisions led to this disaster.[74] Facing a shortage of money, he had levied a number of oppressive exactions (*abvābs*) in the Doab, hoping for an increase of 5 to 10 percent in rates of taxation. This unexpected measure, added to several seasons of scant rainfall, broke the back of the poor peasants (*ri' āiyā-i ghanī*), leading eventually to the devastation of the entire country and an eventual decline of the population. Peasants were reduced to beggars. Those who could took up arms and rebelled. Cultivation almost came to a standstill. Other peasants across the kingdom, alarmed at the news of this debacle in the Doab, began to flee to the forest. The Banjara caravans that brought grain to the marts of Delhi were nowhere to be seen. A raging drought further spiked the price of essentials. The ill-fated measures in the Doab tarnished the reputation of the sultan and rattled the resolve of his administration. Barani gives a vivid description of instinctive, age-old peasant behavior in the face of rack-renting and oppression: flight and desertion (*uftādagī*). The Mughal emperor Babur remarked in his memoirs on how quickly a city or a village could be abandoned and resettled in India.[75]

When the dreaded Iranian conqueror Amir Timur invaded northern India, fording the river Jumna in December 1398 and blockading all the crossings of the river, his soldiers went about sacking markets for grain and taking thousands of captives. Much of the Doab peasantry fled once again, heading for the jungles and mountains. Timur, in one of his letters, described the encounter with the fierce Jat peasantry of the Ganges valley, whom he saw as an insolent, "robust race" of thieves and plunderers who attacked travelers and caravans on the highway, comparing them to an infestation of ants or locusts.[76] They were infidels that needed to be put to sword. As Timur's army approached, the Jats disappeared through the sugarcane, scrubland, and forests of the valley and were hard to pursue. During Timur's invasion much of the Ganges valley was laid to waste. Delhi was sacked ruthlessly, a blow from which it would not recover for a long time, and it led to the eventual downfall of the Delhi Sultanate.

The Blessings of Water

Harnessing water was key to agriculture and the rise of population in the greater Islamic world. From Mesopotamia to the plateaus of Iran, from Egypt to the plains of Anatolia, every regime sought to improve the technology of aqueducts and canals—one of the most valuable legacies of ancient Rome. The search for efficient use of water produced the device known as the *shādūf*, a counterweighted lever for lifting water out of one irrigation channel and pouring it into another; the ubiquitous *noriā*, essentially a vertical water-powered wheel; the *sāqīya*, the Archimedean animal-powered water wheel that employed gears; and the Iranian *qanāt*, or foggara, used more commonly in the desert, which employed subterranean channels tapping into the underground water table and carrying water on gradients draining into agricultural irrigation. We do not know exactly how much of this technology was available in India during the period of the Delhi Sultanate, but there is ample evidence of the use of canals and a local version of the Persian water wheel (*araghaṭṭa*), described in detail by Emperor Babur in his notes on the countryside near Lahore, Dipalpur, and Sirhind.[77] It was far superior to the method of using an ox to pull water out of a well by a bucket with a rope running around a pulley, which he saw in Agra and Bayana.

Sultan Firuz Shah Tughluq had canals dug from the River Sutlej in the Punjab and from the river Jumna (Yamuna) in the greater Ganges plains. The western Jumna canal served the new fort west of Delhi in the town of

Hisar, named after the ruler (Hisar Firozah). The canal named Rajiwah in the east brought water from the northern parts of the Jumna directly to the city of Delhi. Barani is ebullient in his praise of these projects. He had witnessed the ravages caused by drought and famine, setting off peasant outbreaks during the reign of Muhammad ibn-Tughulq. People died from the dearth of water (*be-ābī va tashangī halāk mi-gushtand*), crops failed, trees shed their leaves, and birds even died from the want of rain. When the canals (*jūihā*) were dug, the "beaks of birds became moist again" (*nauk-i parindah tar shavad*). Much like the Rivers Ganges and Jumna, they brought back the trees and crops, and the sultan was able to resettle soldiers and peasants. After these measures the population of Hindustan began to rise again. Where nothing but thorny acacia and wild inedible pods would grow for miles on end, crops like legumes and sesame could be planted.[78] The sultan also took care to repair roads, dig wells, plant shady trees along the way, build resting places and shelters for travelers, and erect mosques for wayfarers throughout his kingdom. The historian Afif notes that during the height of the monsoons, when such canals were susceptible to flooding, Firuz Shah appointed special officers, including Afif's own father, to report on embankments, levees, and the levels of inundation.[79]

The situation in Bengal was quite different from the rest of the Ganges valley. Located directly in the path of the monsoon, with a high mean annual rainfall (about 2.79 meters), much of the delta lay under dense canopies of subtropical forests and swampy mangroves. The hydrological record suggests that for more than five hundred years the River Ganges has been inching eastward, contributing to a gradual shrinking of its principal distributaries and the drainage channel of the Hugli-Bhagirathi, which once flowed much farther west. Early British naturalists who studied these coastal areas observed a restless and dynamic landscape, marked by centuries of volcanic and seismic activity.[80] From the south of the Rajmahal Hills to the shores of the Bay of Bengal they found alternate bands of peaty and vegetal deposits, with remains of trees buried in the soil at considerable depths, suggesting alternate elevation and depression of the surface caused by land erosion, tidal bores, cyclones, and swells. They also found ancient settlements abandoned to the sea and salt water. The cartographic and navigational charts of João de Barros and Giacomo Gastaldi dating to the sixteenth century C.E. show a continual breakdown and accretion of the coastline—the salty mudflats of the Sundarbans, replenished by massive alluvial deposits, where debouching rivers meet the inflow of brackish water inundating vast forests of mangrove.

It is difficult to form a reliable picture of the lives of ordinary inhabitants of the inner delta during this period, except for scattered references to pioneers who cleared brush and planted rice. These were a different breed of peasants who lived side by side with tribal hunter-gatherers, fishermen, boatmen, and ferrymen of the Patni, Malo, Chunari, and Nikari castes famous for their boats and fishing nets. One of the oldest agricultural castes of deltaic Bengal were the Kaibartas, a branch of whom (Jalia-Kaibarta) still make a living from fishing. During the reign of the tyrannical Pala king Mahipala II, there was a rebellion spearheaded by the Kaibarta fishermen rallying under their leader Divya. Mahipala II was killed, and his kingdom based in Varendra in northern Bengal was taken over by the Kaibarta chiefs Divya, Rudak, and Bhim.[81] The poet and court biographer Sandhyakara Nandi's *Ramacharita*—a prolix literary composition in honor of King Ramapala—deprecates the Kaibartas as *dasyus* (robbers) and lowly pretenders who fomented an ill-fated uprising, ultimately put down by the prince Ramapala.[82] The uprising of the Kaibartas gives us a brief view of the hunting, fishing, and agrarian occupational groups of the greater Bengal delta, who do not fit the description of typical medieval peasants.

A geographical distinction must be made here between the ancient settlements along the Bhagirathi-Hugli portion of the Ganges—Anga, Varendra, and Rarha in the western parts of Bengal, and Vanga and Samatata farther east in the delta of the River Padma, which proceeds from the union of the westerly branch of the Ganges with the river Brahmaputra flowing down Assam. Ramparts of the city of Gaur, the ruins of which now stand near the India-Bangladesh border in the Malda district of West Bengal, is also the site of Lakhnauti, the early Turkish capital of Bengal, located in the valleys of the Mahananda River and the Ganges. This Pala Dynasty capital became the seat of the Senas, a Hindu dynasty that was ousted by the Delhi Sultanate in the early thirteenth century. During this time the main channel of the Ganges flowed directly through this region; it has now shifted more than thirty-five miles to the west.

The Turkish incursion did not necessarily lead to the expansion of Islam. In fact, much of the recent scholarship on the rise of Islam in Bengal shows that it tended to spread far away from the seats of military and revenue administration. No large-scale conversion was attempted by orthodox, Sharia-minded clerics, but lone Sufi adventurers who ventured into remote parts of Bengal began to spread the word. The busiest port of Bengal at the time was Saptagram (Satgaon). Arab and European travelers confirm that it had taken the place of Tamralipti (or Tamluk), which since the eighth century C.E. had been the most important center of trade in the delta.[83] Saptagram stood on the Saraswati River flowing southwest, connected to the Rivers Damodar and Rupnarayan branching off from the Ganges headed to the Bay of Bengal. These waterways sustained an extensive network of trade, raising Saptagram to commercial prominence and an important seat of administration for the sultans of Bengal.[84] The renowned Moroccan traveler Ibn Batuta in the fourteenth century saw a burgeoning Saptagram surrounded by seven major villages and large markets.[85] Bipradasa's *Manasabijaya*, written in praise of the goddess Manasa of the serpents, further confirms the flourishing state of contemporary Saptagram.[86]

This is where Zafar Khan Ghazi and his band of faithful warriors seem to have marched against the infidel kings of Bengal. They conquered the port of Saptagram and destroyed a large, ancient temple in nearby Tribeni, and used the spoils to build an imposing mosque. He was conferred the title of Ghazi and established one of the first schools for Qur'anic and Arabic learning (*madrasa*) and charity (*dār-ul-khairāt*). Legend has it that he converted a local raja, Man Nripati, and fought Raja Bhudeb near Hugli, where Zafar Khan was decapitated with his torso buried in Tribeni. There is little agreement among historians about the exact identity of this Ghazi. Some have claimed that he was truly the first spearhead of Islam in lower deltaic Bengal and the founder of the oldest Sufi order in the region. The story of Zafar Khan's martyrdom closely resembles the legend of other pioneer Sufi saint, Hazrat Shah Safi (a nephew of Sultan Firuz Shah Khalji), who had battled the raja of Hugli.[87] It suggests a possible convergence of various legends of militancy and martyrdom, the actual historical details of which have been largely lost over time. Zafar Khan's mosque and tomb still stand overlooking the Hindu pilgrimage of Tribeni, atop the plinth of what was most likely a Vishnu temple, constructed from not only the ruins of Hindu, Jain, and Buddhist monuments but also fragments of older mosques.[88] Zafar Khan's personal tomb is also a part of this complex and one of the main Sufi sites in Bengal. Not only was he a man of exemplary piety known for his charity to poor Hindus and Muslims, he is credited with some of the most evocative Sanskrit hymns to the river.

> *Jāphara khāna gājī roila tribenī sthāne*
> *Gaṅgā jāre dekhā dila ḍāka śuni kāne*[89]
>
> Jafar Khan was the Ghazi in the place called Tribeni
> to whom Ganges revealed herself after having heard his call

The tomb of Zafar Khan Ghazi, Tribeni, West Bengal, early thirteenth century C.E. Photograph by author.

Zafar Khan might have been the first to declare that only the water of the Ganges was pure enough for the ritual ablution of Muslims (Arabic *waḍū*, Bengali *oju*), before prayer or before entering a mosque.

It is difficult to assess how the conflicting images of the iconoclast and the Ganga-worshipper came together in history and folklore. Bengal during this period was a relatively unknown geographical frontier that attracted the early, intrepid emissaries of Islam such as warring Ghazis and the faithful *mutaṭawwi'* (from Arabic *ṭau'*, suggesting voluntary obedience) who dedicated their lives to the cause of their faith. Such inspirational militancy or exemplary piety to dispel the ignorance of infidel rulers and their subjects, or the punishment of unjust tyrants, is typical of narratives chronicling the exploits of early Sufis who brought popular forms of Islam to the interiors of Bengal.[90]

Much more can be said about the tradition of verse poetry venerating Ghazis in Bengali Pir literature, remembered alike by Hindu and Muslim followers, whose graves, tombs, and hospices are scattered across greater deltaic Bengal—their deeds and miracles a common currency in the folk traditions. Thus, for example, Pir Badaruddin, or Badar Ghazi, has long been the spiritual guardian of boatmen and fishermen, and Zinda Ghazi, or Bara Khan Ghazi, a safeguard against tigers. The feats of Ghazis battling denizens of the forests and the deep closely resemble deeds performed by local Hindu deities such as Chandi or Manasa, replicating the standard forms of narrative devotion evident in medieval Bengali *maṅgalakābya* verse poetry—the *Gaji-Kalu-Campabati* story, the *Raimangala*, or the *Banbibi-Jahurnama*.

In the tale of Champabati, Gaji (Bengali for *ghāzī*), son of Sikander Shah, ruler of Vairat, and his foster brother Kalu leave home because of his father's tyranny. During their travels they tame crocodiles and tigers, turn King Sriram to Islam, and erect a golden mosque in his capital. They defeat King Mukutrai and his commander, Daksinarai, converting their subjects. There are frequent references to the river Ganges as a goddess. Gaji's mother, Ajupa, is shown as Ganga's sister, and time and again she turns up when he and his brother are in danger. When a poor woodsman gives Gaji and Kalu shelter in the forest but there is no food, Ganga dispatches one of her serpents (*nāga*) with lavish gifts and provisions.[91] Gaji and Kalu, toward the end of the story, arrive at the ocean and alarm three hundred powerful yogis who have gathered to meditate.[92] Gaji asks them about their penance by the sea, and the yogis explain that they are worshippers of the River Ganges. Hearing this, Gaji summons the mighty river, and she appears in person seated on a lotus leaf. Seeing this, the yogis cannot believe their eyes:

Yogīgaṇa bale dhanya dhanya shāha gājī
Jāhara adhīna gaṅgā morā jāke pūji
Yabanera tulya āra nāhi āche jāti
Jāhāke karena mānya gaṅgā bhāgīrathī[93]

The yogis said praise be to Shah Ghazi
Who commands the Ganges that we worship
There are no other people like the Muslims
Even Ganges Bhagirathi obeys them

Overwhelmed by the appearance of the goddess, they accept Islam as their new faith. Gaji and Kalu build a new mosque on the spot with the blessings of the River Ganges. The Gaji-Kalu story, full of miracles, gives us an intriguing window into the ways in which Ghazi cults spread. It echoes the familiar narrative plot of military conquest, performance of miracles, building of mosques, and the spread of Islam. The figure of the Ghazi is an iconic element, and snakes, crocodiles, tigers, and the river are all part of his appeal.

Stories of this genre can be roughly traced to a period of about three hundred years beginning with the rise of the sultanate of Bengal during the fourteenth century C.E., and they dovetail with the history of the conversion of forest tracts to wet-rice cultivation by peasant colonists who embraced a distinctive Islamic and Sufi identity, but whose practices, true of archaic forest and river cults, remained outside the pale of the Sharia and known orthodoxy. By the mid-fifteenth century, many parts of lower Bengal had been reclaimed by such settlers, edging their way into southern Jessore and northern Khulna in present-day Bangladesh.

The gradual southeasterly drift of alluvium and the deepening of the second eastward channel of the Ganges are natural elements of the dynamic geomorphology of tracts between the Bhagirathi-Hugli and the Padma-Brahmaputra systems.[94] The rise of agrarian settlements and the emergence of grassroots Islam, especially in remote, outlying tracts, has been the subject of much discussion and debate following the lead of Richard Eaton, who first suggested that changing river drainage and exposure of new alluvial deposits abetted forest-felling, draining of marshes, and the extension of rice cultivation suited to the abundance of water.[95] In this instance, the popularity of the fear-dispelling, forest-taming Ghazi might well have been the prelude to what Eaton sees as the opening up of a complex frontier of Islam and agriculture.[96] Eaton suggests that not only did the rise of popular Islam dovetail with the long-term ecological transformation the delta, but conversion of large sections of people outside the fold of Hindu caste society after the sixteenth century C.E. was tied directly to the expansion of rice cultivation—which also led to surplus production and population growth.[97] No matter how typical or pervasive this process was, the rivers and waterways of the lower Gangetic Bengal provide a contrasting picture to the drier uplands, where the incidence of poor rainfall, drought, and peasant rebellions were more frequent.

Sedentary Agriculture

Timur put the prosperity of Delhi to the sword and left a countryside unsettled and impoverished.[98] At the same time, the delay in the restoration of political order encouraged the rise of smaller, independent states. It is said that the writ of Delhi extended only as far as the village of Palam, the site of the present-day airport (*az-dehlī-tā-pālam*). During the tenure of the Sayyids, regular taxation seems to have been largely replaced by periodic raids and forced tribute from places like Jaunpur, Kol, and Tirhut (Bihar) and from the Bengal sultans who ruled from Lakhnauti and Sonargaon.

During the reign of Khizr Khan, founder of the Sayyid Dynasty, the western valley of the Ganges saw a series of outbreaks.[99] The vizier Tajul Mulk led a force against rebel Rai Viram Singh, who had set up autonomous rule along the cities of Katihar, Bulandshahr, and Moradabad, as did Mahabat Khan of Badaon.[100] For a while, Delhi lived on forcible collections from the Doab and Mewat, subjecting the local peasantry to arbitrary and stringent demands.[101] The Sayyids were challenged by new powers emerging in Malwa and Jaunpur, which was the center of Sharqi power, and also Sirhind under the command of Afghan migrants united behind Bahlul Lodi, who repeatedly besieged Delhi. Bahlul would eventually take over the reins of power in northern India (1451 C.E.), terminating Sayyid rule, and finally ending the bid of the Sharqis (1479 C.E.) in an exhaustive, decades-long struggle that did not leave much room for regular affairs of the state. Much of this fight took place over the forts, towns, and villages on either side of the Ganges.[102]

Under Bahlul's successor Sikandar Lodi, the Afghans were finally able to quell the uprisings that had shaken Delhi and its environs, retake Bihar, and bring to heel petty kings of the southern bank of the river such as the raja of Bhata. Bengal was no longer a Sharqi stronghold but under the firm grip of the Ilyas Shahi Dynasty that would hold on to power through most of the fifteenth century C.E. It is difficult to gauge the extent to which Bahlul's successor Sikandar Lodi was able to establish a stable revenue base, given the strident demands of the Afghan nobility for a greater share of power and spoils. In his period the Bahloli coin rose as prime currency, and one Bahloli could apparently buy both ghee and textiles. A couple of verses from the historian and poet Rizqulla Mushtaqi give us a rare and affirmative view of Sikandar's reign, where highways were made passable again and no upstart dared dream of trouble (*khayāl-i fitnah*). It is clear from contemporary accounts that the Afghan officers of the Lodi court were paid less in salary and more in revenue assignments, with the details of collection from village headmen left largely to their discretion.[103] Mushtaqi also says that unruly Hindu landlords and chieftains were humbled, and deserving Muslims were raised to a position of dominance (*muṣalmān cherah dast, hinduvān rām*).[104] Some chroniclers also claim that Sikandar tried to discourage Hindu pilgrimage in Mathura, Allahabad, and Varanasi, but there is little basis for the view that he had the means to pursue such a task.[105] During this period of Afghan rule, the great seer and thinker Kabir, born into a Muslim weaving family, grew up in Banaras, was initiated by his preceptor Ramananda, and attained his first meditative bliss (*samādhi*) on the banks of the northward-flowing Ganges. Ananta Das in his memorable *Kabir Parchai* recounts how Sultan Sikandar and his die-hard jurists tried to kill Kabir the apostate with fire and war elephants, finally flinging him into the Ganges in chains, but failed to harm him.[106] According to legend, Sikander quickly realized his folly and duly humbled himself in the presence of Kabir, who did not hesitate to bless him. There is scant evidence that an actual encounter took place between the Sikandar and the legendary mystic when the Afghan ruler visited Jaunpur in 1495 C.E.[107] The capitulation of the Afghan ruler, apocryphal or not, is still an important historical landmark for the followers of Kabir. Not only did Kabir shun both Islamic and Hindu platitudes, he stood up against all egregious tyranny, rejoicing in the simple devotion of the downtrodden—the weavers and tillers of rural northern India.

The Chagatai prince and conqueror Babur describes in his memoirs how he took over the revenue-rich principalities in northern India from the Afghans—Mewat, Bayana, Agra, Kalpi, Kanauj, Lakhnaur, Buxar, Awadh, Bahraich, Jaunpur, Saran, and most important, in the region of the Doab.[108] Babur did not care much for the climate or geography of Hindustan, which, unlike the mountains and valleys of his native Fergana or Samarkand, was mostly flat. He was puzzled by the fact that people never fashioned aqueducts or viaducts for running water, relying instead on large and untamable rivers or reservoirs for rainwater. During torrential monsoons the rivers flooded, forming gullies that were difficult to pass. He was struck, however, by the natural abundance and lush vegetation of northern India. Crops were effortlessly watered, and saplings had to be tended only at the beginning of the cycle by water wheel or bucket.

During the Tughulq period, as Moreland points out, a key to the prosperity of Delhi lay in the security and produce of the "river country," but undue exactions threatened chronic peasant rebellion and flight, which had to be put out like forest fires.[109] To study the problem closely, Sultan Muhammad ibn-Tughluq had resolved to camp for a few years by the side of the Ganges near Kanauj. He came up with a plan to sow expensive wheat and sugarcane along large stretches instead of the humble barley and millet, handing out money loans to peasants. Corruption, negligence, and malfeasance doomed the project, but Moreland suggests that the real problem lay in the dispersal of the peasant households and a lack of resources in Delhi to rein in assignees and farmers of revenue.[110] The key to Firuz Shah's success in reviving the revenue base was not just his canal projects but a strenuous assessment of the produce in each region, and the strategic redistribution of specific villages to members of the military based on new valuation—who did not have much incentive to torment peasants. Babur's distribution of landed assets to his henchmen after the Mughal conquest did not disturb such old unwritten arrangements between the provincial chiefs, village headmen, and peasants, and his successor Humayun seems to have largely ratified these old assignments.[111]

The real breakthrough in the assessment and exaction of land revenue was made by an enterprising young Afghan chief called Farid from the principality of Sasaram in Bihar—soon to be known as Emperor Sher Shah Suri. The historian Abbas Khan Sarwani tells us that having inherited two districts (*parganā*s) from his father, Hasan Khan, Farid experimented with the resumption of the revenues assigned to his soldiers (*jāgir*s).[112] He summoned the assignees, village headmen, cultivators, and village recordkeepers and let the peasants choose the mode of payment. Some of the headmen asked for written agreements of rent in cash; others preferred payment in kind (*qismat-i ghallā*). Realizing the wide range of produce per square area according to soil and other natural factors, he insisted on setting aside some of the money for the measurement of fields (*jarībānā*) and fees for the collectors (*muhāṣilānā*). He also declared that it was the solemn duty of the ruler to protect helpless peasants from tyranny and exploitation.[113]

After Sher Shah ousted the Mughal emperor Humayun and became the emperor of Hindustan, these early experiments became the basis of a new dispensation for the taxation of produce across the country. In addition to the hereditary office of the recordkeeper, or Qanungo, Sher Shah added the services of the *shiqdār* (collector) and *āmīn* (recorder), who were given an independent staff of experts.[114] The *Ain-i Akbari*, Abul Fazal's administrative compendium of Mughal emperor Akbar's reign, reproduced the lists of revenues affixed to different provinces during the time of Sher Shah and described the basic rule of thumb for calculating estimated taxes by taking the market price of good, middling, and bad staples, averaging them, and assessing one third of it as taxation.[115] Akbar's land revenue arrangements closely followed the precedent set by the Afghan ruler.[116]

These reforms in revenue collection had far-reaching effects on the relationship between chieftains, landlords, headmen, and peasants across the valley. The man responsible for tabulating the variety and quality of agricultural produce and enacting a long-lasting assessment of revenue in the valleys of the Indus and the Ganges was Raja Todar Mal—who came from a family of *khatri* scribes from Lahore. He had distinguished himself in the Gujarat campaign, put down rebellions in Bengal and Bihar, and led a risky expedition against the Yusufzai chiefs of Afghanistan. Todar Mal's regulation of land tenure in many ways underwrote the fiscal foundations of Akbar's great empire. Todar Mal had served with Sher Shah and brought his knowledge of revenue assignments (*jāgir*), peasant holdings, and deeds (*paṭṭā, qabūliyat*) to Akbar's decennial settlement (*dahsāla*). All arable land, from Punjab to Bihar, was classified according to four broad categories: land cultivated throughout the year (*polāj*), land periodically left fallow to regain its fertility (*parautī*), land previously unsown for three to four years (*chāchar*), and land lying untilled for an indefinite period of time (*banjar*). Rope was discarded in favor of a chain made of bamboo tied with iron rings, which was less susceptible to moisture, to measure standing crops for autumn (*kharīf*) and spring (*rābī*) yields.[117] A simple arithmetic of the crop division and redistribution of resources (*ghallābakshī*) was now replaced by a regular system of assessment known as the *ẓabṭ*, a word that has many connotations—possession, control, administration, and governance.[118] It is clear from the *Ain* that not every village or district could be brought under this writ. For example, just under 70 percent of the divisions (*mahal*s) in Bihar and 74 percent in the region of Allahabad were brought under *ẓabṭ*, while all of the divisions in Multan were bound by this regulation. Revenue was no longer dependent on area tilled or yield per unit, but fixed with a schedule of rates (*dastūr-ul-' amal*) and rates of crops and their prices (*raqamī*). It was difficult to keep such detailed inventories current; however, as the empire expanded, average estimates were left to the discretion of local revenue officers, *qānūngo*s.[119]

This was also a valiant attempt to formulate a system of checks and balances against undue extortion. The signature of the humble village accountant (*paṭwārī*) now carried some weight against the claim of stalwart landed intermediaries. It also led to a proliferation of paper records and an army of registrars and accountants, aimed at binding the peasant to his plot, plow, and oxen, bringing obstreperous and remote intermediaries in line (*zamīndārān-i zur ṭalab*), and keeping a close eye on the nobles who were revenue assignees (*jāgirdār*s). Over the years, the original revenue rolls of Akbar's time became largely customary, adjusted to local conditions and political realities. In many parts of the Mughal Empire, the system of *ẓabṭ* could not be enforced and the revenue was assessed by a fixed rate (*nasaq*) without a detailed survey of the price and yield of crops, such as in all of the districts of Mughal Bengal, where a fixed amount seems to have been levied per plow, irrespective of area of cultivation or amount of produce.[120] There is evidence for this in Bengali ballads, such as the songs of Gopichandra, where one finds land measured by how much one plow can till (*ekhāl*), as well as limits placed on interest from loans to given to peasants.[121]

The revenue records of the Mughal Empire are a testament to the uneven nature of its political dominance. Intermediaries taking advantage of lapsed rulings were as baneful to the imperial order as remote and inhospitable fastnesses that sheltered seditious rebels. The ultimate object, however, was access to peasant surplus. In this case the Mughals seemed to follow the dictum of Sher Shah that one must be lenient during assessment but merciless after the harvest. During the best of times the revenue demand was moderate in Mughal India, with periodic revisions of expected yield and added incentives for improvement, as fallow or indifferent land was transformed into valuable farmland through the sweat of enterprising peasant households. Despite such assurances, peasants, especially the landless and subcaste, remained destitute and vulnerable. This is evident not only in the periodic peasant uprisings and desertion of villages but also in the strenuous solicitude shown in Mughal documents urging headmen and collectors to keep track of want and indebtedness and acts of clemency during crop failure and famine. Abdul Qadir Badayuni, writing toward the end of Akbar's reign, noted that it was difficult to weed out rapacious collectors (*karoṛī*s), who not only ruined the countryside but put the poor peasants in such desperate straits that they were forced to sell their women and children.[122] Todar Mal brought many such offenders to justice, subjecting them to imprisonment, beatings, and torture. Undue suffering of the peasantry remained a grave concern for the Mughal government. Aurangzeb's well-known order (*farmān*) of 1665–1666 to Rasik Das Krori makes an explicit injunction not to release plots from which peasants have fled (*uftādah*) for at least one year, just in case they decided to come back.[123]

The abundance of peasant labor ultimately sustained the affluence of the Mughal elite. Badauni remarks with characteristic chagrin that in every prefecture, in dry or wet land, in plain or hilly tract, in city or village—by desert, jungle, lake, river, or well—every bit of land was measured by the order of Emperor Akbar. Hilly, marshy, forested, and stony terrain had for centuries sheltered the indigenous kings, chiefs, and warrior elites of the Ganges valley. Many of these forts and strongholds were broken up by Mughal cannon and matchlock. Expansion of the arable sustained the *jāgir*s, commanded by a handful of military elite, at the apex of wealth and power, leaving the bulk of the subject population in a state of meager subsistence.[124] The tireless extension of settled cultivation in the Ganges valley over the last four hundred years has had a cumulative effect on forests, wetlands, and grasslands, through the extension of what John F. Richards has termed "settlement frontiers."[125] Such overhaul of the natural terrain, noticeable in the Mughal period, would be further accelerated during the period of British colonial rule.

Imperial Traffic

The fort of Chunar, twenty-odd miles from Banaras, stands atop a precipitous ridge overlooking the Ganges where the river takes a sharp bend by the Kaimur Hills—an eastward extension of the Vindhya massif honed from gneiss, hornblende schist, and granite. It affords a commanding view of the river, with turrets from where a fusillade can be directed below to discourage approaching boats. Chunar, with its citadel, is a prime example of strategic holdouts that proved vital to the success of contending regimes during the long period of transition from Afghan to Mughal rule. In order to take a fort such as Chunar, an invading army required clear logistical advantages. It needed to secure the countryside for the sequestration of local boats and boatmen, ford the river at strategic places by laying makeshift pontoon bridges, and ensure an abundant supply of grains and provisions. This is amply illustrated by the tussle over the lower Ganges valley between the Mughals and Afghans during the early fifteenth century C.E.

Emperor Babur had set out in 1529 C.E. on an expedition against the defeated Afghan ruler Ibrahim Lodi's brother Mahmud Lodi, aiming to smoke out rebel Afghans from their strongholds in Bihar and Bengal. After taking Banaras, Babur laid a siege to the fort of Chunar, putting the Afghans to flight across the left bank of the Ganges.[126] As he was settling the *jāgir*s of Bihar, Babur met the young Sher Khan. Impressed by the sprightly, ambitious young Afghan, he supposedly asked his generals to keep an eye on him.[127] Toward the end of his reign, however, Babur left Bengal and Bihar untended, and Sher Khan was able to consolidate his hold over the strategic points along the lower Ganges plain, including the fort of Chunar.[128] A deadly game of political chess had begun in Bihar and Bengal. Babur's successor, Humayun (1532 C.E.), wanted to stamp out the Afghans, which was evident in his cruel punishments meted out to captured Afghan garrisons after Chunar was taken.[129] However, he was beleaguered on many fronts, especially in Malwa and Gujarat, where ruler Bahadur Shah was threaten-

The Ganges from the Chunar Fort, Uttar Pradesh. Photograph by author.

ing to carve out an independent principality. Humayun was forced to make a deal with Sher Khan. It was a mistake that would cost him dearly, because Sher Khan found the temporary withdrawal of the Mughal forces an invaluable opportunity. Guileful and decisive, Sher Khan quickly captured the forts of Rohtas, Ghari, and Munger.[130] Then he carefully laid his battle plans, taking into account the monsoon that made key roads from Agra to Kanauj virtually impassable for the Mughal army, who remained encamped for months in Bengal, while the Afghan advance guards outflanked them, laying siege to Chunar and Jaunpur, and sweeping up districts as far west as Kanauj and Sambhal.[131]

Large swaths of the lower Gangetic plain were prone to flooding, as Humayun found out during his hasty march to Chausa to confront the Afghan menace. Sher Khan at the head of the Afghan forces harassed the discomfited Mughal army as it tried to lay pontoon bridges and unload boats carrying large field artillery, hiding behind temporary mud redoubts.[132] Between Chausa and Buxar, the two armies were encamped on the same bank of the Ganges. Sher Khan's army evaded the Mughal pickets and disrupted the Mughals crossing the river, which was in spate.[133] A sudden attack on the Mughal camp caught Humayun completely unawares.[134] Trying to flee across the Ganges on horseback, he was thrown and would have drowned had he not been rescued by a water carrier, who led the Mughal emperor to safety by letting him cling to his animal-skin ewer. The retinues of highly ranked nobles fell to the Afghans, and almost eight thousand soldiers and camp followers were killed or drowned as the Afghans pursued them in boats. Sher Khan crowned himself as ruler and secured both flanks of the river. After the battle of Chausa, Sher Shah ruled territories in the west along the bend of the Ganges from Kanauj to Chunar, in the south along the hills of Jharkhand, Rohtas to Birbhum, and in the east as far as Assam and the Chittagong Hills. This was a further advantage to the Afghans when they faced the Mughal-Chagatai forces across the Ganges near Kanauj for a final showdown.[135] This last skirmish, which is also known as the battle of Ganges, after Mirza Haidar's account in the *Tarikh-i Rashidi*, was disastrous for the Mughals. Of Humayun's thousand-plus retainers, only eight came out of the river alive.[136] The Mughal camp had been set up on low ground, and it flooded, creating a great confusion in which the center of the army collapsed.[137]

Sher Shah realized the urgent need to pave and reconnect parts of the ancient highway across the Ganges valley for the movement of troops and market-bound produce. After the settlement of the main *jāgir*s, especially in Bengal, he began to lay the foundations of what the British would later rechristen as the Grand Trunk Road, which would ultimately connect Sonargaon near Dhaka in eastern Bengal to lands beyond the River Indus, across the Khyber Pass and Peshawar, into Afghanistan. During Sher Shah's time, about 1,500 *kos* of this road was finished, which, taking into account variations of the Indian measure of distance, would be close to 3,000 miles. Circumventing parts of the monsoon-prone old route skirting the Ganges, and threading through the hilly and forested tracts of Chota Nagpur, the new highway connected the garrison towns of Bayana, Kalpi, Rohtas, Bharkund, Ghari, Chunar, and Rajmahal, punctuated regularly by the typical *kos minār*—an orotund squat pillar, about thirty feet high, atop a masonry base of lime-plastered bricks to mark distances between cities. Sher

Shah took care to build inns for both Hindu and Muslim travelers along the way, plant shady trees, and dig water wells.[138] The strategic advantage of such a highway cannot be gainsaid; it made it very difficult for rebels and enemies to regroup at forts such as Rohtas on the River Son or Chunar on the Ganges.[139]

Military campaigns during the monsoons proved hazardous in the rain-soaked valley, as both Babur and Humayun found out, trying to mount caissons and mortars on boats, make new roads for the army, and travel by litter along the course of the Ganges through Kara, Fatehpur-Hawsa, and Sarai-Munda.[140] For example, it took two days for the early Mughal army to lay a bridge and cross the Tons River, a tributary of the Ganges, which Babur decried as a "muddy morass."[141] These were dangerous roads during monsoon squalls. Once, when Babur was riding on the edge of an embankment on the Ganges, a mud ridge broke and he almost fell into the water with his horse. He also wrote about how during his march against Bengal he had trouble commissioning local boatmen to carry horses, soldiers, arms, and equipage, with hostile forces in sight camped on the opposite bank.[142] The construction of the great highways across the valley provided a clear and decisive logistical advantage to northern Indian regimes that came after Sher Shah and the Afghans.

Imposing Prospects

In the winter of 1555 Humayun fell to his death coming down the steps of his observatory in Delhi, leaving his young heir, Akbar, in the custody of the trusted Iranian commander Bayram Khan. By this time, the last of Sher Shah Suri's descendants, Sikandar Sur, and his large and undisciplined Afghan army had been routed at Sirhind. Sikandar fled to the Himalayan foothills and then to Bengal.[143] Thus it fell upon Hemu, the Hindu vizier of the deposed Afghan ruler, to rally the Afghan forces and force a showdown with the Mughals on the battlefield of Panipat. Hemu was injured in battle, captured, and decapitated, and his head was sent as a trophy to the Mughal headquarters in Kabul. Akbar, the teenager, was installed on the throne of Delhi, and the plains of northern India came under the writ of the Mughal Empire for many centuries to come. The maverick Sher Shah's busy interregnum, which lasted just for a few years before his fatal accident at the fort of Kalinjar, had come to an end. Akbar not only reclaimed the great forts of the valley from the Afghans but also began construction of a new fort at the great Hindu pilgrimage of Prayag.

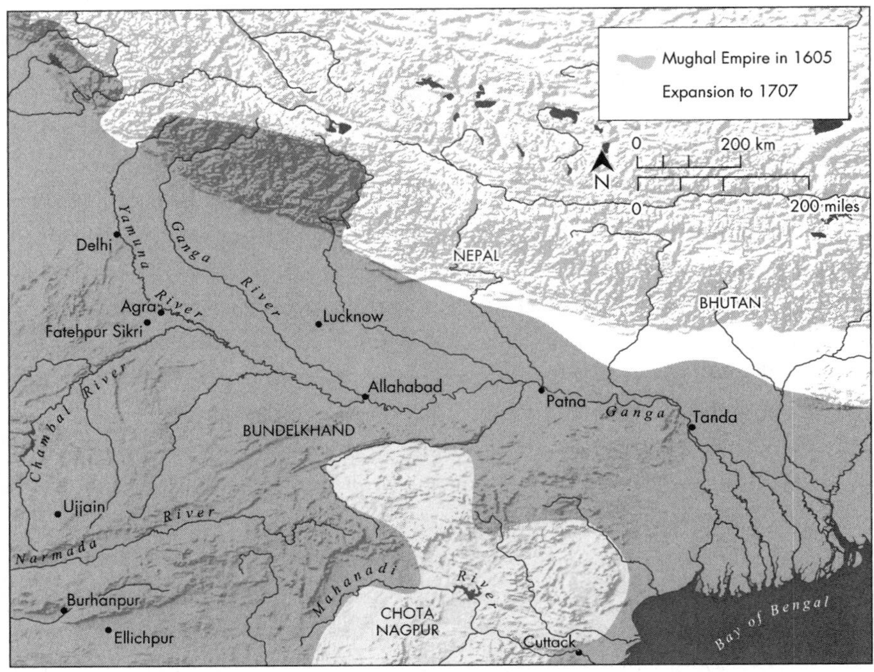

The Mughal Empire in northern India, ca. seventeenth–eighteenth centuries C.E. Map by M. Roy Cartography.

Akbar considered the confluence of Prayag one of the most bracing and peaceful prospects in the Ganges valley. He had visited the Hindu pilgrimage during his military campaigns against rebellious Uzbeks between 1752 and 1757 C.E. and requisitioned the old sultanate-era precincts of Kara-Manikpur, named after the cities of Kara and Manikpur on either side of the Ganges. Here Akbar founded a new district, which he named Illahabas—the Hindustani suffix *bāsa* ("home," "abode") being added to the Arabic word for "divine."[144] The name of the city was later changed to Allahabad during the reign of Shah Jahan. Akbar not only commissioned the sandstone fort but also raised the embankments on two sides of the river to prevent flooding. He also sought to intervene in certain Hindu rituals of sacrifice that he considered inhumane. Historian Badauni, a contemporary witness, remarks that during this time penitent Hindus came to Prayag in large numbers for penance and ritual suicide—some by decapitating themselves with iron saws and others by leaping right into the torrents of the Ganges.[145] These

William Hodges, *A View of the Fort of Allahabad* (1787). © The British Library Board 11/14/17. Print. Shelfmark: X307(20).

remarks confirm observations made by the Chinese pilgrim Xuanzang, who visited Prayag (ca. 644–645 C.E.), about the mortification of pilgrims at one of the main shrines of Prayag, especially the custom of hurtling themselves from the top branches of a giant banyan tree.[146] Some versions of the *Matsya Purana* on the pilgrimage of Prayag also mention this imperishable tree known as the *akṣaya vaṭa*.[147] The tree seems to have overlooked a deep well known as the *kāmya-kūpa* ("well of desire"), into which many devotees jumped and ended their lives. Akbar made it a point to build his fort directly around this notorious landmark, putting an end to the morbid practices of collective suicide. At the same time he saved the venerable old tree held sacred by his Hindu subjects. He closed off the jumping well and placed the banyan tree in an underground chamber, which subsequently became known as the temple of Patalpuri, or the "underground temple." A miracle was now attached to the tree itself. Despite its being cut down and shorn, it refused to die, continuing to sprout new leaves and branches. A flight of steps within the fort opened into a square, pillared courtyard, which now welcomed multitudes of devotees who came to visit the fort as part of their pilgrimage of Prayag.[148]

Akbar's display of authority at the confluence of Prayag, the site of the greatest gathering of Hindu orders in the empire at the festival of Kumbh, is hardly surprising. What is remarkable, however, is that the Brahmins of Prayag claimed that his birth and ascension to the throne of Delhi had been foretold. An early version of the story can be found in the book of geography and curiosities *Hadiqa-al-Aqalim* [Garden of Climes] penned by Ilah Yar Usmani, who also went by the name of Murtaza Husain Bilgrami. Ilah Yar writes that devout Hindus of Prayag confided to him that Akbar in his previous birth was an ascetic known as Mukund Brahmachari, who lived on the banks of the Jumna at Prayag with his three disciples.[149] Once by accident he drank some milk that had not been strained properly and ingested the hair of a cow, which contaminated his body. Forty long and difficult years of fasting and penance having come to naught, in a state of abject resignation, he arranged to take his own life, writing down the year, month, and day of his death and cremation in the form of a Sanskrit verse on a copper tablet, which he buried near the foundation of his residence. He then proceeded to have his body cut up in pieces and his remains consecrated in a sacrificial fire (*homa*). His chief disciple, Biran, also killed himself by suffocation to be with his teacher in the next life. Mukund was reincarnated as the son of Emperor Humayun and Hamida Begam. Biran was born under the name of Birbal, in the house of a Brahmin, and later become one of Akbar's most favorite courtiers.

It is said that Akbar was born with the prior knowledge of Mukunda's prognostication and could recite lines from the original Sanskrit verse, the last of which was *"sakala vratahārī, Brahmacārī Mukunda"*—the ascetic who has renounced all his vows.[150] When his courtier Raja Birbal heard the emperor utter these words for the first time, he immediately responded with the three preceding lines of the original *śloka*, which was in the form of a riddle. It indicated that in the year 1598 of the Vikram Era, in the month of Magha, when the sun enters the house of Capricorn (*makara saṅkrānti*), during the twelfth day of the waxing moon, Mukund Brahmachari had his body immolated, to be reborn as the ruler of Hindustan.[151] This story is also recorded in the compilation of *Darbar-i Akbari* by Muhammad Husain Azad. It seems that before taking his life Mukund had indeed inscribed a few verses on a copper plate, attested by the leading Brahmins of Prayag, stating that he would be reborn in the house of a Muslim, and that the time of Mukund's death would match exactly with the conception of Akbar in his mother's womb.[152]

Akbar was deeply interested in theories of rebirth and came to believe that the doctrine of transmigration was common to all religious creeds.[153] Badauni states with characteristic disapproval that the emperor was often engaged in discussion on this matter with men of diverse religious faiths, including Christians and Zoroastrians. Akbar may have come to believe that he had been a Hindu yogi in his past life. The noted historian and translator of Persian texts Henry Beveridge, who conducted inquiries in Allahabad during the 1890s, found that the legend of Akbar and Mukund was quite well known and even recorded in some later versions of the *Prayag-Mahatmya*. He also discovered a stone image of Mukund Brahmachari in the possession of the leading Brahmins in Allahabad.[154]

Remarkable as Akbar's ecumenical views of popular piety and deep empathy toward Hindu devotional practice may seem, it is the imprimatur of Mughal imperial authority on one of the most important centers of trade and pilgrimage on the Ganges that stands out. Allahabad rose under Akbar as a major city and military outpost along the Grand Trunk Road. The sight of Akbar's great sandstone fort at the confluence of the Ganges and Jumna Rivers was a reminder of Mughal command over the entire valley. Only during the latter years of Emperor Aurangzeb's rule was this established and largely peaceful order rudely disrupted.

As the young governor of Gujarat, Aurangzeb earned a reputation for iconoclasm and the persecution of orders that he saw as an explicit affront to the basic tenets of Sunni orthodoxy.[155] In the first year of his reign (1659 C.E.) he issued the Banaras Order, that no new temples should be built at Hindu pilgrimages such as Banaras, and that Islamic law prevented him from sanctioning any such new construction. The order also noted that temples of long standing should not be demolished, and Brahmins who had the right and obligation to maintain their ancient temples should not be harassed or forcibly removed from their places of worship.[156] He also enjoined his officers not to disturb the Hindu community of Banaras. It is difficult to sort out the exact motivations behind Aurangzeb's subsequent destruction of temples at Banaras, which directly contravened this order. One of the most distinguished historians of Aurangzeb and his reign, Jadunath Sarkar, simply attributed these actions to his innate religious bigotry, while there is also evidence that some of his sanctions had a manifest political agenda.

Aurangzeb resented the wealth and patronage of notable Rajput chiefs and other Hindu potentates at Banaras. He also suspected that in 1666, the rebel Maratha leader Shivaji, after his escape from Mughal custody—fleeing through Mathura, Allahabad, Banaras, and Gaya—had money sent clandestinely to the *pāṇḍās* of Banaras temples. To Aurangzeb, the acceptance of such gifts was an act of insubordination.[157] He also believed that charismatic Brahmin teachers at Banaras were attracting Hindus and Muslims from afar and preaching nonconformist doctrines. In April 1669, he resolved to dismantle the most famous shrine of Banaras, the old Vishwanath temple,

James Prinsep, *Madhoray Ghat and the Minarets of Benares* (ca. 1831). From J. Prinsep, *Benares Illustrated in a Series of Drawings*, Calcutta: Baptist Mission Press, 1831.

adjacent to the fabled Well of Wisdom, the Gyan Vapi cistern, leaving some of the ruins, broken walls, and old columns so that his idolatrous subjects would appreciate the intended sting of his imperial order. He also razed the great Beni Madhav temple that stood overlooking the Pancha Ganga *ghāṭ*, raising in its place the towering Alamgir Mosque intended to dominate the view of the city for miles around. The mosque loomed over the bend of the Ganges, with a long flight of stone steps leading down to the water, its richly carved minarets rising 147 feet from the ground.[158] They were meant to be 50 feet higher in the original plan, but the structure could not support them. Aurangzeb had tried to change the name of Banaras to Muhammadabad, but the people of Banaras did not let that happen, and the new mosque became widely known as Madho Das Ka Daura or the Gates of Madho Das—a nod to the lingering memory of the temple whose site the mosque had usurped.

Akbar's fort in Allahabad and Aurangzeb's imperial mosque in Banaras are visible reminders of the Mughal attempt to cast their authority over the cities, markets, and pilgrimages along the valley of the Ganges. Insubordination and intrigue were Aurangzeb's grave preoccupation, especially toward the end of his long reign. While the extent and depth of his intolerance are the subject of much scholarly debate and controversy, there is no doubt that by the selective persecution of prominent religious groups—Jains, Sikhs, Shias, and Hindus—whom he repeatedly accused of treachery, he tried to drive wedges between different groups of political detractors. Aurangzeb's irate orders and acts of cruel whimsy, as we know, created further distrust and rancor among his subjects, hastening the eventual dissolution of his empire. The later Mughals barely held on to the immense bounty of the Ganges valley. Peasants who had toiled for generations to clear forests, expand arable land, harvest grain, and pay taxes that fed the increasingly idle Indian landed elite began to rise up in rebellion in the heartland of the Mughal Empire. Many threw in their lot with rebellious *zamīndārs* and chieftains, or joined the ranks of religious dissenters such as the Jats, Sikhs, and Satnamis. By the third decade of the eighteenth century, Akbar's vision of a truly authoritarian empire and his latitudinarian approach to its religious orders had all but come to pass.

CHAPTER 9

THE GANGES IN THE AGE OF EMPIRE

The greater Bengal delta is the drainage basin of two formidable rivers, the Brahmaputra and the Ganges, that unite on their journey to the Bay of Bengal. It is estimated to be more than twice the size of the delta of the Nile in Egypt. Closer to the coastline, the rivers break up into a vast network of brackish runs, their low-lying parts forming the Sundarbans, one of the largest mangrove forests in the world. In the late eighteenth century, the pioneering East India Company geographer James Rennell compared the size of this delta roughly to that of Wales. It is a 200-square-mile stretch of sparsely inhabited lowland, with geological evidence of at least eight major channels through which the Ganges and her distributaries have debouched their water into the bay for millennia.[1] Here the ebb-and-flow oceanic tides are interrupted by long, torrential monsoons, and water flowing through multiple drainage channels collides with inrushing tidal bores from the sea. During the months of downpour, parts of the delta are prone to repeated flooding. In summer and winter months the ocean advances farther inward along deepening fissures, breaking away entire sandbanks and seasonal islands, large and small. The Ganges empties a prodigious quantity

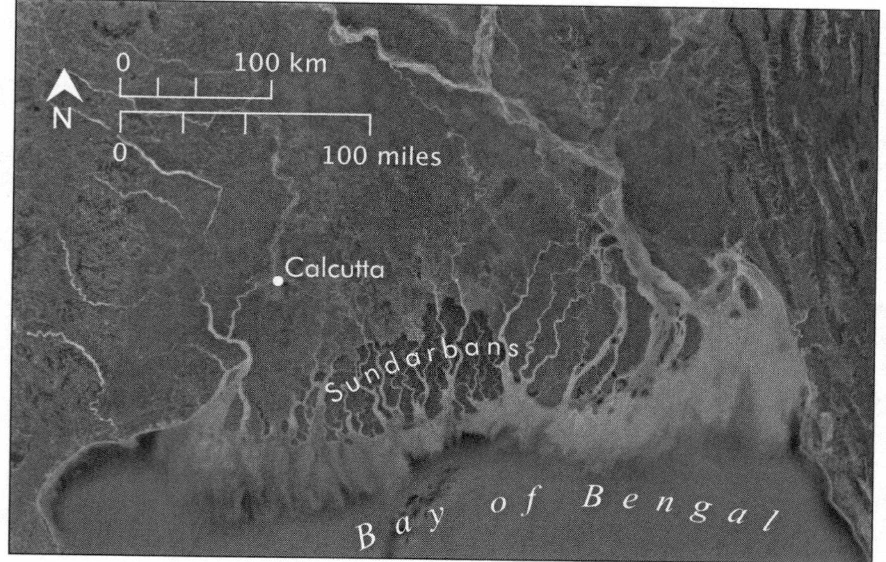

Aerial map of the Sunderbans and the Ganges delta. Adapted from Google satellite imagery by M. Roy Cartography.

of mud and sand, so much so that satellite images show a cloudier band of coastal waters almost sixty miles into the ocean. The ocean floor runs deep near the delta, nearly 60 to 100 fathoms at a distance of about 100 miles from the coastline.[2]

The thick, green belt of the Sundarbans is a perennial forest supported by shifting mudflats periodically inundated by restless ocean currents, drenching tidal mangroves that have adapted to high levels of moisture and salt. Here trees and shrubs with stilts, lenticels, and "breathing" roots (pneumatophores) draw their nutrients from mud and water. The estuaries are dominated by unique plant species (*gāen, kripā*) that can withstand repeated submergence and salinity. The narrow creeks are crowded by the more robust *garjan* (*Rhizophera* spp.), the *garān* (*Cerops* spp.), and the fragrant keora (*Sonneratia apetala*), while higher grounds are crested by genwa, short hental palms, and the ubiquitous sundari (*Heritiera fomes*) from which the Sunderbans gets its name. This ancient wilderness once sheltered the now-extinct dwarf rhino, water buffalo, Gangetic dolphin, swamp deer, and tiger, and an abundance of bird species, including the white-bellied sea eagle, grey-headed fishing eagle, heron, egret, sandpiper, curlew, and stork.

Great egret and mangroves, Sunderbans, West Bengal. Photograph by Debal Sen.

As early British geographers and hydrologists discovered, the relentless erosion of soil caused by the river and the ocean could create and break down islands in the channels well within a single human life span. James Rennell, during the late eighteenth century, saw newly emerged islands that were many miles across, originating from sandbanks formed at the meander of rivers or obstructions such as a large tree trunk or the hull of a sunken boat. Such islands were quickly covered in reeds, grass, and shrubs, creating dense thickets. Human habitation in the heart of the Sundarbans, compared to the rest of the lower delta, was sparse in what Rennell described as a vast "labyrinth of rivers and creeks, all of salt" overgrown by swathes of tiger-infested jungle, where attempts to clear the undergrowth to till the soil was a futile exercise.[3] Spread out across the delta, sluggish with vast quantities of mud and sand, very few channels of the Ganges were deep enough for oceangoing ships. The only major outlet lay along the mouth of the Hugli River, on the western branch of the Ganges, where the British port of Calcutta would be settled during the late seventeenth century. Large sections of this delta had remained for centuries at the margins of the kingdoms and empires of Bengal. A plethora of natural impediments seems to have discouraged the settlement of large populations and the cultivation of land, including inhospitable and saline terrain, infestation of wild beasts, ferocity of cyclonic storms, and the continuous raids of Arakanese pirates—joined in the sixteenth century by outlying bands of Portuguese fortune hunters.[4] These elements contributed to the rising importance of Hugli, which during the Mughal period began to outrival the old port of Saptagram or Satgaon as the busiest hub of coastal and oceanic trade for Bengal and the eastern Ganges plains.

This chapter explores how the advent of Europeans—the Portuguese, followed by the Danes, the Dutch, the French, and the English—opened the delta to new currents of oceanic trade and commerce, transforming the contours of its demography and economy. Overseas trade in the age of mercantilism had already begun to impact trade and exchange, buoying the circulation of currency and rendering the subtropical rice paddies and textiles of Bengal an indispensable asset for a cash-strapped Mughal Empire, which during the time of Aurangzeb gave this land the name "paradise of nations" (*jinnat-ul bilād*). A massive annual tribute sent by the Bengal Nawabs helped keep Delhi in splendor during the waning years of Mughal power. It was also during this period that cities such as Dhaka, Murshidabad, Patna, and Banaras, along with markets and trading networks, began to command the attention of rival overseas traders and corporations.

By the turn of the seventeenth century C.E. European mercantile companies had set up factories and trading posts connecting Bengal, Bihar, and Assam, and their sloops and boats scoured the waters of the main branches of the Ganges near Hugli. Despite treacherous sandbanks and shallows, the depth of the Ganges reaching thirty feet or more in her main channels during monsoons made it relatively easy for heavier sailing vessels to ply their wares farther upriver. This accelerated the trade in relatively cheap manufactures of Indian artisans, especially silk and cotton textiles, indigo, and saltpeter, destined for the distant markets of Europe. Over the course of the eighteenth century the Bengal delta and the plains above became subject to the profits and politics of oceanic trade. Factors and agents of competing European trading companies also began infiltrating local "country" trade that had flourished for a long time on the rivers of eastern India. European merchants and their native agents, aided by local boaters, plied these waters with old-fashioned country boats decked in bamboo, thatched with straw, and fitted with primitive sails, with intimate knowledge of its banks, currents, eddies, floodplains, bluffs, and ravines; the dangers of nor'easters and monsoon deluges; and the various kings, chieftains, and landlords who levied their taxes and tolls on passing merchandise. These vessels ranged in size from the little flat-bottomed swift rafts of Fatehgarh in the west to the plodding 32-ton salt-laden boats of Tamluk in Bengal, along with the ever-present narrow, deep-hulled boats of Dhaka designed for the narrow and convoluted creeks of the vast estuary.

The English East India Company succeeded in edging out its European rivals, scuttling French imperial ambitions in India in the wake of the Seven Years' War (1760) and, in capturing Bengal, turning its Nawab into a cipher. The British traders and their Indian agents lodged themselves in the country trade of the interior, drawn into frequent skirmishes with local authorities. After 1765 they began to treat the eastern seaboard as an imperial acquisition, with exclusive rights to the revenues of Bengal, Bihar, and Orissa in return for one annual payment to the Mughal ruler in Delhi, increasingly powerless and confined to his palace. They set up custom houses, police posts, and checkpoints across much of these tracts, brought to heel unruly zamindars, and secured most of the marketplaces between Dhaka and Banaras.

Temples, mosques, hospices, fairs, and markets along the Ganges had always been subject to political struggle between competing regimes. With considerable forethought, Akbar decided to erect a massive fort at the confluence of the Ganges and the Jumna in Allahabad. In the eighteenth century, before the British had fully curbed the militancy of local regimes, trading and fighting groups such as the ascetic *sanyāsīs* vied for the control of pilgrimages. As the company seized the reins of administration, this volatile landscape began to change once again, not just in the Bengal delta, but beyond Mirzapur and Banaras upriver. Europeans, including company factors, had been floating up and down for decades on the Ganges in native budgerows fitted with both oar and sail, and now they were in charge of the river and its valley, which along with the old Grand Trunk Road was home to the annual traffic of more than 25,000 tons of country cargo.

This chapter describes how the Ganges was mapped and reclaimed in the eighteenth century, how the lower Ganges plains and the Bengal delta formed the core of the East India Company's imperial territories in India, and how, after the great uprising of 1857, the valley of the Ganges became a part of village India of the British Raj. It took a long passage of history to pacify the valley and weed out unfriendly landlords, rebellious chiefs, and hostile tribes. Forests were cleared and swamps drained to make way for an unprecedented expansion of villages and land under cultivation, as steamships and railways followed. A consensus seems to be emerging now among environmental historians of India that the British Empire played an unprecedented role in the transformation of the ecological and environmental landscape of the Indian subcontinent, especially in the manner in which it claimed raw materials and resources for its share of the global industrial economy.[5] The plains of northern India bore the brunt of this upheaval over the course of the nineteenth century. At the same time, in travelogues, etchings, and watercolors, the image of a picturesque and beatific Ganges flowing through the heart of Britain's proud Indian acquisition took hold of the Victorian imperial imagination.

Merchant-Raiders

Sometime during the early years of the sixteenth century, renegade bands of Portuguese and Arakanese descent began to raid the ports and estuaries of the Bay of Bengal from Chittagong in the west to Satgaon and Hugli on the Ganges in the east. Mukundaram Chakarabarti, in his memorable verses dedicated to the goddess Chandi, describing the seaward journey of the legendary merchant Dhanapati Singhala, mentions his fleet passing the country of the *phiriṅgī*s (Franks), sailing all night long for the fear of their armadas (*hārmāda*).[6] It is clear from this description that by the 1530s the Portuguese had extended their maritime activities in Bengal and staked their claim on the banks of the lower Ganges delta, adding to the menace of pirates.[7] The Bengal sultans were well aware of the fact that these waterways to the ocean were difficult to patrol, which is why Arakanese raiders, the dreaded "*mag*s" of the Bengal littoral, had been able to carry on their traffic in local commodities and slaves for centuries.[8] Bengali seafaring vessels traveled with armed mercenaries (*pāik*s) to defend against such sudden raids.[9]

The course of the river Hugli had narrowed over the centuries, but descriptions of Bengali poets from the Mughal era suggest a busy traffic of boats and seagoing vessels up and down the distributaries and channels from Hugli to the great pilgrimage of Sagar Island.[10] These myriad outlets to the Bay of Bengal are marked clearly in Portuguese and Dutch navigational charts and the maps of João de Barros and Mattheus Van den Broucke. De Barros, in his *Décadas da Asia,* describes the Sundarbans, or Cape Segogora, recognized by sailors by its distinctive lines of palm—a coastline known to Portuguese sailors by the beginning of the sixteenth century C.E.[11] De Barros provides a map along with a description of settlements and markets between the mouth of the Ganges and Chittagong. The outlets of the Ganges here, he shows, have dissected the land (*toda terra cortada*) into a myriad of unknown and low-lying (*baixos*) islands. A similar account of these labyrinthine creeks and flats is given in the seventeenth-century compendium of Pedro Barreto de Resende and António Bocarro on Portuguese possessions in the Indian Ocean called the *Livro das Plantas de Todas as Fortalezas, Cidades e Povoações do Estado da India Oriental,* still presenting the Bengal delta after almost a hundred years as a remote and forbidding place (*extremo lugar*).[12] Here the sea cuts the earth in an endless series of islands and lowlands, and rivers shed their waters into the ocean through many mouths of which two are best known (*mais celebres*): Satgaon and Chittagong. By this period these parts of the coast had become the haunts of Arakanese pirates and Portuguese freebooters, along with Malays, Arabs, and other adventurers, who from time to time carried off local inhabitants to be sold as slaves and eunuchs. This may have been one of the reasons why large tracts of the delta between Hugli in the east and the Chittagong Hills in the west remained unsettled, covered with swamps and marshes, subject to periodic flooding, and infested with tigers, crocodiles, and other wild beasts.[13] Accounts of early Jesuit travelers such as Pierre du Jarric, from Toulouse, France, who traveled between Hugli and Chittagong in the late sixteenth century, describe a forbidding terrain of dense vegetation and runny marshes with crocodiles, and tigers with a taste for human flesh that could pursue boats for miles. Father Jarric saw a tiger jump onto a boat with thirty passengers and carry off an African slave from the hold.[14] Such gloomy accounts of the Bengal coastal waters "full of rocks and dangerous shelves" guarding a vast, inhospitable delta of "shallows and flats" is shared by many other travelers and traders, not just Jesuit priests.[15] The letters of Nicholas Pimenta record visitors to Bengal from the Portuguese settlements in Goa and Cochin, especially the mission of Francis Fernandes and Dominick Sosa, who was struck by the extent of this expansive wilderness.[16]

Around the time of Vasco da Gama's historic voyage to India, the Portuguese had already established a string of coastal trading outposts equipped to replenish caravels destined for trade in spices, slaves, and textiles, and other precious cargo across the greater Indian Ocean. Very soon, extensive stretches of the northeastern oceanic rim, including the Arabian Sea and the Bay of Bengal, would be added to the network of islands and enclaves that made Portugal such a formidable amphibious power. This emerging view of the transatlantic Indian Ocean world is evident in contemporary records such as Duarte Pachedo Pereira's handbook *Esmeraldo de Situ Orbis,* written in 1508, which contains not only vivid descriptions of the West African slave coast but also detailed charts of the navigational distances to Indian ports such as Calicut and Cannanore and geographical speculations about the country immediately surrounding the mouth of the River Ganges ("*India rodeando há entrada y foz do rio guanje*").[17]

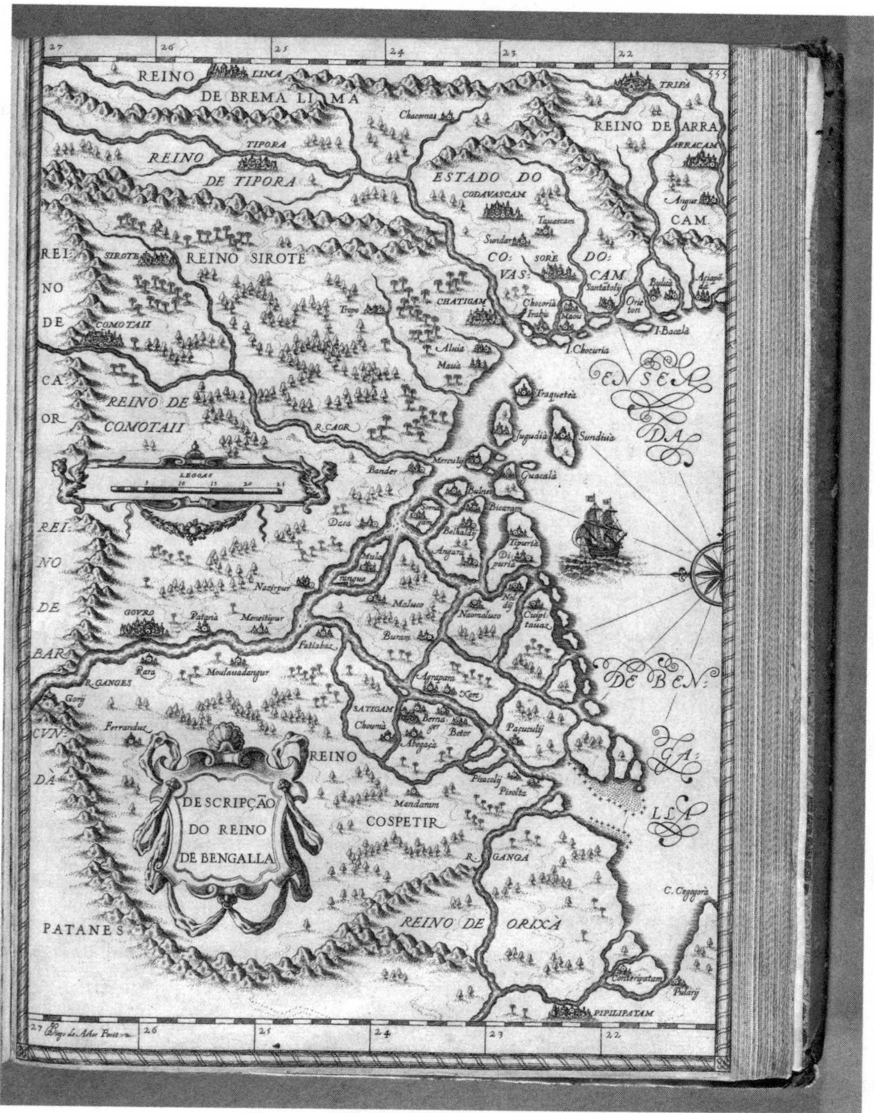

João de Barros's map of the Bengal delta, ca. 1550 C.E., published in 1615 C.E. Courtesy of Houghton Library, Harvard University, Ind. 595.52.10.

Early sixteenth-century accounts begin to take notice of the rapid growth of Portuguese settlements in Chittagong, now rechristened as Porto Grande, distinct from the settlers of the Hugli estuary, known as the Porto Piqueno. This smaller port lay downriver from the old entrepôt of Satgaon. M. Caesar

Frederick, a Venetian merchant, visited the harbor of Porto Piqueno, sailing along the coast and up the Ganges to a city called Satagan (Satgaon), where he saw a long procession of boats. Some had covered a distance of almost 100 miles from the coast, a journey of eighteen hours, and among them were country barks, large *bazārās* (Bengali *bajrā*) laden with merchandise, and smaller dinghies (*pāṭuās*). He was struck by the number of makeshift shacks of straw and mud in a nearby depot known as Betor, a clearinghouse for ships leaving for ports of call farther east. About thirty-five such vessels left Hugli each year during this period carrying rice, cloth, lacquer, and chili peppers.[18] Ralph Fitch, too, during his voyage to Bengal, noted the prosperity of Porto Piqueno, which had become the entry point for boats traveling a league inland to Satgaon.[19] Later British accounts of the ruins of Satgaon, with remains of ground wells and tanks found at depths of 15, 20, or even 30 feet, suggest its gradual decline and the rise of Hugli.[20]

The Legacy of Albuquerque

The Vatican had granted the Portuguese leave to seek out prospects in Persia and India, and they were among the earliest Europeans to settle along the coastline of Bengal. It is important to note that Bengal at this time was largely independent of Mughal Delhi, and the sultans did not have a single fleet to challenge the Portuguese in their coastal waters. By the early fifteenth century C.E. the Portuguese had established themselves as the indispensable agents of trade across the Indian Ocean. Small groups of Portuguese had also begun to arrive in Bengal from their various strongholds across the Indian Ocean, including the spice-rich Malabar coast, Sri Lanka, and the islands of Malacca. News of the bustling trade and fabulous merchandise of Bengal, especially cotton textiles, traveled to Goa and the wider Iberian world, prompting the expeditions of João de Silveira and Fernão Peres de Andrade.

In 1515, following the death of the legendary pioneer of the Portuguese seaborne empire, Afonso de Albuquerque, the young Lopo Soares de Albergaria was appointed as the new governor of the Estado da India, and like his predecessor, he gave a wide birth to Portuguese *casados* (trader-settlers)—not just *fidalgos* (aristocrats)—who sought to expand their commercial operations along the Bengal coast through their own private enterprise. Such *casado*s included not only white (*branco*) married settlers, soldiers, and ecclesiastics but, as Bocarro pointed out, an increasing number of native and mixed-blood Christians or *casados pretos* (blacks).[21] The Portuguese crown at this time was also interested in exploring the possibility of trade across the Bay of Bengal and China and helped finance the voyage of Fernão Peres de Andrade, who reached the coast of China in 1517 and sent an envoy, João Coelho, in a ship belonging to a Muslim merchant sailing to Chittagong. It is not clear whether the Muslim rulers of the Bengal delta saw the Portuguese as a blessing or a menace, but João de Silveira, dispatched by the governor to establish new trading posts in coastal Bengal, did not seem to have succeeded in his mission. There is no doubt that the Portuguese were keen to explore the prospects of coastal and oceanic trade in Bengal as they began to send an annual fleet (*carreira*) of ships. Contemporary Bengali literature, especially the Mangalkabya verses, gives us an impression of a bustling world of coastal and oceanic trade in the Bengal delta and the presence of Chinese and Burmese junks (Bengali *jaṅga-diṅgā*). Common articles of trade included rice, betel-nut, catechu, camphor, spices, dried fish, sugarcane, jute, earthenware, essential oils, clarified butter, vermilion, sea salt, and textiles.[22]

The Portuguese, entrenched in the maritime and coastal trade of Goa and Sri Lanka and in command of the sea lanes frequented by spice-laden Moorish vessels connecting Cambay, Hormuz, Basra, Sumatra, and the Straits of Malacca, were drawn to this burgeoning trade of the Bengal littoral. By the early fifteenth century they had captured Diu, south of Goa, and made Sri Lanka a nominal vassal, weighing in with their armadas and troops in a series of mercantile disputes. Trading and fighting on the high seas were kindred operations for contemporary maritime powers, although J. J. Campos, the pioneering historian of the Portuguese in Bengal, asserts that commercial exploration and evangelization were the main impulses behind these early expeditions.[23] Silveira and João Coelho both found their way to Chittagong, looking for permission to trade and set up manufactories.[24] The governor of Chittagong had taken exception to Portuguese vessels intercepting two cargo ships from Bengal destined for Cambay. Coelho's mission fared better than expected in Chittagong, but he did not have an official mandate. Matters came to a head after Silveira and his men, running out of supplies, seized a local boat and snatched its cargo of rice—an act for which the Bengal ruler decided to expel them. Fugitives in the coastal waters, Silveira and his men were trapped in a long Bengal monsoon, and to avenge this banishment, he set up a temporary blockade of all shipping in the delta.

Despite such mounting mistrust and hostility, the Portuguese continued to send their yearly fleet of ships to Bengal. In 1519 the Turkish immigrant from Tirmiz, Alauddin Husain Shah, who had established himself as the sultan of Bengal, passed away and his son Nusrat Shah succeeded at the capital of Gauda, where the Portuguese sent their first embassy under Antonio de Brito and Diogo Pereira (a private trader from Cochin). They too were treated with due suspicion. When Nuño da Cunha (son of Tristan da Cunha, an Atlantic pioneer) became the capitão-general of the Estado da India, he badly wanted to win back the trust of such local rulers. However, to make matters worse, in 1526, just as the Portuguese envoy to the sultan's court, Dom Martim Affonso, and his men were being taken into custody, Pereira seized a ship owned by the Persian merchant Khwaja Shihabuddin. Cunha promised to release the ship, and Affonso was finally freed along with his companions in 1529. Affonso's envoy to the court of Bengal, accompanied by gifts such as horses and brocades, also included—alas—bottles of rose water taken from the plunder of a Moorish vessel.

Such incidents make it clear that there was never much love lost between the Portuguese and the Bengal sultan, who in an act of caprice captured and killed Cristovão de Mello, nephew of Governor Lopo Vaz de Sampaio, and ten other Portuguese men. João de Barros, in his account, decries this as a dark day in the annals of the Portuguese in Bengal, when the rest of the survivors, who were wounded and starving, were treated like animals while they were in custody at Gauda. To avenge this nefarious act, a fleet of nine ships was dispatched under Antonio da Silva Menezes, and the Portuguese attempted to force Sultan Mahmud Shah to release the prisoners, who were about to be forcibly dragooned for the campaign against the Afghan upstart Sher Shah, threatening to annex Bengal. Menezes's men began to kill, plunder, and torch the port of Chittagong. They also set up a blockade in Satgaon. The Portuguese eventually made peace with Sultan Mahmud and assisted him in delaying Sher Shah's advance by cutting off ferries to mountain passes in the north, fighting valiantly alongside Mahmud's troops in a losing cause. Before Mahmud died from wounds sustained in his last battle against the Afghans, he entrusted the customs houses of Satgaon and Chittagong to the Portuguese in recognition of their assistance and bravery. The Afghan interlude in Delhi, however, came to a sudden end with the accidental death of Sher Shah, weakness and disunity in the Afghan ranks under his successors, and the return of the Mughal emperor Humayun. In 1579 Pedro Tavares traveled to Emperor Akbar's court in Agra and obtained a much-coveted imperial order, or *farmān,* allowing his people to trade free of duty throughout India. Jahangir would renew these rights in the early seventeenth century. By this period, the Estado da India had been firmly established as a major center for the spice trade between Asia and Europe.[25]

From the point of view of Lisbon, the coastlines of Bengal and Arakan remained distant backwaters of the far-flung Portuguese seaborne empire, places where the Estado da India could never exercise adequate control over informal and clandestine networks of trade. These places were not usually inhabited by honorable *fidalgos*. Recruits who worked the galleys of the Portuguese armadas often came from among the poorest sections of the empire, and many were prisoners of war or ex-convicts.[26] Resende and Bocarro's *Livro das Plantas*, while strenuously affirming the fidelity of the Portuguese scattered across the various forts and settlements of the Bay of Bengal to the Estado and the crown, notes nevertheless that the *filho*s of Portugal in the various forts and settlements (*bandéis*) were often beyond the reach of the viceroy's writ.[27] In such parts, remote from the home country (*partes tão remotas de Portugal*) and in a state of constant warfare (*sempre em continuas guerras*), it was easy to disobey the authority of the crown.[28] Crews and musketeers were poorly paid, and many who went to India never came back. The descendants of the Portuguese who settled in these parts, Resende and Bocarro inveighed, were of common stock and lacking in self-esteem and honor. They engaged in coastal traffic, were drawn into local political intrigue, and enlisted in rival armies. They also ended up marrying local women and leaving behind progeny of mixed blood.[29] Father Pierre du Jarric of Toulouse, in his account of the Jesuit missions in Bengal undertaken in 1598, noted the extensive *peuplade* (settlement) of Portuguese along the River Hugli and their church dedicated to St. Mary, frequented by native Christians and Portuguese sailors who dedicated foresails of their ships to her for good luck.[30] While in Hugli, the fathers also received an appeal from the Portuguese Christians of Satgaon, who had lost their priest and could not hold Mass. Du Jarric notes that these denizens of the interior were in a state of "great discord and dissention," having long settled down with local women and at the head of large families, attended by slaves and native servants.[31] The Jesuit fathers persuaded many of these women to enter into lawful Christian marriages and gladly baptized them and their children.

The Portuguese of the Bengal littoral and their mixed-blood, Christianized, Portuguese-speaking descendants have not been treated kindly in the travelogues and histories of the subsequent period. Duarte Barbosa, in his account of Chittagong and Satgaon, described them as castoffs who teamed up with the rogue Arakanese to set up an extensive traffic in slave boys and eunuchs.[32] Sebastien Manrique, the noted traveler who visited Bengal and Arakan during the 1630s, observed that they were seen by local people primarily as raiders and slavers and noted how he and other priests in his entourage alarmed the natives, who mistook them for soldiers who had come to capture and sell their young.[33]

The Age of Mercantile Rivalry

Notwithstanding their depredations, the Portuguese had opened up the coastal trade of Bengal to the markets of the greater Indian Ocean. Sandalwood, prized as a ritual commodity among Hindus, came by Portuguese vessels from the distant islands of Malacca, Sumatra, Borneo, and Timor, as did brocade, velvet, satin, and damask from the Persian Gulf. Bengali and Coromandel taffetas and muslin made their way to new European markets. Cinnamon and pepper from Sri Lanka and the Malabar Coast, as well as red chili peppers from the New World, became part of the established culinary repertoire. Sugar, silk, woodwork, and boxes came from China. Pearls, conches, and sea shells came on Portuguese caravels plying through the Maldives.

Emperor Akbar's investiture of 1579 supporting Jesuits and Iberian merchants gave further legitimacy to the Portuguese settlements of the lower Ganges delta. A visible reminder of this history is the present-day Basilica of the Holy Rosary, or the Bandel Church, overlooking the River Hugli, which is perhaps the most significant Portuguese monument in Bengal—a symbol of pious Augustinian resolve to uphold the faith in this distant corner of the Estado da India, administered through the dioceses of Cochin and Saint Thomé. Hugli was roughly 40 miles downstream from the Mughal garrison of Rajmahal. Kafi Khan, following the account of Abdul Hamid Lahori, provides a colorful account of the outbreak of hostilities between the Portuguese and the Mughals.[34] Emperor Shah Jahan had received word that the *firang*s had erected a fort in Hugli with lofty towers and walls, amassed cannons and muskets, and built an imposing church where they displayed their gaudy idols.[35] They were reported to oppress travelers and to use their leave for trade to run a lucrative monopoly in tobacco. They were also suspected of converting native orphans to Christianity and trading in slaves. Shah Jahan dispatched Kasim Khan at the head of a large army and a fleet that made a surprising appearance on the Hugli. Bandel was attacked in June 1632, and after a siege of about three months it fell to the Mughal forces.

The old church was destroyed, and the Augustinian fathers who survived were captured and taken to Agra. One of them, Father DeCruz, regained the favor of Shah Jahan and obtained permission to rebuild the Bandel Church, which in subsequent years would proudly display the Mughal investiture alongside the veil of the blessed Virgin Mary.[36]

For the native population of Bengal, the Portuguese were the archetype of the hat-wearing, musket-toting, brash European overseas traders. Scattered along remote outposts of the Mughal Empire, they are described by contemporary court chroniclers such as Abdul Hamid Lahori as insolent troublemakers rather than a direct political threat, committing acts unbecoming of their station as traders: exacting unjust duties from boats that passed their factories, stifling the commerce of Satgaon, trafficking in poor children, and consorting with the Arakanese pirates.[37] Just like on the west coast of Africa, the Portuguese of the Bengal delta were feared as musketeers, pirates, slavers, and ruthless mercenaries. They infiltrated the coastal trade and purchased leases for the collection of tolls and duties especially from salt and tobacco, in Satgaon, Hijli, Chittagong, Sandip, and Sripur. Many were involved in the trade between Bengal, the Coromandel Coast, and parts of Southeast Asia, including Sumatra and Malacca. People of mixed Portuguese and Arakanese extraction frequented the slave markets of contemporary Hijli and Tamluk. Others, marked by their distinctive hats (*topas*), were people of mixed Portuguese descent who joined the ranks of mercenaries known as the Topasses. The Bengali word *kinṭāl*, denoting people of mixed Indian and Portuguese blood (Luso-Indians), is derived from *quintal*, the Portuguese word for a garden house. Not only did a population of mixed-blood Eurasians emerge from the Portuguese presence, but Portuguese soon became the most common European tongue and, through regional variations, the lingua franca of the trading world of the Indian Ocean littoral. Portuguese-speaking Luso-Indians held high positions of trust as mercantile brokers and keepers of ledgers.

The Portuguese grasp on the trade of the Bengal delta did not go unchallenged. John Huyghen von Linschoten, a Dutchman who visited in Goa during 1583–1589, took careful notes on the state of Portuguese settlements in India. His *Itinerario*, translated into several European languages, quickly became a guide for all aspiring European traders and adventurers, because it spoke plainly of the great and rich provinces of Mughal India, while noting its poorly defended cities and the weakening hold of the Spanish-Portuguese crown over its far-flung outposts. A similarly lavish picture was also circulated through the account of Ralph Fitch from England, who had the opportunity to visit Emperor Akbar's durbar in 1591. Using Linschoten's charts, a Dutch fleet left for the Indies under Cornelis Houtman, circumnavigating the Cape of Good Hope in 1595 and sailing up to western Java. Houtman returned without much to show in his cargo and having lost about two thirds of his original crew, but he had succeeded in sealing a treaty with the king of Bantam, which would soon open up the Straits of Malacca and the Indonesian archipelago to Dutch oceanic trade. In 1598 five more Dutch squadrons set sail for the East Indies, leading to the triumphal return of naval officer Jacob Corneliszoon van Neck who had pioneered the exploration of Madagascar and Bantam. The Dutch now began to threaten the Portuguese sea lanes of the Indian Ocean, blockading Mauritius for a period of thirteen years, supplanting the spice trade in Malacca, setting up factories and garrisons in Java and Sumatra, and establishing precedence along the sea lanes of the Far East and the Pacific. In 1597 they had founded the Society for Trade to Distant Countries, which was followed by the establishment of the Dutch East India Company (Vereenigde Oostindische Compagnie) in 1602. By 1625 Dutch ships and factories became a common sight along the lower course of the Hugli and the coastal waters of the bay. The Dutch would soon become the most formidable sea power on the open ocean. They expelled the English from the Indonesian archipelago after the slaughter of Amboina in 1623—a piece of infamy that would live long in the annals of English naval history. After an extended maritime conflict ending with the Battle of Malacca (1641), they finally succeeded in edging out the Portuguese from the Straits of Malacca.[38]

After their expulsion from Hugli the Portuguese had begun to settle in Chinsura, farther down the Hugli River, in 1632.[39] François Bernier, the French physician from Anjou who visited Mughal India and Bengal in 1665–1666, noted extensive Dutch settlements across the Bengal delta. Bernier found that Bengal was the great emporium of cotton and silk textiles not only for the Mughal Empire but for many of the trading nations of Europe, led by the Dutch. "I have been sometimes amazed by the vast quantity of cotton cloths, of every sort, fine and coarse, white and coloured, which the Hollanders alone export to different places, especially to Japan and Europe," he wrote.[40] He also gave a description of the Dutch silk factory at Kasimbazar, near the capital of the Bengal Nawabs, Murshidabad, which seems to have employed close to eight hundred native artisans and weavers during this time. Chinsura and the nearby village of Baranagar on the eastern bank of the Hugli also attracted the Dutch. Diaries of the English factor Streynsham Master reveal that the Dutch had gained an upper hand in the riverine and oceanic trade of Bengal at this time, dealing in saltpeter, muslins (known also as *malmal*), taffetas, ginghams, and raw silk.[41] The Dutch built warehouses to protect silk textiles from humidity and exposure and looms for silk weavers.[42] Their ships could navigate the treacherous sandbanks and shoals of the Ganges better than others.[43]

Alexander Hamilton, in his account of Bengal, describes the sprawling houses and gardens of the factors of Chinsura, fitted with balustrades and pavilions.[44] Many had Moorish-style terraces with floors of pulverized red stone and walls plastered with lime. There were similar establishments at Kasimbazar and at Malda, a small town conveniently located at a branch of the Ganges. Malda was a place of "great resort" for weavers, dyers, and dealers of cloth that arrived from all parts of the country, providing for the European factors located as far away as Agra and Gujarat, and also agents dispatched by the merchants of Banaras. By the mid-eighteenth century, the Dutch ran factories at all the major centers of trade and commerce in Gangetic Bihar and Bengal.[45] Not only did they dominate the trade in bullion from Europe, Dutch ships carried tea, porcelain, and finished and raw silk

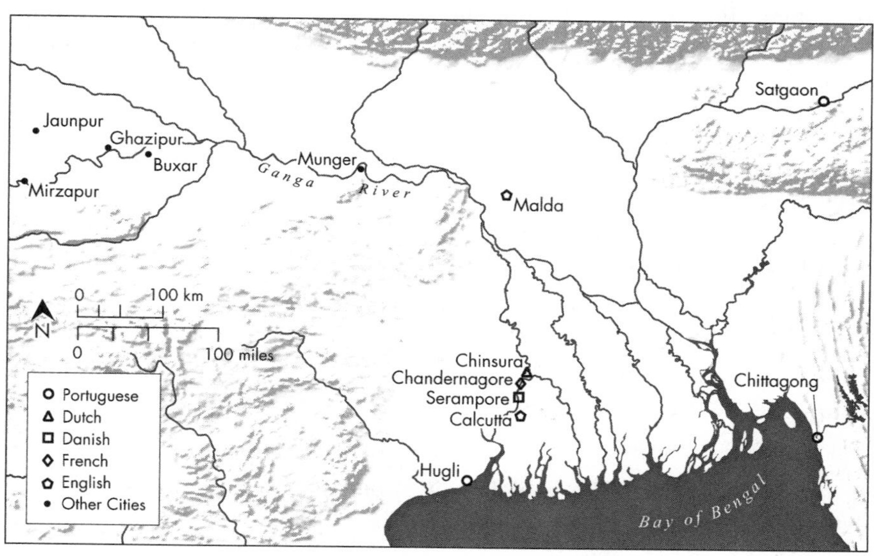

European maritime empires in the Bengal delta.
Map by M. Roy Cartography.

from China; lacquered ware and gold dust from Japan; cloves from Malacca; nutmeg and mace from Banda Aceh; pepper and opium from Sulawesi; textiles, indigo, and diamonds from the Bengal delta; calicoes, muslins, cotton, and flax from the Coromandel Coast; and shellac, rosewood, saltpeter, and dyes and cinnamon from Sri Lanka.[46]

Hugli flourished for a long time as the hub of European trade in the lower Ganges delta, with the English, French, Danes, and Dutch having carved out their respective enclaves in Calcutta, Chandernagore, Serampore, and Chinsura. Europeans became familiar figures in the markets and towns of Bengal, typically described as "hat-wearers" (*kulā-poshān*) in Persian accounts. Poet Bharatchandra, in his ballad *Bidyasundara*, which is a section of his celebrated composition *Annadamangala* written in the early eighteenth century, mentions hatted European mercenaries (Bengali *kolāpoṣa*) including Englishmen, Dutchmen (*olandāja*), Frenchmen (*farāsa*), Danes (*dinemāra*), and Germans (*elemāna*).[47] While Aurangzeb reigned in Delhi, and Bengal was in the hands of his able deputy Shaista Khan, the European companies did not dare indulge in any armed conflict. The early company pioneer and Madras agent Streynsham Master noted in his aforementioned diary that the English too refrained from raising forts and garrisons, seeking instead the goodwill of the nawab of Bengal and his magistrates.[48] The rivalry between the Dutch, English, French, and Danes, however, was difficult to contain as they fought over markets, weavers, brokers, credit, currency, and bids to obtain favorable grants from the Indian powers in exchange of gifts and embassies. "Bengal abounds with every necessary of life," Bernier noted in his travels.[49] This abundance had lured so many Portuguese and their half-caste progeny to the banks of the Hugli River. Jesuits and Augustinians vied with Dutch Calvinists and English Protestants for merchandise and profit. There were close to nine thousand Christians in Hugli alone, and perhaps more than twenty-five thousand in Bengal during the late seventeenth century. Bernier recorded a common saying among Europeans who resorted to Bengal that the country had "a hundred gates open for entrance, but not one for departure."[50] For nearly two hundred years the burgeoning commercial world of the lower Ganges delta assimilated European newcomers into its extensive networks of trade and exchange, absorbing the influx of New World silver in exchange of manufactured goods. This part of the East Indies was, as one contemporary put it, a "bottomless pit for bullion."[51] The tide turned in the eighteenth century, when the French and the English, through their respective trading companies, began to intervene directly in Indian regional politics and local military conflict to hedge their imperial rivalry.

Spoils of Trade and War

Duties levied on inland trade had always been a bone of contention between the Indian regional powers and the Europeans. This was particularly true in Bengal, where, during the heyday of Mughal rule, the nawabs of Bengal exacted duties from all merchants, Indian or European, not only at the major custom houses such as Sultanganj in Patna, Pachottera in Murshidabad, or Bakhshbandar in Hugli, but also at a myriad of smaller posts and toll houses along rivers and roads of the countryside. Grand *farmāns* of the Mughal emperor, which gave a general leave to European companies for their overseas trade, did not guarantee exemption from all local duties and tolls. At Hugli the English East India Company paid a handsome annual tribute (*peshkash*) of 3,000 rupees to trade free of added duties, while the Danes, the Dutch, and the French paid 2.5 percent on their total invoices.[52] The nawab's customs officials and clerks furnished inventories (*ravānās*) of each boatload of goods carried up and down the Ganges and issued handwritten clearances (*dastaks*) inspected at smaller toll posts at crossings and ferries. The post (*chaukī*) of Jaldanga, situated between the Dutch factory of Chinsura and the French factory of Chandernagore, was set up to stop boats trying to pass under one flag to evade these duties. A contemporary chronicler, Munshi Salimullah, noted that while Murshid Quli Khan, the nawab of Bengal, was thankful that his province benefitted handsomely from duties levied on the profits of European trading companies and did not wish to impose added exactions on their overseas ventures, he also wanted to make sure that they did not set up forts, bastions, and trenches.[53] This was one of the reasons why Job Charnock, former head of the Kasimbazar factory who had also served as a factor in Patna, sought a remote site for a fortified settlement for the British farther downriver toward the tidal mouth of the Hugli.[54]

He would eventually settle for the brackish, low-lying marshes of Sutanati, Gobindapur, and Kalikata, where he signed a lease with the family of Sabarna Chaudhuri, local zamindars, for a string of villages that would eventually grow into the city of Calcutta, destined as the capital of the British Empire in the East. Alexander Hamilton, who was part of the early English settlement between 1688 and 1723, notes in his account that Charnock "could not have chosen a more unhealthful place" because of the proximity of the great saltwater lake to the northeast of the English colony that overflowed after each monsoon, exuding what was then believed to be miasmal vapors afflicting local inhabitants with fevers and agues and taking a hefty mortal toll every year.[55] Large tracts of tiger-infested forest were cut down for the building of Fort William and the esplanade, and over time, the "straggling villages" of Chowringhee were converted into a bustling thoroughfare for carriages, lined with elegant brick and whitewashed mansions, and jungles were cleared between the fort and the native settlement of Chitpur. However, contrary to Rudyard Kipling's niggling epithet for Calcutta as a city "chance-directed, chance-erected, laid and built on the silt," the site for the fort and settlement of Calcutta was chosen carefully.[56] It was defended to the west by a tortuous and deep stretch of the river that was difficult to ford with large armies; it was near multiple outlets to the Bay of Bengal; and Sutanati was a burgeoning market for thread, cotton yarn, and textiles. The defenses of Fort William were intended for strategic and military protection, not just from the corrupt and rapacious native officers of custom but from rival European companies, especially the French and the Dutch. In 1717, the British sent an embassy to the Mughal court under John Surman; it returned in triumph with a new and extensive *farmān* under the seal of Emperor Farruksiyar for trade free of duty. Shielded by this new document, the British pressed their advantage in Bengal during a period when the Mughal Empire had been considerably weakened by succession struggles, disloyal nobles, palace coups, and peasant rebellions. During these politically fraught times, many provinces, such as Bengal and Awadh, while remaining nominally loyal to Delhi, began to exert a far greater degree of autonomy in their internal affairs. While this created an uncertain climate for trade, it also offered new opportunities for the company, its officials, and their dependents.

The East India Company was a chartered, joint-stock corporation. The Scottish theorist of political economy Adam Smith disparaged it famously as a despotic and oppressive monopoly protected by parliamentary fiat.[57] It was controlled by a small group of affluent merchants who served on an exclusive court of directors, whose annual meetings, according to the British prime minister William Pitt the Elder, were run like "little parliaments."[58] While it commanded the major sea lanes and shipping to India and its vast profits generated an average return of around 8 percent for its major shareholders, it did not pay its factors, ensigns, and cadets much in terms of salary. Instead, company servants in India were indirectly rewarded by the perquisites of their station, which included not only gifts and bribes but lucrative deals in inland or country trade.[59] Private trade yielded high margins of profit, as employees were able to conduct business transactions free of local duties along with those of the company, claiming protection under the Mughal permission against which the chiefs of English factories could issue handwritten passes (*dastaks*). Armed with such licenses, they outbid local traders and merchants in the traffic of commodities such as salt, betel-nut, and tobacco. Later, the court of directors of the company would acknowledge that the letter of the Mughal emperor was intended exclusively for the trade of the company and not for the enrichment of individual members who had been for years conducting an illegal trade, profiting from a spurious claim, occasionally supported by violence.[60] Many of these items of everyday use, such as salt, saltpeter, tobacco, betel leaves, and betel-nut were seen as prestige goods coveted by local merchants, and trading in them required a special favor or dispensation from the Bengal nawab.[61] The English intrusion into inland trade was particularly irksome to the deputies and officials of the nawab, as well as local *zamīndārs* and landlords, and it led to repeated conflicts over the passage and examination of boats laden with East India Company merchandise plying up and down the River Ganges. Boats passing under company papers (*dastaks*) were subject to attack and looting, which is why they often carried armed retainers and muskets.[62] The news of such outbreaks and the increasing militancy of the English worried the Bengal nawabs, and the lengths to which the company was prepared to go to defend their trade threatened to infringe the accepted hierarchy of ruler and merchant in late Mughal Bengal. Nawab Alivardi Khan, who supported the English trade, reportedly asked why their merchants needed a fort and an army.[63] In 1756, when Alivardi passed away, his young grandson, Siraj-ud-daulah, took charge. Accusing the company of having violated the terms of trade and sheltering fugitives, he ransacked the English factory and settlement in Kasimbazar, ordering Roger Drake, the governor of Calcutta, to bring down all fortifications. When this was refused, the nawab marched into the city. Fort William had a small garrison of artillerymen and not enough supply to last a siege, and Drake decided to abandon his post and seek shelter in sloops off the coast, leaving its inhabitants to the mercy of the invaders. Siraj stormed the fort and took the garrison, allegedly throwing 146 captives into an ill-ventilated, small cell—which now lives in infamy as the Black Hole of Calcutta—where apparently all but 23 perished, according to the eyewitness account of surgeon John Holwell.[64] The veracity of this event, which Karl Marx once called a "sham scandal" invented by the "English hypocrites," has been questioned by many historians, but it gained quick and widespread notoriety as clear evidence of the barbaric perfidy typical of Oriental despots.[65] The incident also stoked the outrage of a young Robert Clive, kept away at Fort St. David in the south, who wrote to the directors in London of the loss of the city having caused the "greatest mortifications to the Company and the most barbarous and cruel circumstances to the poor inhabitants."[66]

Clive promised to recover Calcutta and its losses and restore its former rights and privileges, as he left for Bengal with five armed ships from the Royal Navy under the command of Admiral Charles Watson. What followed is one of the most well-known chapters in the history of the British Empire in India: Fort William was recaptured, General Mir Jafar was induced to desert the nawab's army with the promise that the British would install him as the new ruler, and Bengal fell to the British at the battle of Plassey—fought near the shady banks of the Ganges in 1757. The East India Company's military success in Bengal and its creation of a string of puppet nawabs on the throne of Murshidabad in exchange for unprecedented sums of money was not just a great victory and a terrific windfall that gave the company the keys to the richest province of the Mughal Empire, but a decisive punctuation to the extended Indian theater of the European wars that lasted from 1744 to 1763, which put the Dutch and the French competition to rest. More than textiles, indigo, or opium, the control of the lower Ganges delta enabled the English to edge out the Dutch and the French and secure a monopoly over a much-coveted substance in Europe—potassium nitrate, found in its natural state as saltpeter—a crystal that is produced in soil from the action of bacteria on animal droppings and urine. Along with carbon and sulfur, saltpeter was an essential element in the manufacture of gunpowder. The environs of Calcutta produced copious quantities of high-quality saltpeter, processed by an entire caste of indigenous saltpeter harvesters known as Nuniyas. After they conquered Bengal, Britain would control almost 70 percent of the world's production of saltpeter.[67]

The Prize of Bengal

European trading settlements in the lower Ganges delta had always vied for favor and advantage at the courts of Bengal and Delhi, but they also acted in common interest, rooting out interlopers who tried to claim a share of the trade in contravention of established monopolies and rights. This fragile neutrality fell apart during the outbreak of the War of Austrian Succession in Europe (1740–1748), a tortuous conflict over the Austro-Hungarian

Empire's claim over Silesia that upended the old alliances between the European powers, pitting the English against Austria and France. The 1748 Treaty of Aix-la-Chapelle could not prevent the drift toward further conflict that spilled over into the French and Indian Wars in North America and the Anglo-French hostilities that engulfed Bengal and the Carnatic. The East India Company appealed for support from the Royal Navy, whose warships began to frequent Indian waters by 1745 alongside those of the French national fleet. Naval skirmishes took place near the Coromandel Coast, where the English settlement of Fort St. George in Madras was less than 90 miles from the French stronghold in Pondicherry. The Dutch Republic had so far remained neutral in the crisis brewing in Europe; the Dutch and the English in Bengal signed a joint truce with the French at Chandernagore. In 1749, however, the French brazenly seized the Dutch gardens and factories at Chinsura. J. Huyghens, the director of the Dutch settlement, wrote to the English in Calcutta that the French by this act had broken the "neutrality of the Ganges."[68] Rival trading companies were gearing up for armed conflict: fortifying factories, conscripting native mercenaries into standing armies, and stockpiling muskets and cannons. This looming threat of war and the political uncertainty in Bengal also contributed to the military confrontation that took place between the company and the nawab of Bengal.

In southern India the French outrivaled the British in creating a network of alliances with new and upstart regimes that had reared their heads after the decline of Mughal power: the nizam of Hyderabad, the Maratha ruling houses, the kings of Mysore, and the nawabs of Arcot and the Carnatic. English troops stationed at Madras were critically outnumbered in Pondicherry. Joseph François Dupleix, son of a former director of the French Campagnie des Indes, was the chief architect of the French empire in India. He took charge of the struggling factory of Chandernagore, put it on a profitable footing, revived the fortunes of the settlement, and became the governor of Pondicherry in 1741. Political turmoil in the Carnatic gave an opportunity for the French to attack Madras, which they took in 1746, burning down several English factories. A truce was declared in 1748 with the Peace of Aachen, and the French gave Madras back to the British. Dupleix wanted to establish himself as the undisputed nawab of the Carnatic, vehemently opposed by "the Company's Nabob," Muhammad Ali Wallajah of Arcot, who was propped up by the British. French political affairs in India, however, became very chaotic because of the rivalry and quarrel between Dupleix and Bertrand-François Mahé de La Bourdonnais, a commander in the French navy and the former governor of Mascarenes. When La Bourdonnais's fleet was ruined in a major storm, Dupleix refused to render help, and he was taken prisoner by the British, who sent him in ignominy to Paris for trial. He was consigned to the Bastille in 1749, which was a banishment that broke his health, and he died within two years of his acquittal. Dupleix also fell afoul of the members of the French court for his flamboyant lifestyle and the huge debts he incurred in setting up the French as an Indian power. His recall and replacement by the more prosaic Charles Godeheu stymied the French political designs in India, clearing the path for Clive and the English.

During the siege of Calcutta the French were allied with the nawab of Bengal; they plied him with 250 chests of gunpowder and did not help the English who were under attack. While the English company gained power in Bengal, the French Compagnie des Indes wielded power in the Deccan. In June 1758, French troops, under the command of the Irish-born Comte de Lally, attacked and burned down the English outpost of Fort St. David, followed by an unsuccessful siege of Madras. British troops from Bengal engaged the French forces in Hyderabad under the command of Marquis de Bussy. By January 1760, the British general Eyre Coote defeated de Lally and took Bussy prisoner in a skirmish that has made its way into the annals of British conquests in India as the short but significant battle of Wandiwash. Pondicherry fell to the British in 1761, spelling the end of French imperial prospects in India.

In 1749 William Forth, the Dutch political agent at Chinsura, attempted to form an alliance with the newly installed ruler Nawab Mir Jafar, promising to procure ships and reinforcements from Batavia and help him expel the British from Bengal. It is doubtful that the nawab was serious, but a Dutch vessel with Malayan soldiers appeared at the mouth of the Hugli, which Clive and the English promptly intercepted.[69] The Dutch pretended that the vessel was bound for the south and had docked near Hugli to avoid monsoon squalls, but their designs became clear in October 1759, when seven more warships arrived. The provocation was defused by a timely trade deal offered by Clive, but Holland remained a thorn in the side of the English in Bengal, helping the nawab increase his troops in Patna and Kasimbazar with seven hundred European and eight hundred crack Malayan mercenaries. Matters soon came to a head when the English began to conduct searches and seizures of Dutch vessels. In the naval battle that took place at the mouth of the Hugli, seven Dutch ships were destroyed and over a hundred Dutchmen were killed or wounded. The British seized

Dutch factories in Baranagar and Chandernagore and drove the Dutch back to Chinsura. A renewed conflict ensued on land. Colonel Francis Forde engaged the Dutch, led by the French adventurer Jean-Baptiste Roussel. The Dutch forces fought back valiantly, but they did not have adequate artillery support. In a quick, bloody, and decisive battle that took only half an hour, 120 Europeans and 200 Malay soldiers were killed across a ditch used as a line of defense, with Dutch officers and 350 European soldiers taken as prisoners.[70] The Dutch debacle at this lesser-known but important battle of Bedarrah, in effect, ended Dutch territorial ambitions in Bengal and the Indian subcontinent.

Despite the end of the French and the Dutch threats, the restoration of Calcutta, the deposition and murder of Nawab Siraj-ud-daulah, and the setting up of Mir Jafar on the throne of Bengal, political affairs of the British in the lower Ganges valley were far from settled. The East India Company had to secure its factories and weaving districts of Murshidabad, Patna, Dhaka, Lakshmipur, and Balasore, as well as the safety of its goods passing up and down the markets and riverside landings (*ghāṭ*s) in the west from disgruntled and hostile elements in the nawab's administration, zamindars, and rural chieftains. At least three uprisings against Mir Jafar, in Midnapore, Purnea, and Bihar, were put down with the help of company troops. The assassination of the Mughal emperor Alamgir II and the usurpation of power by his minister, threatening to march on the English settlements in the east, did not assuage English fears. In this climate of intrigue and uncertainty, Clive's replacement as governor-general of Bengal, Henry Vansittart, decided to remove Mir Jafar and replace him on the throne of Bengal with his son-in-law, Mir Qasim, who in exchange for a handsome subsidy ceded the provinces of Midnapore, Burdwan, and Chittagong to the Company.

Mir Jafar, installed in 1757 by Robert Clive as the first "puppet" nawab, had turned out to be not only corrupt but feeble and ineffective. Mir Qasim, on the other hand, was capable, ambitious, and willful. Rather than being a tool in the hands of company officials, he sought to pursue his own political interests. Unlike the traditional nawabs of Bengal, he did not attempt to reconstitute dynastic rule in Murshidabad. Instead, he worked tirelessly to restore the fiscal solvency of his provinces. He did not hesitate to exercise a measure of coercion to realize arrears in rent from the largest estates. While he agreed to meet increasing financial burdens placed on him by the company, he also began to petition company officials about the interference of English factors and their Indian agents in the internal trade and local markets of Bengal, Bihar, and Orissa. One of the main items in contention was the trade in salt, a lucrative commodity that yielded considerable duties and tolls for the nawab's treasury.[71] Along with salt, company servants as private traders had infiltrated the networks of trade in saltpeter, tobacco, betel-nut, and opium, where they wielded the handwritten *dastak*s, or passes, to ply their goods free of duty. They also abused this authority in local bazaars buying goods below market prices and selling them for high profit. Clive, who would later set up a society of traders to oversee the manufacture and trade of salt, acknowledged that the private investment of company factors was deeply resented by local traders and rural landlords, and an enormous number of boats and mercenaries had to be maintained to protect the trade of salt along the Ganges.

The abuse of internal trade by the factors and agents of the company remained one of the key bones of contention between Calcutta and Murshidabad, straining the alliance between Mir Qasim and the company. As Clive discovered during his second visit to Bengal, the tyranny and oppression of private traders had thrown the entire internal commerce of the country into chaos.[72] Brandishing the *dastak*s and company standards, European traders and their agents along with armed militias roamed the manufactories, markets, and riversides of Bengal, cornering trade in the most valuable items—salt, saltpeter, opium, areca nut, and tobacco. Mir Qasim repeatedly complained about the arbitrary and tyrannical nature of such conduct and the financial loss it inflicted on his treasury.[73] He pointed out that native servants and brokers (*gumāshtā*s) went about misappropriating and abusing the papers and other insignia of the company while forcing themselves into trade in everyday goods such as oil, fish, straw, bamboo, sugar, ghee, rice, and betel-nut, acting like insolent landlords, disregarding the officers of customs and tolls at the markets and warehouses, and unleashing violence on local villagers and traders.[74] Warren Hastings, writing to the governor in 1762, noted that almost every boat he saw on the Ganges, whether it belonged to the company or not, was flying English colors.[75] He also rued the fact that in many parts of the Bengal countryside, traders masqueraded as agents of the English or paraded around in the uniform of company troops.

Mir Qasim eventually decided to abolish all duties on inland trade for native traders and merchants, putting them at par with the servants and middlemen of the company. He also sought to end his puppet regime, abandoning the old capital city of Murshidabad on the northern reaches of the Bhagirathi and refortifying the city of Munger in Bihar farther west

Old ramparts overlooking the Ganges, Munger, Bihar.
Photograph by author.

up the Ganges and away from the reach of the company. He also began to strengthen his musketry and artillery. In the end, however, Mir Qasim's ambition of creating an independent state in eastern India eluded his grasp. The company had already decided to remove him from power after he had abolished the duties on trade. Mir Qasim tried to find a last-minute military alliance with the Mughals and the nawab of Lucknow, Awadh, which proved to be insufficient. At the battle of Buxar in 1764, the English East India Company crushed the combined military forces of Mir Qasim, the Mughal emperor Shah Alam II, and Shuja-ud-daulah, the ruler of Awadh. Mir Qasim became a fugitive, never to return to Murshidabad. The importance of the English victory at Buxar could not be gainsaid. Following the defeat of the Dutch in 1759 and the rout of the French in 1760, it gave the company an opportunity to snuff out the political ambitions of Awadh and to humiliate the Mughal emperor, Shah Alam II—significant steps in the elevation of the English East India Company as the undisputed political masters of Bengal. It presented the prospects of a new arena of empire to Britain that would eventually assuage the loss of the American colonies. In 1765, the company pressed its advantage with the Mughal emperor, who had taken temporary refuge in an English camp, holding him to the terms of the Treaty of Allahabad, which included the transfer of the office of revenue collection (Diwani) of its three great eastern provinces—Bengal, Bihar, and Orissa—to Clive and the East India Company. For the next few years the company, having reinstalled old Mir Jafar on the throne of Bengal, reaped the immense spoils of agrarian revenue without having to assume direct responsibility for the administration of Bengal.

A Delta Famished

The East India Company's acquisition of the Diwani would be later heralded by the English statesman Edmund Burke as a "great act of constitutional entrance of the Company into the body politic of India."[76] The passing of this mantle was recognized by all native powers, but the day-to-day collection of revenue in Bengal, Bihar, and Orissa remained in the hands of Reza Khan and Shitab Roy, while the nawab was placed under the watch of the British resident stationed at Murshidabad. Mir Jafar died in 1765, and his teenage son, Najm-ud-daulah, was installed on his father's seat in 1765. Mir Jafar had squandered most of his fortune, and his son ruled for only a year, dying in a state of near penury from a severe bout of fever. He had entrusted the entire burden of revenue collection to Reza Khan, who proved to be both ruthlessly efficient and corrupt at the same time.[77] As the new regime sought to ratchet up the collection of land revenue, eastern India was lurching toward a severe drought. Revenue officers began to detect widespread crop failure in 1768, and a sharp spike in the price of staples followed. To add to the miseries of the peasants, the monsoon rains stopped prematurely in September 1769. Governor John Verelst; his deputy, Henry Cartier; and the members of the council, however, did not heed these early warning signs.[78] Taxes were being collected without let from the peasantry as late as January 1770, when rice was scarce but barley and wheat seemed plenty along the banks of the River Ganges. Despite the rising clamor of food shortages, a further one-tenth rise on the assessment had been sanctioned for the coming year.

William Hunter, in his vivid account of the Bengal famine in the *Annals of Rural Bengal*, suggests that the scarcity of 1769–1770 appeared without warning, and the company was caught woefully unprepared to address the widespread and abject starvation of its new subjects. One third of the population was rumored to have perished in the district of Purnia alone. The native officers who knew what was happening did not inform the council in time to prepare for relief. The Bengali peasant, writes Hunter, "bears existence with a composure that neither accident nor chance can ruffle." The complaints of victims, Hunter seems to suggest, did not reach the ears of the officials, because the starving masses accepted their fate with silent forbearance.[79] There is not enough evidence to support or discredit this impression of multitudes starving and dying without a whimper, but the devastation wrought by the famine was quick and massive, with millions perishing within the space of a few months. To quote from Hunter's grisly account:

> The husbandmen sold their cattle; they sold their implements of agriculture; they devoured their seed grain; they sold their sons and daughters, till at length no buyer of children could be found; they ate the leaves of trees and the grass of the field . . .[80]

Exhausted from burials and cremations, people began to throw their dead into the Ganges. The English resident stationed at the nawab's seat of power in Murshidabad reported that the living were feeding on the dead along with vultures, dogs, and jackals.[81] Pestilence followed hunger. The smallpox epidemic that broke out across Bengal did not spare the life of Nawab Saif-ud-daulah. Ghulam Husain Khan, in his contemporary account, writes that entire villages and towns "suddenly disappeared from the face of the earth."[82] In 1772, at the end of the famine, Warren Hastings reported to the court of directors that close to one third of the population of greater Bengal had disappeared over the course of two to three years.[83]

Hunter describes the famine of 1769–1770 as the most significant event in the Gangetic delta, one that shaped the course of the history of Bengal over the next forty years.[84] Edmund Burke, in one of his lengthy speeches during the trial of Warren Hastings, described the Bengal famine as the singular episode that "dishonored and disgraced" the government of the East India Company in Bengal in the eyes of both England and Europe at large.[85] The French belletrist Abbé Raynal in his widely read *Histoire philosophique des deux Indes* pointed the finger directly at the policies and conduct of the company government for this "appalling famine in this the most fertile of lands"; they saw the warning signs of scarcity and still deliberately stowed away a large portion of the grain harvest for their own consumption and that of their native Indian soldiers.[86] Notwithstanding the flurry of recriminations, company officials understood that the entire organization of land revenue had become corrupt and moribund. Young English supervisors of revenue such as Gerard Gustavus Ducarel appointed to the district of Purnia or Boughton Rous in Rajshahi, both in their early twenties, pleaded that it was nearly impossible for them to penetrate the labyrinth of arcane practices of assessment and collection of land revenue left over from Mughal times and the obtuse forms of Persian account keeping.[87] While some attributed the failure of custodianship to the inexperience and youth of the English officials, others blamed the suffering of the poor on the rapacity of native moneylenders, middlemen, and dealers of grain (*pāikār*s and *dālāl*s).[88] Ghulam Husain indicts the conduct of Reza Khan himself, who was found guilty of stopping merchant boats for extra levies and of running corrupt monopolies in rice and other necessities of life for illicit profits that "increased the calamities of the poor during the height of the famine."[89]

Richard Becher, the English resident at Murshidabad, observed at the onset of the famine in 1769 that the East India Company's accession to the management of land revenue had significantly worsened the living conditions of the common people.[90] It was, according to Becher, the very nature of the company's investments, its export of currency, and most of all, its connivance at the pitiless exactions of native functionaries that had intensified the crisis. Colonial historians examining the origins of the famine in the breakdown of the prevailing system of land tenure concluded that the taxation of Mughal Bengal had been a top-down affair, with too much leeway given to individual officials (*'āmil*s), zamindars, and revenue farmers, who often imposed additional cesses (*ābwāb*s) for personal gain.[91] The famine, paradoxically, destroyed the old land revenue order of the lower Ganges valley and gave the company a free rein to overhaul the entire structure of land tenure and taxation.

Pax Gangetica

Famine and pestilence wreaked havoc along the agrarian heartland of the lower Ganges valley through the year 1772. There had been a widespread desertion of villages and pastures, with multitudes of starving and destitute people flocking to Murshidabad and Calcutta for survival and accepting abject forms of servitude, including chattel slavery and the meanest forms of wage labor. Large tracts of formerly cultivated land had been turned to waste, leading to a crisis in the collection of taxes. In many districts British officials remained in the dark about the extent of arrears and acreage that could be reclaimed from waste.[92] Edmund Burke inveighed in Parliament that corrupt officials, both English and native, had profited handsomely from the Bengal famine, aggravating the distress of the famished for their own financial gain, which "dishonored and disgraced" the company government in the eyes of both England and Europe. The dishonesty and bribery of the East India Company servants, according to Burke, was the original sin and "ruinous distemper" of British affairs in India.[93] The calamity, historian and statesman Thomas Macaulay later concluded, utterly ruined Bengal, the richest province of the Mughal Empire—a veritable garden of Eden. "No part of India possessed such natural advantages, both for agriculture and for commerce," he wrote; Bengal was blessed by the Ganges, which, "rushing through a hundred channels to the sea, has formed a vast plain of rich mould . . . that rivals the verdure of an English April," and where rice, spices, sugar, and oils could grow with little effort. Nonetheless, in an unexpected way, the famine had given the company a second opportunity to reclaim the fields of plenty and overhaul the old fiscal-military regime.

The passage of the Regulating Act of 1773, which was a formal declaration of the company's intent to assume direct governance of its Indian territories, confirmed Warren Hastings as the new governor-general of India; he took over the charge of revenue collection from the hands of Indian intermediaries. In the previous year the office of the naib diwan, held by Reza Khan and Shitab Rai, had been discontinued, and the entire revenue apparatus (*khālṣa*) transferred from Murshidabad to Calcutta.[94] The act, promising greater vigilance of Parliament over Indian affairs, forbade all gifts and presents from native subordinates, intent on rooting out the corruption rife among all ranks of company servants in India, especially Bengal. The famine of 1769–1770, disruption in trade and manufacturing following the conflict with the nawabs, and the chaotic disorganization in the collection of land revenue compelled the company to assume a much more direct and responsible administration.[95]

As the famine receded and the everyday economy of the Bengal countryside came back to life, the company found its investments on very insecure footing. There was an urgent need to resettle parts of the famine-stricken country where millions of peasants had either perished or been uprooted from their fields. The future of British territorial possessions now hinged on a quick restoration of the agrarian economy of Bengal. Governor-General Hastings pushed through an extensive inquiry into the fallen state of land revenue collection, attempting to repair and secure the tenure of primary producers whose welfare he thought "ought to be the immediate care" of the British government in India—who needed to be protected against the ruthless exactions of despotic and rapacious landlords.[96] The reports submitted by the Amini Commission (1776–1778) on the condition of agricultural production in the eastern provinces pointed out that neither the appointment of English collectors nor the delegation of native officers overseen by the provincial councils had been able to stem the decline in land revenue. For years, instead of taxes, these officers had reported returns of deficit, along with details of defaulting zamindars, absconding farmers, and fleeing agricultural laborers. The Committee of Circuit in its reports on the revenue and commerce of Bengal stated that one of the major impediments to the free movement of "goods and necessaries of life passing by water through the interior parts of the country" was the petty tyranny of the myriad rural zamindars and farmers of revenue whose estates lay along the Ganges and its various tributaries. They collected a plethora of dues and tolls on boats and caravans as part of their traditional claims over the passage of mercantile goods, listed under the head of *sā'ir* in Mughal accounts as separate from land revenue (*mālguzārī*).[97] Along key bends of the river and points along the overland trading routes, merchants, pilgrims, and travelers were accustomed to paying a repeated number of small duties—most commonly known as *rahadārī* (highway tolls) and *zakāt* (contributions) in exchange for safety of their goods and persons.

The scarcity of grain and its poor distribution had made it clear to officials of the company that the Ganges and the vast network of waterways was essentially the pulse of life in the lower provinces of the valley. During the height of famine, Shitab Rai had been able to commission special boatloads of grain arriving daily from Banaras to relieve the starving population of the city of Patna and its environs, a measure that saved thousands of lives.[98] A similar measure in Murshidabad did not succeed because desperate fights broke out over boats transporting foodgrains, and armed militias of the elite at the various *ghāṭ*s wrested provisions away from the hands of the needy. Pursuant to the end of the Bengal famine, a new set of rules was passed to ease the passage of trade and commerce along the main waterways. District collectors were directed to stop all levies and tolls along the *ghāṭ*s of the Ganges and its branches on boats plying between markets stretching from Banaras in the west to Dhaka in the east. These rules overturned the ancient and time-honored rights of landlords and chieftains to claim a share from the profits of merchandise passing through their territorial jurisdictions. These petty and customary duties were replaced by a string of custom houses set up by the company, with major establishments in Calcutta, Hugli, Murshidabad, Dhaka, Patna, Manji, and Banaras to monitor and tax all internal trade as the exclusive prerogative of the company-state.[99]

This was not an easy task. For many years to follow, the outlawing of such privileges led to repeated conflict between company forces and local landlords along the banks and bends of the Ganges. Over time, however, the custom houses of the company, supported by police outposts, were able to exert greater authority over the passage of trade along the river, inspecting and stamping every boatload of goods after payment of dues. After 1781, when Hastings, having extracted large sums of money in lieu of troops from the raja of Banaras, Chait Singh, finally had him deposed, similar regulations of police and customs were extended along the course of the Ganges at Mirzapur, Banaras, Jaunpur, and Ghazipur. Here, as in the Bengal delta, such measures were obstructed by zamindars, chieftains, and merchants who stood to lose the most in terms of power and profit, leading to widespread incidents of defiance including armed confrontations. G. H. Barlow, who was sent by Governor-General Charles Cornwallis to inquire into the state of manufactures and commerce in Banaras in 1787 and to report back to the court of directors about prospects of investment in that province, found that there were still myriad restrictions placed on the free movement of commodities due to the residuum of petty dues and exactions levied on trade.[100] Once these impediments were removed, he wrote in one of his reports, "The navigation of the Ganges from the sea to its source would be freed from all obstructions, and the manufactures from Bengal and the exports from Europe would be transported to the heart of Hindustan at a trivial expense."[101] Toward the end of the eighteenth century, along the 400-mile stretch of the Ganges plains between the cities of Calcutta and Banaras, the company government enacted and pushed through a series of economic reforms to put an end to the ancient and hereditary claims of local regimes on the passage on goods and market exchange.

The overhaul of the entire system of land revenue collection in Bengal, Bihar, and Orissa has been the subject of extensive historical research, especially changes that took place in the 1790s during the tenure of Lord Cornwallis, who initiated a new regime of collection, validating by law the estates of the major *zamīndār*s of the Bengal presidency who were seen as amenable instruments in the creation of a colonial agrarian order.[102] This compelling vision of a new and improved landed society based on secure agricultural tenures and a propertied class of rural landlords, as historian Ranajit Guha argued in his definitive study on the subject, emerged from two decades of contentious debate among framers of the company's land-revenue policies.[103] With the passage of the Permanent Settlement in 1793, Cornwallis sought to reinstitute the *zamīndār*s of Bengal by stripping them of their old quasi-feudal privileges and inculcating the values of thrift, progress, and capitalism as exemplified by the rural gentry of England.[104] A parallel set of reforms was passed during the same period integrating the various species of internal revenue, which included the resumption of rights over the various internal marketplaces—weekly *hāṭ*s, residential *ganj*s and *qasbā*s, and markets for bulk produce such as *katrā*s, *manḍī*s and *golā*s—aimed at rescuing market exchange from what company policy makers saw as the predatory and corrupt influence of native landed elements and religious institutions.[105] Market duties were thus taken away from the traditional class of landlords and made a part of company government's taxation; only those who could provide documents from the Mughal era supporting their claims were compensated with small deductions in land revenue. Thousands of markets set up between 1765 and 1790 were declared unauthorized, and their levies resumed.[106]

Between 1790 and 1815, revenue collectors of the company took on the exhaustive and painstaking task of securing control of marketplaces, pilgrimages, and fairs along the roads and waterways following the course of the Ganges from Mirzapur and Banaras in the west to Calcutta and Sagar Island in the east. How effective these proved to be in the long run is subject to debate, and some historians of the period, notably C. A. Bayly, have argued that the company, despite its unquestioned political and military mastery, was still financially hamstrung and beleaguered, heavily dependent on native intermediaries and financiers, its dictates compromised by the direct and indirect resistance from the various ranks of indigenous society.[107] There is little doubt, however, that the Ganges valley came under the direct purview of the company government. Captain James Rennell's military sketch surveys conducted over a period of seven demanding years, during which he nearly lost his life in an encounter with Sanyasi rebels, led to the comprehensive *Bengal Atlas* of 1781. Rennell wanted to plot the precise route of the Ganges from the Himalayan foothills to the Bhagirathi delta, improving upon the work of Jesuit Father Tieffenthaler, who had visited Allahabad and Banaras in 1765 and prepared a gargantuan 15-foot-long chart of the river. Rennell, a military engineer and the first surveyor general of British India, wanted more fastidious details. He was fascinated with how after winding for almost 800 miles through the cavernous and rocky slopes of the Himalayas, the Ganges "issues forth a deity to the superstitious, yet gladdened, inhabitant of Hindoostan" taking the shape of a "smooth and navigable stream through delightful plains."[108] As a geographer, Rennell valued not only its alluvial bounty but its eminent navigability, especially as it served the interests of the company "in the capacity of a *military way*."[109] Rennell's *Atlas* yielded a commanding prospect of the Ganges and its surrounding territories. Its accompanying inventory of maps is testimony to the enormity of the task he undertook, attempting to track every meander along the course of the river, the elevation of its banks, the frequency and size of its seasonal islands and sandbanks, the forces of its

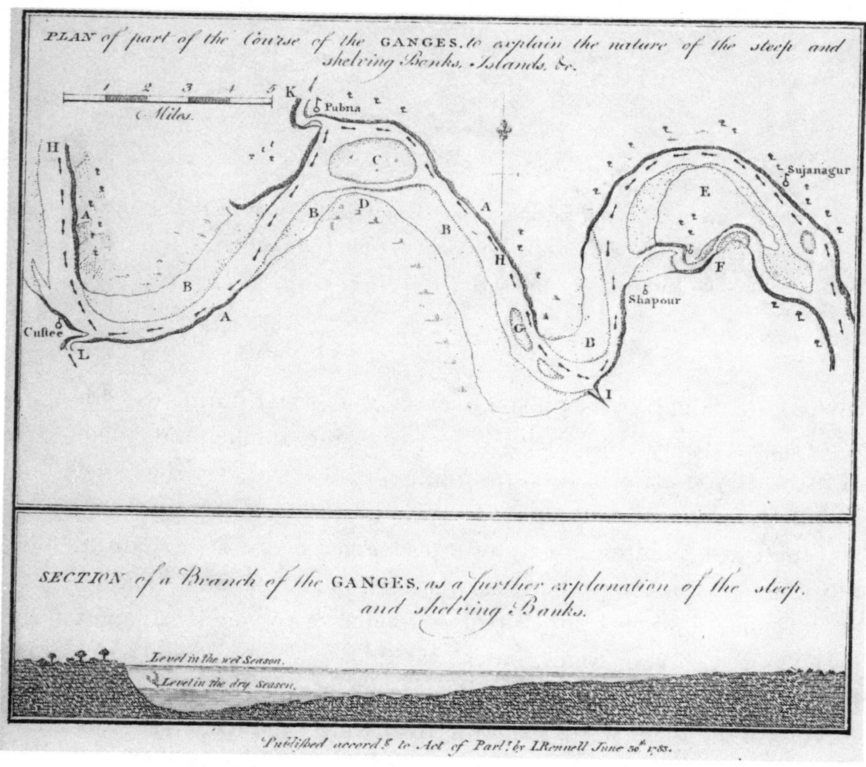

James Rennell, *Plan of Part of the Course of the Ganges* (1788).
From James Rennell, *Memoir of a Map of Hindoostan, or the Mogul Empire*, 3rd ed., London: W. Bulmer, 1793.

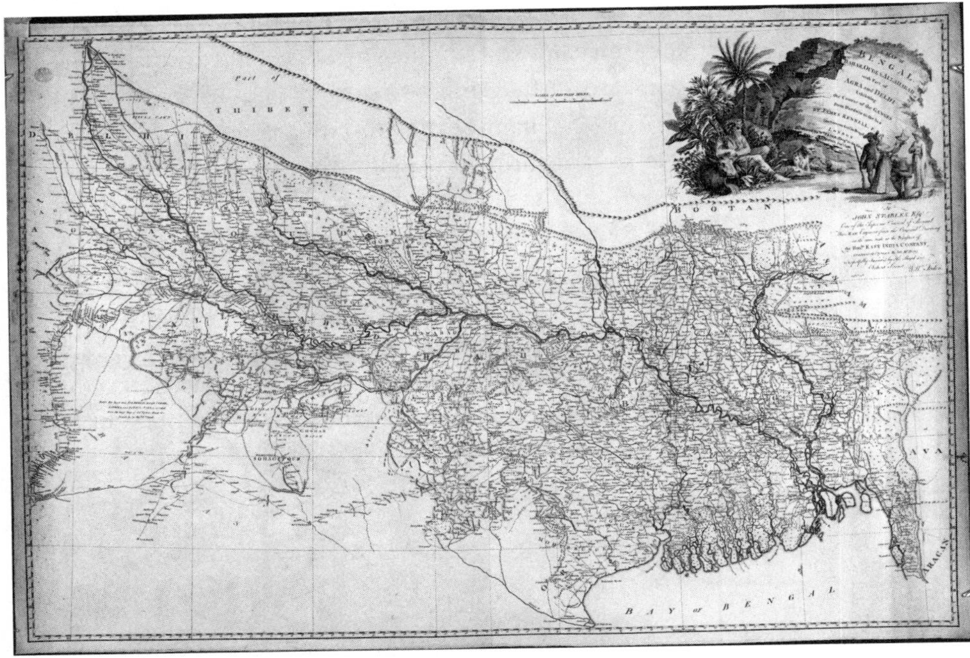

James Rennell, *A Map of Bengal, Bahar, Oude & Allahabad, with Part of Agra & Delhi, Exhibiting the Course of the Ganges from Hurdwar to the Sea* (1786). © The British Library Board 11/14/17. Maps 52335 (1).

current, the length of its gulfs, the levels of its water during the driest and wettest months, the soil quality of its banks, the number and convenience of its fords, the areas of its surrounding jungles, the size of its creeks and channels, the extent of its monsoon floods, the state of its inland navigation, and the exact quantities of water it discharged every season. His memoir, furnished with a comprehensive atlas, contains a detailed table of routes and distances upriver from Calcutta intended as a guide for inland navigation, along with 510 "water routes," taking into account the accessibility of ferries and fords during months of drought and monsoon floods.[110] Further surveys of the main stretches of the river were conducted by his assistants William Richards, Luis Felix de Gloss, and Henry Huygens. These charts, along with a compendious tabulation of cities, towns, and markets lying along the main routes between Calcutta, Murshidabad, Patna, and Dacca (Dhaka) in the *Description of the Roads in Bengal and Bahar* are proof of the visibility and ascendancy of the company across the great plains of northern India toward the end of the eighteenth century. During the course of the next century these passages would be further subdued and secured under the aegis of the Geographical Survey of India and the East Indian Railway Company, which laid down rails to connect Calcutta to Delhi through Rajmahal and Mirzapur.

A New Landscape

During the eighteenth century, with the unraveling of the Mughal administration, travel via the markets, towns, pilgrimages, quays, ferries, and roads along the Ganges—visited repeatedly by traveling armies—had become increasingly perilous. W. H. Sleeman's widely read *Report on the Depredations Committed by the Thug Gangs* of 1839, at the end of the company's long and bloody Thugee campaign, provides abundant details of the various denominations of stranglers and looters operating on the pilgrim routes along the river between Allahabad, Mirzapur, and Banaras, and farther north at the sacred precincts of Haridwar, serving large masses of pilgrims and traders—many of their victims carriers of holy Ganges water.[111]

Before law and order was reasserted by the East India Company, pilgrimages along the river were claimed by rival powers such as the Sikhs and the Marathas. Captain Thomas Hardwicke, who stopped at the fair of Haridwar on his journey to Srinagar, Kashmir, witnessed the clout of the Hindu ascetic and militant Gosains, wielding their swords to maintain a vigilant order at the bathing steps, collecting duties, imposing sanctions, and administering lashes to offenders.[112] When fourteen thousand Sikh horsemen arrived nearby for ablutions, and Sikh Udasi ascetics were sent to mark a spot by the river with their flag, the Gosains attacked them and stripped them of money and belongings. Hardwicke saw the Sikhs dispatching an agent to recover damages and negotiate for a portion of the riverbank for their rituals. The dispute, however, had not been fully resolved, and the Sikhs ultimately exacted their vengeance on the Gosains. The bloody carnage that followed not only wrecked the pilgrimage but left five hundred Hindu ascetics and fifty Sikhs dead.

The scene described reveals the continuing power and prestige invested in the patronage of fairs and pilgrimages along the Ganges. In Banaras, for example, where the great Rajput noblemen once underwrote the upkeep of the *ghāṭ*s and temples of Banaras during the reigns of Mughal emperors Akbar and

Shah Jahan, the Marathas now asserted their preeminence during the greater part of the eighteenth century, striking an alliance with the nawab of Awadh and lavishly endowing the riverfront of the city that had once sheltered their leader Shivaji with many of its lasting structures. As the eminent historian of Indian antiquity A. S. Altekar aptly remarked, modern-day Banaras was "largely a creation of the Marathas."[113] Peshva Bajirao I (1720–1740) endowed the Manikarnika and Dashashwamedh ghats, and Queen Ahilyabai of Indore built the present Vishwanath temple (1775–1776). The city flourished under the patronage of the rajas of Banaras, Balwant Singh and Chait Singh, until the kingdom was annexed in by 1782 by Warren Hastings for the East India Company. The patronage of Indian potentates continued even during the early period of company rule. Baijabai, the widow of Daulat Rao Scindia of Gwalior, built the pavilion around the holy well of Gyan Vapi in 1828, and Maharaja Ranjit Singh of Lahore donated the gold filature covering the spires and domes of the Kashi Vishwanath temple in 1839.[114] In 1765, as signatory to the Treaty of Allahabad ending the battle of Buxar, the East India Company acquired the confluence and pilgrimage of Prayag, which had been under the Mughal imperial charter and enjoyed the lavish patronage of heads of the Rajput princely states, the rana of Nepal, and kings and chieftains of Kathiawad and Kashmir.[115] Thus a vast concourse of trade and pilgrimage now came under the power and protection of the company raj.

Through its many custom houses the company attempted to keep a reliable account of the annual traffic of boats and the thirty-thousand-odd boaters plying their trade between the markets and ferries from Dhaka to Banaras. Having declared that the inhabitants of India must be allowed a free exercise of their religious practices, the company resumed and consolidated all major duties levied on pilgrims by Indian local rulers and petty landed elite, adding to both their coffer of taxes and their administrative burden, paying no heed to missionaries and other critics who felt that a Christian government should not be in the business of collecting revenue from the mass of poor pilgrims or interfere with the working of idolatrous Hindu charitable endowments.[116] As these passages became secure for travel, prospects along the Ganges began to attract the attention of missionaries and tourists from England and Europe, including romantic travelers and professional artists in search of novel vignettes of Indian antiquity inspired by contemporary ideas of the naturalistic sublime and the picturesque.[117]

Back in England, artistic tastes of the time were moving away from classical formalism to the more inadvertent pleasures of the picturesque, which was a new form of gentlemanly, artistic pursuit popularized by enthusiasts such as Uvedale Price and William Gilpin—an emotive depiction of the natural landscape that derived, as Christopher Hussey explained in his classic work on the subject, from the cultivation of "feeling through the eyes."[118] Just as they would in the English countryside, a traveler in search of the picturesque in India was expected to discern the aesthetically pleasing visual properties of the Indian rural and natural landscape, in which the River Ganges was one of the most eagerly sought subjects. From the late eighteenth century through the early Victorian era, amateurs and professionals armed with pen, pencil, sketchbook, and watercolors labored tirelessly on the idylls and prospects of the river. Their prints brought alive exotic imperial locales for viewers at home, contributing to a distinctly Gangetic view of England's northern Indian possessions. William Hodges, a draftsman from the South Pacific expedition, led by example, traveling up and down the river between 1780 and 1783, depicting ruins against quiet and contemplative backdrops of nature. Hodges's landscapes allude to a timeless arcadia where human subjects are diminished to enhance expansive and majestic

Thomas Daniell, *View from the Ruins of the Fort of Currah, on the River Ganges* (1803). © The British Library Board 11/14/17. Print. Shelfmark: X432/3(21).

views of the Ganges. His travelogues are littered with descriptions of the river flowing through the rocks of Kahalgaon (Colgong) in Bhagalpur or descending through the passes of the Rajmahal Hills, "meandering and glittering through an immense plain, highly cultivated, as far as the extent of the horizon, where the eye is almost at a loss to discriminate the termination of sky and land."[119] Sailing from Munger to Calcutta, Hodges describes how the "inexpressible grandeur" of the Ganges filled his imagination, with its perennial fleet of boats, its crumbling ancient temples with their flights of steps leading to the water, along with its woods, forests, and hills reminiscent of the verdant parks of Old England.[120]

From 1786 to 1794, Thomas Daniell and his nephew William, following in the footsteps of Hodges, scoured the breadth of the river from the foothills of the Himalayas to Calcutta, using the camera obscura to great effect for the accuracy of scale and perspective. Their aquatints and engravings resulted in lavish and costly publications such as the series *Oriental Scenery*, released in six parts between 1795 and 1808, and *A Picturesque Voyage to India*, published in 1810.[121] Along with natural landscapes, the Daniells also had a keen eye for ruins of ancient buildings. They seized the opportunities presented by the military exploits of the company to travel to the interior parts of India and indulge in what they saw as the "guiltless spoliations" of artists capturing the beautiful and the picturesque aspects of India and taking them back to Europe as trophies.[122] The Ganges was often at the center of such landscapes, not just in albums but also in travelogues and romances, especially in popular accounts written by women such as Fanny Parks, Maria Graham, and Emma Roberts. Roberts in particular wrote evocatively about native life unfolding by the banks of the river. She found the old city of Patna particularly appealing, with its tall riverfront *haveli*s, its giant fig trees, and the remains of its "Gothic" gateway in dark red stone redolent of feudal Europe. The river here, brimming to its banks, became a mirror for the turrets, spires, and domes of the temples and mosques, making an "exceedingly imposing" coup d'oeil.[123] And there was also the timeless Banaras, the spirit of which she thought could not be adequately conveyed in writing. Only paintings—as the Daniells had shown in their aquatints and engravings—could convey its unique blend of the beautiful and the grotesque. The confused mass of the structures of Banaras that made up the "stupendous wall" spreading along the bank of the Ganges captured for Roberts an element of Gothic wonder in the alchemy of nature and art in India.[124] Seen

Bishessur Nath Temple, Benares. Photograph by John Murray (1858). © The British Library Board 11/14/17. Shelfmark: Photo 52/(42).

through the romantic tint of antiquarianism, such riparian vistas offered succeeding generations of Britons who visited, worked, and died in India a glimpse of something akin to an imperial sublime, with the Ganges and the land through which it flowed viewed as a tribute to a timeless India.[125] A popular traveler's guide in its rapt musings on the panorama of palaces, temples, mosques, pinnacles, minarets, shrines, and canopies declared that Banaras was "without question the most picturesque city in India."[126]

The mutiny and upheavals of 1857, which saw some of the bloodiest episodes of carnage in cities and villages along the length of the upper course of the Ganges, including the much-reviled Cawnpore Massacre in which a mob attacked British women and children, could not in the end dispel this long-term view of the river. It is fascinating to see how the first photographs of Banaras, taken by Dr. John Murray, a British surgeon, shortly after the uprisings of 1857, tried to recapture this very ambience of contemplative serenity.[127]

Steamships, Canals, Railroads

Coal and steam changed the experience of travel along the Ganges beyond recognition. Robert Fulton's steam ferry had begun to brave the Hudson River in 1811, and within a decade the company proposed similar steamboats for the Ganges capable of maneuvering its capricious bends, shallows, and banks. In 1828 the very first steamer, *Hooghly*, went up from Calcutta to Allahabad, able to reach a speed of four knots against the tide. Throngs of natives filled the banks of the Jumna to catch a glimpse of the boat moving without men or oars, causing "great astonishment," in the words of travel writer Fanny Parks.[128] Within a few years the river would be abuzz with deep-hulled steamers trailed by draft tugs, busily carrying officers, troops, and horses to the various cantonments, followed by regular passenger service connecting the cities and towns along the course of the river and its navigable tributaries.

During the first half of the nineteenth century, India under British rule had become a testing ground for some of the most pioneering advances made during the Industrial Revolution. With the advent of steam, the company government turned its attention to the greater Ganges valley and the Ganga-Yamuna Doab to extend irrigation, navigation, and investment capital by taking over the remnants of Mughal-era aqueducts, causeways, reservoirs, and bunds, and reengineering an entire system of canals. They took over, for example, Muhammad Abu Khan's canal from the Ganges north and northwest of the city of Meerut, which ran only for twelve and a half miles and was not more than fifteen feet wide, and also the old Doab, or Eastern Jumna, Canal near Meerut.[129] Such old waterways were seen as having little benefit to poor cultivators whose fields often were placed at a higher elevation. A reliable system of canals was a key to the irrigation required for lucrative crops produced in bulk such as sugarcane and rice. A number of proposals had already appeared on the desk of Governor-General William Bentinck, who was contending with a spate of violent monsoon floods. As Sir Proby T. Cautley, engineer and early architect of the Ganges Canal project, took up the post of supervisor of canals in Rohilkhand and Delhi during 1837–1838, a devastating famine raged across the Central Provinces and the lower districts of the Doab, and the company government became increasingly wary of the rising cost of famine relief for its starving subjects.

Famine was a visible object, but the prospect of capital and investment in a regenerated valley struck the industrial ambition of the company-state, and soon the work of excavation started right next to one of the pilgrim spots, Har-ki Pauri ("steps of Shiva"), near the holy city of Haridwar in the Himalayan foothills. Colonial engineers aided by an army of cheap native laborers dug through the centuries-old alluvium of the floodplain (*khaddar*), charting a course from Kankhal to Kanpur, 325 miles downstream.[130] They set up gauges to measure the discharge of the Ganges at Haridwar, Garhmukteshwar, Fatehgarh, Kanpur, and Allahabad.[131] The object of the great Ganges Canal was to connect with the main river jetties of Allahabad, opening up the entire waterway to steamer traffic. The East India Company saw canals as a means to ameliorate the impact of famines, which in India, much like in Ireland, were seen not so much a result of the scarcity of produce as the pitiable purchasing power of the masses living at the edge of subsistence. The Ganges Canal, instrumental in opening up the traffic of boats and steamers, was also seen as the conduit through which grains and foodstuffs from Mirzapur, Banaras, and Calcutta could be brought to the inland towns and bazaars of the greater Doab.[132]

A new faith was placed on the ability of machinery to harness the power and energy of the Ganges. Canals would employ the starving poor in the undertaking and maintenance of masonry and earthworks.[133] The canal authorities also took over sacred sites, promising to ease the difficulties of millions of poor pilgrims beholden to their ritual ablutions, especially during spates and droughts, and to keep the bathing ghats of pilgrimages such as Haridwar safe for traffic. The canal was inaugurated on April 8, 1854 at Roorkee, the site of the first college dedicated to engineering in British India, which was now the new headquarters of the Ganges Canal project. Charles Eliot Norton, writing for the *North American Review*, was greatly impressed by the grand opening, writing that the word had now spread all across India and pilgrims were arriving from distant parts of the country to see for themselves that "the revered Ganges was about to leave her ancient and hallowed channel for one formed for her by the hand of strangers."[134] Lieutenant-Governor John Colvin, in his inaugural toast lauding Colonel Cautley, declared that the British had finally left a "permanent mark on the soil of India to attest the power, the wealth, and the munificence of their nation."[135] The British in India thus became the architects and custodians of one of the largest canal systems of the world, which with its main channel, branches, and terminal lines totaled 898 miles. Contemporary engineers like J. S. Beresford saw the canal as a "machine composed of many parts" that connected canals, distributaries, rural watercourses, and individual

Head of Ganges Canal, Hardwar (Haridwar). Photograph by Samuel Bourne (1860). © The British Library Board 11/14/17. Shelfmark: Photo 15/1(28).

cultivators.[136] Beresford also noted that the success of irrigation lay not just in the nature of absorption of water into the soil but also in the industrious "habits of the cultivators," and for the successful implementation of engineering plans he stressed the importance of the acquisition and use of the detailed village maps of India.[137] The irrigation engineer, as Anthony Acciavatti points out in his comprehensive study of the hydrological past of the river, was thus thrust into the prime position of the architect of the colonial rural landscape.[138]

Canals, aqueducts, and causeways transformed the waterscape of the Ganges. The introduction of the railways, with culverts, masonry brick, and iron bridges, changed the views of the river forever. R. Baird Smith, the chief engineer and a main architect of the siege of Delhi during the mutiny of 1857, later noted that the orderly network of canals and railways had redrawn the surface of northern India, removing the "unsightly marshes and swampy tracts of land," purifying the "air they taint," and renewing the promise of cultivation in the Ganges basin.[139] The railways reinforced the desire to reorder the industrial landscape of the valley, its culverts, level crossings, and masonry bridges working in tandem with locks, escapes, and viaducts of the Ganges and Jumna Canals. The Haridwar branch of the northerly extension of the Oude and Rohilkund Railways constructed a bridge over the Ganges at Balawali in 1887, an impressive feat of imperial engineering.[140] The Dufferin Bridge across the Ganges at Banaras followed, undertaken by the Oude and Rohilkund Guaranteed Railway in 1881–1887, which, anticipating the erosive power of the river at the height of the monsoons, erected piers that were 140 feet below the lowest water level and still stand as some of the deepest in the world.[141] The Jubilee Bridge was built between 1883 and 1887, carrying the East Indian Railways over the lower Hugli, 28 miles north of Calcutta.[142] Such constructions, proclaiming the triumph of steam and iron, bear witness to the might of Britain as an industrial powerhouse, which imposed its will on the relationship between her imperial subjects and their beloved and venerated river. To the ritual almanacs of Hindu pilgrims traveling to Banaras, Allahabad, or Haridwar was added the railway timetables of Bradshaw and Company, which also published the annual *Railway Manual, Shareholders Guide and Official Directory*. Such speed and facility of travel, reducing the sense of distance between sojourner and river, quickened the flux of historical time. Somewhere through the long apprenticeship of culture under the watchful eyes of a modern industrial empire, the Ganges began to recede from its living, universal, empyrean presence, and resemble—to borrow the words of India's first prime minister, Jawaharlal Nehru—"a symbol and memory of the past of India."[143]

EPILOGUE

THE TWO BODIES OF THE RIVER

The body of the Ganges in certain respects reminds us of the king's two bodies celebrated in traditional European political theology.[1] One is undying, everlasting, and the other vulnerable—flesh, blood, and bone. The bodies of gods and goddesses in Indian mythology and popular culture often exhibit a strangely similar duality, both playful and brutal. Krishna's final exit in the *Mahabharata* in his human, mortal form, mistakenly struck in the foot by an arrow from the quiver of the hunter Jara while asleep, is a well-known instance—especially as *jarā* is also a name for old age. Ganga, the river goddess, is likewise endowed with a twofold character: one as the immaculate and eternal deity of the flowing waters, the other as a mundane river, repository of accumulated human misdeeds. During her descent to the world, an anxious Ganga asks Bhagiratha whether if she appears on earth—the exact passage in the *Bhagavata Purana* uses a word derived from the root *mṛja*, to wipe, cleanse, purify—would not people wash off all their sins in her waters? And then, where would she go to cleanse *herself* of such accumulated poison?[2] Bhagiratha gives an unexpected reply. The purest and the most upright of all mortals who have forsworn all action and thought of gain, and have conquered all forms of sensory desire, he says, also have the power to expiate the sins of this world. Vishnu himself, the destroyer of sins, dwells within them. Along with the wicked and corrupt, they too shall bathe in the Ganges, and their bodies touching the waters of the river will wipe out her burden of the world's collective sin. The river of last resort, refuge of the wretched of the earth, is thus not immune to the mounting burden of human depravity and needs the sacrifices of the virtuous to be able to carry out her unenviable task. In our times, the river's second body is not just a receptacle for the sins of the unredeemed; it is also a vast depository of toxic waste.

In this brief epilogue, I want to return to this eternal refrain of the pure and the polluted plainly manifest in the myth and history surrounding the Ganges, its sacred confluences, and its forbidding high-Himalayan pilgrimages described in the first chapter of this book. This beguiling duality has exerted a profound and fundamental effect on the way Indic culture, the Indian bureaucratic nation-state, and the millions of people of India, especially Hindus, have approached the question of the use and custodianship of the river and its water. The river's two opposing bodies have allowed it in a strange way to be worshipped and neglected at the same time, regardless of its worth as a finite and tangible resource that has been held in an unstated and common trust for thousands of years. Consider these two well-known verses of the fifteenth-century weaver, songwriter, and radical mystic Kabir, who was born and spent his life in the sacred city of Varanasi, that offer a tribute to the waters of the Ganges.[3] In the first couplet Kabir likens a pure state of the mind to the undisturbed waters of the Ganges:

Kabīrā mana nirmala bhayā, jaise gaṅgā nīra
Pāche lāgā hari phirai, kahata "Kabīra Kabīra"

Kabir says I have made my mind pure like the waters of the Ganges
Lord Hari now runs after me, calling: "Kabir, Kabir"

In another couplet he laughs at the gullibility of the common folk:

Kabīrā khāyi koṭa kī, pānī na pīvai koi
Jāye mile jaba gaṅgā men, gaṅgodaka hoi[4]

Kabir says, no one drinks water from the roadside gutter
But when it mingles in the Ganges, it becomes Ganges water

In the first instance Kabir insinuates that purity itself is a matter of attitude, a state of mind. In the second, in his signature upside-down (*ulaṭbāṃsī*) language—what Linda Hess has called the language of "enigma and paradox"—he undermines the simple, misplaced faith in the physical characteristics of water itself.[5] Kabir's barbed examples outline an important paradox: where does the purity of the Ganges reside except in the heart and mind of the believer? Kabir says bluntly in one of his songs: "*tīrtha*s are nothing but water [*saba pānīhai*], I have bathed and seen no such results."[6]

Such rhetoric of feigned puzzlement is common in the songs and verses of the radical devotional ferment that swept across India of the late Sultanate period, led by figures such as Nanak, Raidas, Dadu, and Kabir, all of whom questioned the blind following of prescriptive rituals in normative and orthodox forms of Hindu worship and piety. In the body of legends surrounding Nanak's life and teachings recorded in the Adi Granth, a similar lesson is taken about the folly of believing that simply bathing in the Ganges makes one pure. When Nanak visited pilgrimages on the River Ganges, he saw tens of thousands of people bathing and offering Ganges water to their forefathers.[7] Nanak did not think that this solemn act was creating any lasting effect on their state of being. Nanak also got into the water, and while people were worshipping with their faces turned to the direction of the rising sun, he proceeded to sprinkle water in the opposite way. Surprised to see this, the people asked Nanak why he was acting in such a strange manner. Nanak asked them, whom were they casting Ganges water toward? They replied that it was for their forefathers, whose departed souls were in heaven. Nanak said that he too was watering the field near his house in the village of Kartarpur in the Punjab. If your offerings can reach your ancestors in heaven so far away, Nanak asked, why could not these waters reach my field in the Punjab?

The litany of references to the Ganges in the teachings of the Indian revival movements are further testament to the iconic status the river has occupied in the popular culture of the Indian subcontinent, especially its ideas of purity and pollution, cleanliness and filth, sanitation and contamination. They are still very much part of the everyday *habitus*, attitude, and worldview within which present-day scientific knowledge dedicated to the long-term health and survival of the holy river—from the related fields of hydrology, irrigation, chemistry, forestry, environmental studies, and conservation—are received. Decades before India's official independence, the Ganges had already secured its place as the national river of the new nation. Iqbal in his *Song of India* (*Tarana-i-Hind*), written in Lahore in 1903, had asked the river whether she remembered the day when the first caravans of his people, presumably Muslims from afar, alighted on her shores. The poet Akbar Allahabadi in his patriotic tribute in Urdu paid to Mahatma Gandhi in the *Gandhi Nama*, imagined that the holy waters of the well of Zamzam in Mecca, once locked in bottles, had now been freed to mingle with the torrents of the mighty Ganges.[8]

Rabindranath Tagore put both the Yamuna and the Ganges in his patriotic reverie *Bharata Vidhata*, penned in 1911, a part of which is the present-day Indian national anthem. Jawaharlal Nehru in his magisterial *Discovery of India* wrote that the story of the Ganges was indeed the "story of India's civilization and culture."[9] In his *Will and Testament* written in June 1954, Nehru asked that he be allowed to pay his homage to the icon of India's cultural inheritance by having a handful of ashes from his cremated body cast into the Ganges near his childhood home at Allahabad, to be carried by the great river to the ocean.[10] While the river was seen unquestionably as the living embodiment of the nation's past, it was also seen as a natural conduit of technological progress and economic development. Despite Mahatma Gandhi's deep suspicion of technology as the portent of human arrogance, the Working Committee of the All India National Congress recommended that future members of the party running for elections in the provinces consider a committee of experts who would look into "comprehensive river surveys" as the first step toward national reconstruction and planning, and to outline plans for the prevention of floods, expansion of irrigation, arresting of soil erosion, eradication of malaria and other diseases, and development of hydroelectric power through the construction of dams and reservoirs.[11] Today, the discharge of the Ganges is impeded by two of the most disputed dams in India. One is the Farakka Barrage at the point where the eastern branch of the river enters Bangladesh, originally intended for irrigation and flood control, which has caused decades-long strife between India and Bangladesh over the sharing of water as a vital resource. Since the early 1970s India has defended the viability of Farakka as a safeguard against excessive siltation that has progressively diminished the navigability of the river between Hugli and Allahabad, and the fading prospect of ships of significant tonnage entering the port of Calcutta.[12] Bangladesh, on the other hand, has maintained that the distribution of waters from Farakka is weighted unfairly in India's favor, and it has worsened the effects of flooding, contributing substantially to the degradation of the environment and diminution of the natural habitat for species in the Ganga-Brahmaputra delta. Years of bilateral talks, and the appointment of a Joint Rivers Commission, have not been successful in reaching a mutually acceptable compromise.[13]

Much closer to the sacred sites on the Ganges in the lower ranges of the Garhwal Himalayas, which now comprise the relatively young state of Uttarkhand, the river has been tamed by the high Tehri Dam near the site of its confluence with the Bhilangana River. First proposed in 1949, the project was sanctioned in 1972 and met stiff resistance from activists and local hill folk who feared that it would submerge the town of Tehri along with a hundred villages and displace thousands of people.[14] The construction of the much-vaunted second-highest dam in Asia thus became mired in controversy, provoking fears about serious environmental degradation in the Himalayas.[15] A flood of petitions and public-interest lawsuits followed, resulting in the Anti-Tehri-Dam Committee, which fought the government of India in local and national courts, citing the devastating convergence of geological, seismological, technical, environmental, social, economic, and humanitarian setbacks and casting serious doubts over the purported longevity of the dam; arguing that the exposure of erodible rocks and mounting debris would suffocate the reservoir in the long run.[16]

A brief history of the making of the Tehri Dam is instructive, because it shows the stark contrast between the desire to preserve the landscape of the ancient pilgrim and mountain trails and the lure of electricity and irrigation. The plans were drawn up between 1961 and 1972, and construction began in 1978. The first part of the project was finally completed in 2007, a second phase in Koteshwar has been recently finished, and the project is still under way. This is one of the tallest dams in the world (over 855 feet), crafted out of rock and gravel embankment, intended for flood control, irrigation, and municipal water supply for the pilgrim towns of Hrishikesh and Haridwar, and also meant to power the Tehri Hydro Power Complex with a capacity of 2,400 megawatts. It is also intended to revive the canal system below Haridwar, meeting the insatiable demands of the rapidly expanding outskirts of India's capital. How does a dam take almost a lifetime to build? This stalling of the project for decades can be seen as a tribute to the resiliency of the environmental movement in India. The Tehri Dam had drawn opposition from a number of local Garhwali hill communities from the very beginning, and in the late 1970s it widened into a powerful movement for natural and human rights, gathering further momentum from the Chipko tree-saving struggles of 1978.[17] It was led by the Anti-Tehri-Dam Struggle Association and took the form of rallies, demonstrations, and hunger agitations inspired by the leaders of the Chipko, many of whom were elderly women. They raised their collective voice against the loss of forest, the submergence of land, and the forcible relocation of almost a hundred thousand people. As I finish writing this book, the legal battles over resettlement rights are still being fought in the Supreme Court of India.

The Tehri agitation catapulted activist Sundarlal Bahuguna to the national stage, whose message to his followers was that the dam will destroy the mountain and kill the river, just as the Dakpathar barrage had strangled the Jumna.[18] Bahuguna's struggle resonated with the people of the northern hills and a large section of the less-educated Indian population because the refrain of his movement was simple. The people of India had a *sacred duty* to protect the rivers, mountains, and forests of the Himalayas.[19] One of the primary architects of the legendary Chipko movement, Bahuguna was a follower of the environmentalist Richard St. Barbe Baker and Mahatma Gandhi.[20] He is remembered for his hunger strikes—forty-nine days in 1995 and seventy-five in 1996—and his moving, personal letters to Prime Minister Indira Gandhi imploring her to halt the project. The Tehri movement affected the momentum of the constriction of the dam and drove wedges between different branches of the Indian bureaucracy, complicated by the creation of a new department of the environment. The Supreme Court in 1990 had ruled against the Tehri activists and their claim that the construction of the dam violated Article 21 of the Indian constitution on the right to life, but the agitation persisted. A devastating earthquake in Uttarkashi in 1991 near the dam area raised further awareness of the grave dangers of building dams such as Tehri within zones prone to seismic activity that, in the event of a breach, had the force to drown 6.2 million acres along the Bhagirathi Valley.[21]

A national review committee set up in 1995 to examine the ecological impact of the Tehri Dam stalled its construction for yet another decade, adding further fuel to a national debate that pitted engineers and technocrats against villagers, devout Hindus, Indian leftists, and environmental militants. The Tehri Dam movement was also a beacon for the struggle ahead for the rightful place of a river in the future of the nation's economy and environment. Some scientists have sounded the alarm that the Tehri Dam, added to the minor barrages in Bijnor, Narora, and Kanpur, has so accelerated the siltation rate of the Ganges that its life span is limited to a meager forty to fifty years. Already, the confluence of the Bhagirathi and its tributary the Bhilangana has shifted because of alluvial fans to a distance of 150 meters downstream. Large locked water bodies close to areas with a history of frequent earthquakes, many environmental activists argue, are human and ecological disasters waiting to happen. We have interfered too much with the free-flowing action of the great river, they say, and it can no longer mitigate the effects of sudden tear-faults or landslides.

Environmental historian Richard White, as noted earlier, has likened the Columbia River that flows through Washington and Oregon in the Pacific Northwest to a giant organic machine, which reminds us that throughout much of history human labor and ingenuity have bound the mutual fate of humans and rivers.[22] White shows how the advent of steam, mechanization, irrigation, and industry helped advance the idea that nature itself was a machine that could be harnessed and put to human use. A comparable case can be made about the British in mid-nineteenth-century northern India, who wanted to control and redirect the waters of the Ganges through a busy network of canals. A founding father of the project, the East India Company army engineer Richard Baird Smith, called it a "great work for the good of the people."[23] The colonial irrigation projects, in retrospect, exerted a profound influence on the political and economic mind-set of the young, independent Indian nation-state. The Ganges valley and its water system became a touchstone for the early framers of economic planning and development, spurred by profound anxieties over the metrics of growth and development. The Nehruvian state had always pinned its hopes on experts, economists, and technocrats and the development of multipurpose river valley projects—as every Indian schoolchild of the sixties and seventies would have known from their geography and social studies classes—became a virtual fixation for the Indian Planning Commission in charge of India's future and industrial progress.[24] By the 1970s the battle cry for self-sufficiency in foodgrain production, which accelerated the pace of research on genetically modified, high-yielding varieties of rice and wheat, also led to a pervasive concern with underdeveloped water resources in the great drainage basin of the Ganges. The harbingers of economic reform pointed their fingers at the never-ending trap of poverty and bare subsistence, a result ultimately of the abject dependence of Indian cultivators on the caprice of monsoon rains that ruled the chances of flood and drought. In this context of developmental crisis, the Ganges basin became part of a nationwide hydrological mission. Experts in agricultural science and irrigation engineering warned that poverty, ignorance, and stubborn cultural and social habits were stymieing agricultural output and the optimal use of water resources. The chief object of the new "water machine," as a seminal article proclaimed in 1975, was to "break the chains of tradition and injustice."[25]

Architect and designer Anthony Acciavatti, who has studied the long-term effects of hydrological projects along the Ganges River basin in great detail, calls it "one of the most engineered spaces on the planet," where the manipulation of rivers, canals, tanks, lakes, pumps, and tube wells has transformed the surface and infrastructure of the valley and pushed it to the limits of sustainable practice. It resembles a gargantuan hydrological engine that somehow still sustains almost 40 percent of the nation's billion-plus population. A reminder of this obsession with water is the ubiquitous tube well, which can be as simple as a metal or plastic pipe fitted with a diesel- or electric-powered pump that can tap groundwater. Ever since the 1930s, tube wells have been officially encouraged across India to offset the poor distribution of water. While this may have contributed to the explosion of foodgrain production in the state of Uttar Pradesh, which constitutes about 20 percent of India's entire output, indiscriminate sinking of over 3.5 million private wells, combined with the distribution of plots, fragmented capital, and minimal regulation, have led to a mounting crisis in groundwater contamination and unexpected increases in the levels of fluoride and arsenic.

There are many instances in human history of faith placed in the ultimate efficacy of technology. The story of the settling of North America and the conversion of wilderness to pastures, with machines augmenting the work of hands in a peaceful partnership with nature, was central to Leo Marx's notion of the "pastoral ideal" and his compelling argument that the inroads of industrial capital precipitated a unique and idealized view of nature and the outdoors.[26] The settlement, exploitation, and overcrowding of the Ganges valley is a comparable case. Over centuries, this watercourse has generated a surplus that not only supported the great agrarian empires of the distant past but also underwrote the mighty industrial empire of Victorian Britain. Such mounting extraction took place at the expense not only of forest and scrub but of millions of peasants who starved and perished from famines. This is what the first generation of economic nationalists such as Dadabhai Naoroji and R. C. Dutt wanted to hold the British accountable for. For them, the technological benefits of empire, such as steam and coal, railways and canals, did not in sum outweigh their human cost. The flush of independence, with Nehru at the helm, a socialist who believed in the public good of a bold, technocratic state, lifted old misgivings toward

Western machines. Just as it was during the decades of five-year economic plans, India in the era of global, neoliberal capital seems to be still captivated by the idea that technological progress is a prophylactic that can somehow eradicate structural inequity and the scarcity of resources. It is not surprising that industrialists, policy makers, and urban planners believe that with the implementation of the right technology India's national river too can be cleansed and restored.

The Ganges brings water to half a billion people across twenty-nine major cities for drinking, irrigation, and sewage. It is also a river that carries an astonishing burden of contaminants. In the most dense industrial belt along the middle Ganges plains, between the cities of Kanpur, Allahabad, Varanasi, and Patna, tanneries, chemical plants, textile mills, distilleries, slaughterhouses, and hospitals discharge their organic and nonbiodegradable industrial effluents—including high levels of chromium and mercury—that render the river dangerously toxic. Cities along the Ganges generate 2.7 billion liters of raw sewage daily, of which nearly two thirds enter the river untreated.[27] Pathogenic analysis of bacteria of the coliform group in places presents an equally daunting picture. A study sponsored by the Uttarkhand Environmental Protection and Pollution Control Board between 1996 and 2006 found that the presence of coliform bacteria in Ganges water is at a level that is considered entirely unsafe for agricultural irrigation, let alone drinking or bathing. Consider such indices in the context of the auspicious winter solstice at the Kumbh gathering in Prayag at the confluence of the Yamuna and the Ganges, where upward of sixteen million people congregate and bathe next to each other.

The idea of a sanitized national river is thus the focus of both national policy and electoral capital. Most political parties in India, especially those who profess allegiance to a fundamental Hindu tradition, have at one time or another pledged that they would try their utmost to preserve the river as their national and maternal icon. The cleansing and the purification of the Ganges has been an urgent and much-vaunted national priority for a very long time. Prime Minister Indira Gandhi, who noted that the river from time immemorial "watered and nurtured an entire civilization," was moved by the concern that the Ganges had become contaminated and needed a plan for its restoration.[28] Gandhi wanted to identify sections of the river that were the most affected by urban and industrial effluence and sections that were the most "pristine."[29] In 1980, she asked the Central Pollution Control Board to prepare a report on the pollutants of the Ganges. They came back with a report in 1984 that stated that three fourths of the Ganges's pollution burden was untreated municipal sewage.[30] After Indira Gandhi's assassination, the idea was passed on to her son and successor, Prime Minister Rajiv Gandhi, who launched the Ganga Action Plan (GAP) in January 1986, aimed at reducing the pollution load on the river. These efforts proved largely unsuccessful, even after the expenditure of over 1 billion USD. The plan was officially withdrawn in 2000, and a postmortem was done by the National River Conservation Authority.[31] Another body, the National River Ganga Basin Authority, was established in 2009 under Section 3 of the Environment Protection Act of 1986, officially declaring the Ganges as the "National River" of India, whose drainage basin must be protected. From public-interest litigators to urban ecologists, an unprecedented range and number of experts in India have studied these problems, and they point to a number of stubborn obstacles—corruption, population density, failed sewage plants, lack of public awareness, and red tape—that stand in the way of tangible progress toward a cleaner and more healthy river.

Grand schemes to improve the living conditions of the poor masses undertaken by both colonial and postcolonial states, as James Scott has pointed out, often end up in failure. As top-down exercises they tend to simplify complex and intractable local realities, making them misleadingly obvious for the purposes of governance. Scott argues that this is a function of the "ideology of high modernism," a misplaced devotion to science and technology.[32] Developmental regimes participate in such schemes because they also provide an overarching legitimacy for the mobilization of resources and labor. Most plans devised for the control of pollution, flooding, and drought in the Ganges basin have been victims of their own ambitious scope and unspecified responsibilities. Projects aimed at "cleaning" the river run into the problem of inertia of norms and practices, habits of wasteful consumption, and the vexing problems of everyday realities on the ground. The antipollution drive, for instance, has pitted the government against the poor fishing and corpse-handling untouchable communities around Banaras. More important, as noted repeatedly in this book, for a multitude of Indians, the mythology and deep history of the Ganges makes it difficult to accept that the river may be in imminent danger or that its waters are as contaminated as they are sacred.

Will the Ganges survive its burden of human and industrial contaminants? Will dams and barrages strangle its flow one day with an unbearable burden of slit and detritus? Will it go the way of the great Yellow River of

China, which dried up in 1997, at a staggering distance of 400 miles inland from the delta, sacrificed in the pursuit of industrial progress and in the name of modernity? The physical death of the most cherished river of India would be unthinkable for most people in India, who, despite the evidence of its endangered environment and ecology, still find solace in the idea of the Ganges as the maternal spirit of their civilization. The river, with its water and its valley that have sustained the imaginative life, material culture, and daily subsistence of millions of inhabitants of the subcontinent over so many centuries, is now—alas—facing its most daunting challenges.

NOTES

Introduction

1. M. R. Dhital, *Geology of the Nepal Himalaya: Regional Perspective of the Classic Collided Orogen*, Cham: Springer, 2015, p. 288.
2. Ananda K. Coomaraswamy, *Time and Eternity*, Ascona: Artibus Asiae, 1947, p. 15.
3. Some of the earliest accounts of the legend associate the mythical fish (*matsya*), savior of the world during the great deluge, with Prajapati, an older god of creation, sometimes seen as a version of Brahma. The Puranas, however, see the primordial fish as an avatar of Vishnu. See Sri Vyasadeva, *Matsya Purana*, Lucknow, India: Naval Kishore, 1892, p. 3.
4. Xuanzang, *Si-yu-ki: Buddhist Records of the Western World*, translated by Samuel Beal, vol. 1, Boston: J. R. Osgood, 1885, p. 232.
5. Harisadhana Mukhopadhyaya, *Kalikata Sekalera o Ekalera*, reprint of 3rd ed., 1985, Calcutta: P. M. Bagchi, [1915] 1991, p. 83.
6. M. V. Krishna Rao, *The Gangas of Talkad*, Madras: B. G. Paul, 1936, p. 70.
7. See the Udayendiram Plates of Prithivipati II in E. Hultzsch, *Tamil and Sanskrit [Inscriptions] from Stone and Copper-Plate Edicts at Mamallapuram, Kanchipuram, in*

the North Arcot District, and Other Parts of the Madras Presidency, Madras: Superintendent, Government Press, 1890, p. 387.

8. Narendra Nath Sarma, *Panditaraja Jagannatha: The Renowned Sanskrit Poet of Medieval India*, Delhi: Mittal, 1994, p. 5.
9. Eknath Vishnu Dadape, ed., *Bhaminivilasa of Jagannath Pandit*, Delhi: Motilal Banarsidass, 1994, pp. 5–6.
10. Swami Agehananda Bharati, *The Ochre Robe*, London: Allen and Unwin, 1961, pp. 153–154.
11. R. E. Hume, *The Thirteen Principal Upanishads*, London: Oxford University Press, 1921, p. 389.
12. *The Anguttara-Nikaya: Part II, Catukka Nipata*, ed. Richard Morris, reprint, London: Pali Text Society, 1976, p. 140.
13. David Gordon White, *The Alchemical Body: Siddha Traditions in Medieval India*, Chicago: University of Chicago Press, 1996, pp. 225–226.
14. David A. Pietz, *The Yellow River: The Problem of Water in Modern China*, Cambridge, MA: Harvard University Press, 2015, pp. 2–3.
15. Ibid., pp. 16–17.
16. Philip Ball, *Water Kingdom: A Secret History of China*, Chicago: University of Chicago Press, 2017, p. 5.
17. Rachel S. Havrelock, *River Jordan: The Mythology of a Dividing Line*, Chicago: University of Chicago Press, 2011, p. 26.
18. Bronislow Malinowski, *Magic, Science and Religion, and Other Essays*, Boston: Beacon Press, 1948, p. 78.
19. Nigel J. Thrift, *Spatial Formations*, London: Sage, 1996, p. 4.
20. Eduardo Kohn, *How Forests Think: Toward an Anthropology Beyond the Human*, Berkeley: University of California Press, 2013, pp. 6–7.
21. Eric Voegelin, *Order and History*, Columbia: University of Missouri Press, 2001, p. 19.
22. Jonathan Z. Smith, *Map Is Not Territory: Studies in the History of Religions*, Chicago: University of Chicago Press, 1993, pp. 100–101.
23. Kuntala Lahiri-Dutt and Gopa Samanta, *Dancing with the River: People and Life on the Chars of South Asia*, New Haven, CT: Yale University Press, 2013, pp. 40–41.
24. Samaresh Basu, *Ganga*, Calcutta: Bengal Publishers, 1957.
25. Carl Sauer, "Theme of Plant and Animal Destruction in Economic History," in *Land and Life; A Selection from the Writings of Carl Ortwin Sauer*, Berkeley: University of California Press, 1963, p. 145.
26. William Cronon, "A Place for Stories: Nature, History, and Narratives," *Journal of American History*, no. 78, March 1992, p. 1349.
27. Jamie Linton, *What Is Water?: The History of a Modern Abstraction*, Vancouver: University of British Columbia Press, 2010, p. 34.

28. W. W. Hunter, *The Imperial Gazetteer of India: Volume IV*, 2nd ed., London: Trubner, 1885, p. 471.
29. Charles Lyell, *Principles of Geology*, 11th rev. ed., New York: D. Appleton, 1877, p. 479.
30. Smithsonian Institution, *Smithsonian Contributions to Knowledge*, vol. 9, Michigan Historical Reprint Series, Ann Arbor: Scholarly Publishing Office, University of Michigan Library, 2005, p. 86.
31. Ibid., pp. 256–257.
32. L. S. S. O'Malley, *Bengal District Gazetteers: Monghyr*, vol. 17, Calcutta: Bengal Secretariat Book Depot, 1906, pp. 7–8.
33. J. H. E. Garret, *Bengal District Gazetteers: Nadia*, vol. 24, Calcutta: Bengal Secretariat Book Depot, 1910, p. 9.

Chapter 1. The World of Pilgrims

1. See Sunitikumar Chattopadhyay's excellent essay "Darāpa Khāna Ghājī," in *Sanskrtiki*, vol. 1, Calcutta: Vakya Sahitya, 1962, pp. 156–174. Translation by the author.
2. This is distinct from the river of the same name in northern India and several streams of the same name in eastern Bengal.
3. Chattopadhyay, "Darāpa Khāna Ghājī," p. 161. Translation by the author.
4. See Bhudev Mukhopadhyay, "India's History Revealed in a Dream," trans. Sujit Mukherjee, *Indian Economic and Social History Review*, vol. 32, no. 2, 1995, p. 224.
5. Mirza Asadullah Khan Ghalib, *Ghalib, 1797–1869*, Cambridge, MA: Harvard University Press, 1969, p. 47.
6. From the translation of Qurrutulain Hyder cited in Rafiq Zakaria, *Indian Muslims: Where Have They Gone Wrong?*, Mumbai: Popular Prakashan, 2004, p. 53.
7. Rana P. B. Singh, *Cultural Landscapes and the Lifeworld*, Varanasi, India: Indica Books, 2004, p. 131.
8. *Hymns of Guru Nanak*, ed. Khuswant Singh, Hyderabad, India: Orient Blackswan, 1997, p. 8.
9. Thomas McCormick, "The Jaina Ascetic as Manifestation of the Sacred," in Robert H. Stoddard and E. Alan Morinis, eds., *Sacred Places, Sacred Spaces: The Geography of Pilgrimages*, Baton Rouge, LA: Geoscience, 1997, p. 236.
10. Jim Corbett, "The Man-Eating Leopard of Rudraprayag," in *The Jim Corbett Omnibus*, New Delhi: Oxford University Press, 1991, pp. 431–432.
11. Rudyard Kipling, "The Miracle of Purun Bhagat," in *The Second Jungle Book*, Garden City, NY: Doubleday, Page, 1921, pp. 31–32.
12. Aldous Huxley, *The Jesting Pilate: The Diary of a Journey*, London: Chatto and Windus, 1948, p. 107.

13. Ibid., p. 108.
14. Ibid., p. 109.
15. Ibid., p. 125.
16. Ibid.
17. Ibid., p. 129.
18. Francis Sydney Smythe, *The Valley of Flowers,* London: Hodder and Stoughton, 1938, p. 40.
19. Ibid., p. 33.
20. Ibid., p. 40.
21. *Vishnu Purana,* quoted in Kamalakrishna Smrititirtha, ed., *Tirthacintamani of Vacaspati Misra,* Calcutta: Asiatic Society, 1912, p. 203.
22. Mircea Eliade, *The Sacred and the Profane: The Nature of Religion,* London: Houghton Mifflin Harcourt, 1987, p. 11.
23. Émile Durkheim, *The Elementary Forms of the Religious Life,* trans. Carol Cosman, Oxford, UK: Oxford University Press, 2001, p. 207.
24. See Diana L. Eck, *Banaras: City of Light,* New York: Columbia University Press, 1999, p. 347.
25. My reading of this text is from an edition with Bengali commentary and translation. See *Vedavyasa Pranita Kasikhanda,* Calcutta: Ramakrishna Mission Sevasram, 1916, p. 43.
26. *Kasikhanda,* p. 194.
27. Ibid., p. 45.
28. Diana L. Eck, *Encountering God: A Spiritual Journey from Bozeman to Banaras,* Boston: Beacon Press, 1993, p. 58.
29. *Tirthacintamani,* p. 191.
30. Ibid., p. 44.
31. Ibid., p. 164.
32. See Mahendranath Tattvanidhi, ed., *Sri Sri Gangastakam, Maharsi Valmiki Viracitam,* Calcutta: Diamond Press, 1927, p. 1.
33. Ibid., p. 2.
34. See John Erskine Clarke, *Chatterbox,* vol. 12, London: Wells, Gardner, Darton, 1867, p. 93.
35. Ibid.
36. Library of Entertaining Knowledge, *The Hindoos,* vol. 2, London: Charles Knight, 1835, pp. 10–11.
37. Kelly D. Alley, *On the Banks of the Ganga: When Watershed Meets a Sacred River,* Ann Arbor: University of Michigan Press, 2002, pp. 17–18.
38. Ibid., pp. 24–25, 245–246.
39. See Kelly D. Alley, "Idioms of Degeneracy: Assessing Ganga's Purity and Pollution," in Lance E. Nelson, ed., *Purifying the Earthly Body of God: Religion and Ecology in Hindu India,* Binghamton: State University of New York Press, 1998, p. 305.

40. Mary Douglas, *Purity and Danger: An Analysis of the Concepts of Pollution and Taboo*, reprint, London: Routledge, 1984, p. 7.
41. Ibid., p. 2.
42. See Meenakshi Sharma, "Polluted River or Goddess and Saviour: The Ganga in the Discourses of Modernity and Hinduism," in Helen Tiffin, ed., *Five Emus to the King of Siam: Environment and Empire*, Amsterdam: Rodopi, 2007, pp. 31–50.
43. *Tirthacintamani*, p. 192.
44. Ibid., p. 197.
45. Ibid., p. 200.
46. Ibid., pp. 262–263, 268.
47. This is discussed in compendia of Vedic domestic rituals. See, for example, *The Grihya-Sutras: Rules of Vedic Domestic Ceremonies*, trans. Hermann Oldenberg, Oxford, UK: Clarendon Press, 1886, p. 18.
48. See Arjun Appadurai, "Introduction: Commodities and the Politics of Value," in *The Social Life of Things: Commodities in Cultural Perspective*, Cambridge, UK: Cambridge University Press, 1986, pp. 22–23.
49. See Richard White, *The Organic Machine: The Remaking of the Columbia River*, New York: Hill and Wang, 1996, p. 3.
50. *Transactions of the Epidemiological Society of London*, vol. 8, London: D. Bogue, 1889, p. 154.
51. See Nirmal Dass, *Songs of Kabir from the Adi Granth*, Albany: State University of New York Press, 1991, p. 333.
52. See *Gazetteer of the Bombay Presidency*, vol. 8, Bombay: Government Central Press, 1884, p. 499.
53. Harold Begbie, *Other Sheep: A Missionary Companion to Twice Born Men*, New York: Hodder and Stoughton, 1912, p. 153.
54. Ibid.
55. E. Sachau, ed., *Alberuni's India*, vol. 2, London: K. Paul, Trench, Trubner, 1910, p. 104.
56. K. S. Lal, *Twilight of the Sultanate*, Bombay: Asia Publishing, 1963, p. 275.
57. S. M. Latif, *Agra: Historical and Descriptive with an Account of Akbar and His Court and of the Modern City of Agra*, New Delhi: Asian Educational Services, 2003, p. 223.
58. Françoise Bernier, *Travels in the Mogul Empire A.D. 1656–1668*, trans. and ed. Archibald Constable and Vincent Smith, Oxford, UK: Oxford University Press, 1914, p. 221.
59. *Indian Horizons*, Vol. 30, New Delhi: Indian Council for Cultural Relations, 1981, p. 31.
60. D. S. Bhargava, "Nature and the Ganga," *Environmental Conservation*, vol. 14, no. 4, 1987, pp. 307–318.
61. Edmund Hillary, *From the Ocean to the Sky*, New York: Viking Press, 1979, p. 20.
62. Robert Montgomery Martin, *History of the Colonies of the British Empire*, London: W. H. Allen, 1843, p. 278.

63. See Walter Clemence and Max Recklinghausen, "The Filtration of Public Water Supplies and their Sterilization by Ultraviolet Rays," *Transactions of the Institution of Water Engineers*, vols. 1–16, 1912, p. 81.
64. See *Tropical Diseases Bulletin* (Bureau of Hygiene and Tropical Diseases), vol. 30, 1933, pp. 414–417; see also "Abstracts of Bacteriology," *American Society for Microbiology*, vol. 6, 1922, p. 159.
65. Ramananda Chatterjee, "The Sacred Ganges and the Jamuna," *Modern Review*, vol. 58, 1935, p. 352.
66. Sharma, "Polluted River," p. 40.
67. D. S. Bhargava, "Nature's Cure of the Ganga: The Ganga-Jal," in Rashmi Singh, ed., *Our National River Ganga: Lifeline of Millions*, Cham, India: Springer, 2014, p. 174.
68. J. C. Hollick, *Ganga*, New Delhi: Random House, 2007, pp. 146–147.
69. *Popular Science*, vol. 119, no. 4, October 1931, p. 129.
70. See Alexander Hamilton's account of the East Indies in *A General Collection of the Best and Most Interesting Voyages and Travels in All Parts of the World*, London: Pinkerton, Longman, Hurst, Rees, and Orme, 1811, p. 415.
71. Caleb Wright, *Historic Incidents and Life in India*, Chicago: J. A. Brainerd, 1869.
72. A. K. Ramanujan, *Speaking of Shiva*, reprint, Harmondsworth, UK: Penguin, 1979, pp. 24–25.
73. Ariel Glucklich, *The Strides of Vishnu: Hindu Culture in Historical Perspective*, Oxford, UK: Oxford University Press, 2008, p. 151.
74. Diana L. Eck, "The Imagined Landscape: Patterns in the Construction of Hindu Sacred Geography," *Contributions to Indian Sociology*, vol. 32, 1998, pp. 165–188.
75. See Surajit Sinha and Baidyanath Saraswati, *Ascetics of Kashi: An Anthropological Exploration*, Varanasi: N. K. Bose Memorial Foundation, 1978, p. 158.
76. Diana L. Eck, "Banaras: Cosmos and Paradise in the Hindu Imagination," *Contributions to Indian Sociology*, vol. 19, no. 41, 1985, p. 48.
77. *Kasikhanda*, p. 45.
78. Jonathan Parry, *Death in Banaras*, Cambridge, UK: Cambridge University Press, 1994, pp. 30–32.
79. K. V. Rangaswami Aiyangar, ed., *Krtyakalpataru of Bhatta Laksmidhara: Brahmacarikanda*, vol. 1, Baroda, India: Oriental Institute, 1948, pp. 44, 47.
80. S. C. Mukerji, "Allahabad," *Modern Review*, vol. 8, no. 6, 1910, p. 649.
81. Norman Macleod, *Days in North India*, Philadelphia: J. B. Lippincott, 1870, p. 20.
82. James Prinsep, *Benares Illustrated in a Series of Drawings*, Calcutta: Baptist Mission Press, 1831, p. 16.
83. Ibid.
84. See "India's Ultimate Pilgrimage," *Time*, February 18, 2013, p. 11.
85. Jawaharlal Nehru, *Toward Freedom: The Autobiography of Jawaharlal Nehru*, New York: John Day, 1941, p. 60.

86. Karl Marx, *Critique of Hegel's "Philosophy of Right,"* Cambridge, UK: Cambridge University Press, 1977, p. 131.

Chapter 2. Ganga Descends

1. Various versions of the *Garuda Purana* dated roughly to the fourth century C.E. have this reference to the city of Kanchi. See Manmatha Nath Dutt, ed., *The Garuda Puranam,* Calcutta: Society for the Resuscitation of Indian Literature, 1908, p. 210.
2. Durkheim, *Elementary Forms of the Religious Life,* pp. 64–65.
3. Alexandre Moret, *The Nile and Egyptian Civilization,* Mineola, NY: Courier Dover, 2001, pp. 28–29.
4. Edith Hamilton, *Mythology: Timeless Tales of Gods and Heroes,* New York: Grand Central, 1942, p. 13.
5. Sir James G. Frazer, *The Golden Bough,* Edinburgh: Canongate Books, 2010, p. 28.
6. E. B. Tylor, *Primitive Culture: Researches into the Development of Mythology, Philosophy, Religion, Art, and Custom,* vol. 1, London: J. Murray, 1871, p. 257.
7. Ibid., p. 365.
8. Steven Mithen, *The Prehistory of the Mind: The Cognitive Origins of Art, Religion and Science,* London: Thames and Hudson, 1996, pp. 124–125. See also Howard Gardner's review of the book: "Thinking about Thinking," *New York Review of Books,* October 9, 1997.
9. Ernst Cassirer, *The Philosophy of Symbolic Forms,* vol. 2, *Mythical Thought,* New Haven, CT: Yale University Press, 1955, p. 69.
10. R. K. Narayan, *God, Demons and Others,* Chicago: University of Chicago Press, 1993, p. 4.
11. A. K. Ramanujan, "Three Hundred *Rāmāyaṇas*: Five Examples and Three Thoughts on Translation," in Paula Richman, ed., *Many Rāmāyaṇas: The Diversity of a Narrative Tradition in South Asia,* Berkeley: University of California Press, 1991, p. 46.
12. Ramanujan, *Speaking of Shiva,* pp. 23–24.
13. Marcel Detienne, *The Greeks and Us: A Comparative Anthropology of Ancient Greece,* Cambridge, UK: Polity Press, 2007, p. 29.
14. Valmiki, *Ramayana, Book One: Boyhood,* trans. Robert P. Goldman, New York: New York University Press, 2005, pp. 45–47. See also K. N. Dave, *Birds in Sanskrit Literature,* Delhi: Motilal Banarsidass, 2005, p. 314.
15. Ibid., p. 185.
16. Ibid., p. 201.
17. Ibid.
18. Shiva is referred to as *Śitikaṇṭha* here. He is also referred to as *Nīlakanṭha*—the blue-throated one, because of the poison that he drank from the oceans of the world to save creation.

19. Valmiki, *Ramayana, Book One*, p. 211.
20. For a comparison of the various versions of the story of the birth of Kartikeyas, see Sudhirchandra Sarkar, ed., *Pauranika Abhidhana*, rev. ed., Calcutta: M. C. Sarkar, 1982.
21. Valmiki, *Ramayana, Book One*, p. 219.
22. See Purnendu Narayana Sinha, *A Study of the Bhagavata Purana: Or, Esoteric Hinduism*, Varanasi, India: Freeman, 1901, p. 196.
23. Ibid., p. 225.
24. In other versions of this myth, it was Kapila the sage who told Amshumant about Ganga. See Sarkar, *Pauranika Abhidhana*, p. 96.
25. Valmiki, *Ramayana, Book One*, pp. 230–231.
26. Ibid., p. 239.
27. The Puranas are supposed to have five major identifying characters: *sarga, pratisarga, vaṃsa, manvantara*, and *vaṃsānucarita*.
28. Ludo Rocher, *The Puranas*, Wiesbaden, Germany: Otto Harrassowitz, 1986, p. 101.
29. Romila Thapar, *The Penguin History of Early India: From the Origins to AD 1300*, New Delhi: Penguin Books, 2002, pp. 98–99.
30. Rocher, *Puranas*, pp. 102, 147–148.
31. Ibid., p. 40.
32. I have consulted the text edited by M. Eugene Burnouf. See M. E. Burnouf, ed., *Le Bhagavata Purana: Ou Histoire Poétique de Krichna*, vol. 3, Paris: Imprimerie Royale, 1847, p. 220.
33. See Radharaman Mitra's Bengali version of the *Bhagavata Purana* rendered from the Sanskrit original: R. Mitra, ed., *Mahabhagavatapurana: Maharsi Krishna-Dvaipayana Vedavyasa Pranita*, Calcutta: Vedanta Press, 1893–1894, pp. 162–163.
34. Ibid.
35. Ibid., p. 164.
36. Ibid.
37. Ibid., p. 110.
38. Ibid. The Sanskrit original is as follows: *Uddhārattung mānavān sarvvān jahnavī bārirūpinī, jagāma tadā bhumau bhagīrathena sevita.* In other words, Ganga appears on earth because of the ministrations of Bhagiratha.
39. Ibid.
40. *Brahmavaivartapuranam*, ed. by Panchanan Tarkaratna, Calcutta: Natabara Chakravarti, 1917, p. 81.
41. Ibid., pp. 114–115.
42. See *Vedavyasa Pranita Kasikhanda*, Calcutta: Ramakrishna Mission Sevasram, 1322 (1915–1916).
43. Burnouf, *Bhagavata Purana*, vol. 3, p. 216.

44. *Brahmavaivarta Purana*, p. 105.
45. This has recently been pointed out by Ruth Vanita. See Ruth Vanita, *Gandhi's Tiger and Sita's Smile: Essays on Gender, Sexuality, and Culture*, New Delhi: Yoda Press, 2005, p. 241. Vanita has drawn attention to the "Swarga Khanda" of the *Padma Purana*, Varanasi, India: All India Kasiraja Trust, 1972, pp. 140–141.
46. See the "Sṛṣṭikhaṇḍam" of the *Padmapurāṇa*, vol. 4, Calcutta: Natabara Chakravarti, 1913, pp. 66–67.
47. "Uttarakhaṇḍaṃ," *Padmapurāṇa*, vol. 7, Calcutta: Natabara Chakravarti, 1915, p. 88.
48. See Subodhchandra Majumdar, ed., *Krittivasa Racita Sampurna Ramayana*, Calcutta: Deba Sahitya Kutira, 1958.
49. Ibid., p. 38.
50. Ibid. English translation is by the author.
51. Ibid., p. 39.
52. Ibid., p. 37.
53. Dvija Madhab, *Gangamangala*, Calcutta: Bangiya Sahitya Parishad, 1916. See also Sukumar Sen, *Bangala Sahityera Katha*, Calcutta: Kalikata Visyavidalaya, 1960, p. 53.
54. Ibid., p. 91.
55. Durgaprasad Mukhopadhyaya, *Gangabhakti Tarangini*, Calcutta: Chaitanyachandrodaya, 1855, p. 8.
56. Ibid., pp. 60–62.
57. Ibid., pp. 109–111.
58. Ibid., pp. 113–117.
59. Kapil Deo Giri, ed., *The Gangavataranam of Nilakantha Diksit*, Varanasi, India: Chaukhamba Sanskrit Series, 1985, pp. 26–27.
60. Ibid., p. 26.
61. Ibid., p. 75; see also Pandit Kedarnatha Sastri, ed., *The Gangavatarana of Nilakantha Dikshit*, Bombay: Tukaram Javaji, Niranaya Sagar Press, 1916, p. 56. English translation is by the author.
62. Panditaraja Jagannatha, *Gangalahari: The Flow of the Ganges*, Varanasi, India: Indica Books, 2007, p. 63.
63. Ibid., p. 117.
64. J. A. B. van Buitenen, *The Mahabharata, Book 1: The Books of the Beginning*, Chicago: University of Chicago Press, 1973, p. 216.
65. Ibid., p. 217. For the passages in the original Sanskrit, see Pratap Chandra Roy, ed., *Mahabharata*, vol. 1, Calcutta: P. C. Roy, 1887, pp. 268–269.
66. Roy, *Mahabharata*, p. 270.
67. J. Gonda, *Aspects of Early Visnuism*, Delhi: Motilal Banarsidass, 1969, pp. 212–213.
68. Alf Hiltebeitel, *The Ritual of Battle: Krishna in the Mahabharata*, Albany: State University of New York Press, 1990, pp. 67–68.

Chapter 3. Digging Out of Prehistory

1. See the preface to Alexander Cunningham, *Four Reports Made during the Years 1862–63–64–65*, vol. 1, Simla, India: Government Central Press, 1871, p. iv.
2. See *Imperial Gazetteer of India*, vol. XII, new ed., Oxford, UK: Clarendon Press, 1908, p. 30.
3. Cunningham, *Four Reports*, pp. 268–270.
4. See Ajit Ghosh, ed., *Indian Archeology: A Review, 1963–64*, New Delhi: Archaeological Survey of India, 1967, p. 45.
5. R. C. Gaur, *Excavations at Atranjikhera: Early Civilization of the Upper Ganga Basin* [original publication, Aligarh, India: Centre for Advanced Study, Department of History, Aligarh Muslim University], Delhi: Motilal Banarsidass, 1983, p. 121.
6. Ibid., p. 234.
7. Ibid., p. 239.
8. J. G. Shaffer and D. A. Lichtenstein, "South Asian Archeology and the Myth of Indo-Aryan Invasions," in Edwin F. Bryant and Laurie L. Patton, eds., *The Indo-Aryan Controversy: Evidence and Inference in Indian History*, New York: Routledge, 2005, pp. 93–94.
9. Thapar, *Penguin History of Early India*, p. 117.
10. J. N. Pandey and R. P. Tripathi, "The Settlement and Subsistence Pattern of the Early Farming Cultures of the Middle Ganga Plain," in D. P. Chattopadhyaya and Lalanji Gopal, eds., *History of Agriculture in India: Up to c. 1200 A.D.*, New Delhi: Concept, 2008, pp. 79–80. See also Upinder Singh, *A History of Ancient and Medieval India: From the Stone Age to the 12th Century*, Delhi: Dorling Kindersley (India), 2008, p. 120.
11. Dilip K. Chakrabarti, *Archeological Geography of the Ganga Plain*, New Delhi: Permanent Black, 2001, p. 187.
12. Randi Haaland, "Porridge and Pot, Bread and Oven: Food Ways and Symbolism in Africa and the Near East from the Neolithic to the Present," *Cambridge Archeological Journal*, vol. 17, no. 2, 2007, pp. 172–173.
13. Randi Haaland, "Sedentism, Cultivation and Plant Domestication in the Holocene Middle Nile Region," *Journal of Field Archaeology*, vol. 22, no. 2, Summer 1995, p. 163.
14. Ibid., p. 164.
15. Randi Haaland, "Emergence of Sedentism: New Ways of Living, New Ways of Symbolizing," *Antiquity*, vol. 71, 1997, p. 381.
16. Haaland, "Sedentism, Cultivation and Plant Domestication," pp. 166–167.
17. Haaland, "Porridge and Pot, Bread and Oven," p. 172.
18. See Shereen Ratnagar, "Approaches to the Study of Ancient Technology," in Sabyasachi Bhattacharya, ed., *Approaches to History: Essays in Indian Historiography*, New Delhi: Indian Council of Historical Research and Primus Books, 2011, p. 79.

19. Ballabh Saran, "Technology of the Painted Grey Ware," in B. P. Sinha, ed., *Potteries in Ancient India,* Patna, India: Patna University, 1969, p. 124.
20. Vibha Tripathi, *The Painted Grey Ware: An Iron Age Culture in Northern India,* Delhi: Concept, 1976, p. 42.
21. Saran, "Technology of the Painted Grey Ware," p. 127.
22. A. Ghosh and K. C. Panigrahi, "The Pottery of Ahichchhatra, District Bareilly, U.P.," in *Ancient India,* no. 1, 1946, pp. 38–41.
23. B. B. Lal, "The Painted Grey Culture of the Iron Age," in A. H. Dani and V. M. Masson, eds., *History of Civilizations of Central Asia: Vol. 1,* Paris: UNESCO, 1992, p. 421.
24. A. Ghosh, "A Note on the Homeland of the Painted Grey Ware," in *Painted Grey Ware: Proceedings of Seminar on Archaeology Held at the Aligarh Muslim University, 1968,* Jaipur, India: Publication Scheme, 1994, p. 22.
25. B. K. Thapar, ed., *Indian Archeology: A Review, 1975–76,* New Delhi: Archaeological Survey of India, 1967, p. 51.
26. Irfan Habib, "Unreason and Archaeology: The 'Painted Grey-Ware' and Beyond," *Social Scientist,* vol. 25, no. 284–285, January–February 1997, p. 23.
27. For an engaging discussion on some of the problems in the lacunae between archaeological findings and the Indian epic literature, see Brajadulal Chattopadhyaya, *Studying Early India: Archeology, Texts and Historical Issues,* London: Anthem Press, 2006, p. 23.
28. Hemchandra Raychaudhuri, *Political History of Ancient India: From the Accession of Parikshit to the Extinction of the Gupta Dynasty,* revised edition, Delhi, Oxford University Press, [1923] 1996, pp. 6–10.
29. Chattopadhyaya, *Studying Early India,* pp. 23–24.
30. This was acknowledged quite some time ago by M. Lal, "Date of Painted Grey Ware Culture: A Review," *Bulletin of the Deccan College Research Institute,* no. 39, 1980, pp. 65–77. See also Gregory L. Possehl, *Radiocarbon Dates for South Asian Archaeology,* Philadelphia: University of Pennsylvania Press, 1989.
31. These findings were based on the pioneering work of Mohammed Rafique Mughal, who excavated this area during 1974–1977. See M. R. Mughal, *Ancient Cholistan: Archaeology and Architecture,* Rawalpindi, India: Ferozsons, 1997, pp. 40–56.
32. See Shaffer and Lichtenstein, "South Asian Archeology and the Myth of Indo-Aryan Invasions," pp. 86–87.
33. J. G. Shaffer, "Re-urbanization: The Eastern Punjab and Beyond," in H. Spodek and D. M. Srinivasan, eds., *Urban Form and Meaning in South Asia,* Hanover, NH: University Press of New England, 1993, p. 57; Habib, "Unreason and Archaeology," p. 19.
34. Brian Hayden, "Practical and Prestige Technologies: The Evolution of Material Systems," *Journal of Archaeological Method and Theory,* vol. 5, no. 1, March 1998, pp. 2–3.

35. See Suraj Bhan, "Aryanization of the Indus Civilization," in T. J. Byres, Irfan Habib, K. N. Panikkar, and Utsa Patnaik, eds., *The Making of History: Essays Presented to Irfan Habib*, London: Anthem Press, 2002, pp. 41–55.
36. Ibid., p. 44. See also Michel Danino, *The Lost River: On the Trail of the Sarasvatī*, New Delhi: Penguin Books India, 2010, pp. 38–39.
37. S. Ratnagar, *Understanding Harappa: Civilization in the Greater Indus Valley*, New Delhi: Tulika, 2001, p. 4.
38. Danino, *The Lost River*, pp. 117–118. See also Jane McIntosh, *The Ancient Indus Valley: New Perspectives*, Santa Barbara, CA: ABC-CLIO, 2008, pp. 3–4.
39. McIntosh, *The Ancient Indus Valley*, pp. 64–65.
40. Ibid., pp. 19–20.
41. B. B. Lal and Peter N. Peregrine, *South and Southwest Asia: Encyclopedia of Prehistory*, New York: Kluwer, 2002, p. 8.
42. For an overview, see George Erdosy, "Language, Material Culture and Ethnicity: Theoretical Perspectives," in *The Indo-Aryans of Ancient South Asia: Language, Material Culture and Ethnicity*, Berlin: Walter de Gruyter, 1995, pp. 1–31.
43. For a general sense of some these assertions, see S. P. Gupta, *The Indus-Sarasvati Civilisation: Origins, Problems and Issues*, New Delhi: Pratibha Prakasan, 1996, pp. xii–xiii.
44. J. M. Kenoyer, "Cultural Change during the Late Harappan Period," in Edwin F. Bryant and Laurie L. Patton, eds., *The Indo-Aryan Controversy: Evidence and Inference in Indian History*, New York: Routledge, 2005, p. 23.
45. McIntosh, *The Ancient Indus Valley*, pp. 4–5.
46. See A. Ghosh, ed., *An Encyclopedia of Indian Archeology*, Leiden, The Netherlands: E. J. Brill, 1990, p. xiv.
47. Thapar, *Penguin History of Early India*, p. 89.
48. Kenoyer, "Cultural Change during the Late Harappan Period," p. 43.
49. Ibid.
50. Acheulean "industry" or stone making is attributed to certain groups of *Homo erectus* and early *Homo sapiens*. It derives from the site of Saint-Acheul, in Somme, northern France. Acheulean tools are made out of stones that fracture with sharp edges, such as chalcedony, jasper, flint, or quartzite. The Acheulean period is dated from 1.5 million to 200,000 years ago. Stone tools were likely one of the most crucial factors in the spread and dispersal of early humans out of Africa.
51. John G. Fleagle et al., eds., *Out of Africa I: The First Hominin Colonization of Eurasia*, Dordrecht, The Netherlands: Springer, 2010, pp. 41, 127.
52. Rajeev Patnaik and Avinash C. Nanda, "Early Pleistocene Mammalian Faunas of India and Evidence of Connection with Other Parts of the World," in Fleagle et al., *Out of Africa I*, p. 159.
53. See J. G. Fleagle and J. J. Shea, "Summary and Prospectus," in Fleagle et al., *Out of Africa I*, p. 276.

54. R. Potts and R. Teague, "Behavioral and Environmental Background to 'Out of Africa I' and the Arrival of *Homo erectus* in East Asia," in Fleagle et al., *Out of Africa I*, pp. 74–75.
55. K. N. Prasad, "Observations on the Paleoecology of South Asian Tertiary Primates," in R. H. Tuttle, ed., *Paleoanthropology: Morphology and Paleoecology*, The Hague: Mouton, 1975, pp. 24–25.
56. Robin Dennell, "'Resource-Rich, Stone-Poor': Early Hominin Land Use in Large River Systems of Northern India and Pakistan," in M. D. Petraglia and Bridget Allchin, eds., *The Evolution and History of Human Populations in South Asia: Inter-Disciplinary Studies in Archaeology, Biological Anthropology, Linguistics, and Genetics*, Dordrecht, The Netherlands: Springer, 2007, pp. 41–42.
57. A. P. Khatri, "The Early Fossil Hominids and Related Apes of the Siwalik Foothills of the Himalayas: Recent Discoveries and New Interpretations," in Tuttle, *Paleoanthropology*, p. 44.
58. Dennell, "'Resource-Rich, Stone-Poor,'" p. 63.
59. Ibid.
60. Matt Cartmill and Fred H. Smith, *The Human Lineage*, Hoboken, NJ: Wiley-Blackwell, 2009, pp. 273–274.
61. V. N. Misra, "Ecological Adaptations during the Terminal Stone Age in Western and Central India," in K. A. R. Kennedy and G. L. Possehl, eds., *Ecological Backgrounds of South Asian Prehistory*, Ithaca, NY: South Asia Occasional Papers, Cornell University, 1976, pp. 28–51.
62. Kenneth A. R. Kennedy, *God-Apes and Fossil Men: Paleoanthropology of South Asia*, Ann Arbor: University of Michigan Press, 2000, pp. 202–203.
63. Virendra N. Misra, "Microlithic Industries in India," in V. N. Misra and Peter S. Bellwood, eds., *Recent Advances in Indo-Pacific Prehistory*, Leiden, The Netherlands: E. J. Brill, 1985, p. 118.
64. Kennedy, *God-Apes and Fossil Men*, pp. 239–240.
65. For a critique of traditional approaches to evolution among early hominids and humans, see C. Loring Brace, *Evolution in an Anthropological View*, New York: Rowman and Littlefield, 2000, pp. 123–125.
66. Ibid., p. 229.
67. Misra, "Microlithic Industries in India," p. 118.
68. Petraglia and Allchin, *Evolution and History of Human Populations*, p. 13.
69. Ibid., p. 203.
70. John R. Lukacs, "Hunting and Gathering in Prehistoric India," in Kathleen D. Morrison and Laura L. Junker, eds., *Foragers-Traders in South and Southeast Asia: Long-Term Histories*, Cambridge, UK: Cambridge University Press, 2002, p. 60.
71. Kathleen D. Morrison, "Historicizing Adaptation, Adapting to History: Forager-Traders in South and Southeast Asia," in Morrison and Junker, *Foragers-Traders*, p. 30.

72. Gregory L. Possehl and P. Rissman, "The Chronology of Prehistoric India from Earliest Times to the Iron Age," in R. W. Ehrich, ed., *Chronologies in Old World Archeology*, Chicago: University of Chicago Press, 1992, p. 488.
73. V. Gordon Childe, *Man Makes Himself*, New York: New American Library, 1951, p. 59.
74. Ibid., pp. 67–69.
75. For a substantive discussion of the main facets of Carl Sauer's views on the history of agriculture, see Daniel W. Gade, "Introduction," in William M. Denevan and Kent Mathewson, eds., *Carl Sauer on Culture and Landscape: Readings and Commentaries*, Baton Rouge: Louisiana State University Press, 2009, p. 186.
76. Ibid., pp. 190–191.
77. Carl Otwin Sauer, *Agricultural Origins and Dispersals: The Domestication of Animals and Foodstuffs*, 2nd ed., Cambridge, MA: MIT Press, 1969, pp. 3, 53.
78. For a detailed account of the domestication of sheep and goats as livestock in Europe, see Susan A. Gregg, *Foragers and Farmers: Population Interaction and Agricultural Expansion in Prehistoric Europe*, Chicago: University of Chicago Press, 1988, p. 123. See also Alan H. Simmons, *The Neolithic Revolution in the Near East: Transforming the Human Landscape*, Tucson: University of Arizona Press, 2007, pp. 260–261; and Catherine Perlés, *The Early Neolithic in Greece*, Cambridge, UK: Cambridge University Press, 2001, p. 38.
79. Kennedy, *God-Apes and Fossil Men*, p. 246.
80. Graeme Barker, *The Agricultural Revolution in Prehistory: Why Did Foragers Become Farmers?*, Oxford, UK: Oxford University Press, 2009, pp. 153–154.
81. Lewis Roberts Binford, *In Pursuit of the Past: Decoding the Archeological Record*, Berkeley: University of California Press, 1983, p. 200.
82. Barker, *Agricultural Revolution in Prehistory*, p. 155.
83. These animals' records have been discussed extensively in P. P. Joglekar, *Mesolithic Mahadaha: The Faunal Remains*, Allahabad, India: University of Allahabad, 2003, passim.
84. Petraglia and Allchin, *Evolution and History of Human Populations in South Asia*, p. 13.
85. U. C. Chattopadhyaya, "Settlement Pattern and the Spatial Organization of Subsistence and Mortuary Practices in the Mesolithic Ganges Valley, North-Central India," *World Archaeology*, vol. 27, no. 3, pp. 461–476.
86. Nayanjot Lahiri, "Archeology and Some Aspects of the Social History of Early India," in Brajadulal Chattopadhyaya, ed., *A Social History of Early India*, New Delhi: Pearson Longman, 2009, pp. 4–5.
87. See J. R. Lukacs and J. N. Pal, "Skeletal Variation among Mesolithic People of the Ganga Plains: New Evidence of Habitual Activity and Adaptation to Climate," *Asian Perspectives*, vol. 42, no. 2, Fall 2003, p. 345; and J. R. Lukacs, "Interpreting Biological Diversity in South Asian Prehistory: Early Holocene Population Affini-

ties and Subsistence Adaptations," in Petraglia and Allchin, *Evolution and History of Human Populations in South Asia*, p. 279.

88. Ibid., p. 289.

89. K. D. Morrison, "Foragers and Forage-Traders in South Asian Worlds," in Petraglia and Allchin, *Evolution and History of Human Populations in South Asia*, p. 330.

90. See Nayanjot Lahiri, "Ganges Valley," in the *Encyclopedia of Archaeology*: Vol. 3, Oxford, UK: Elsevier Press, 2008, pp. 683–694.

91. Sacha C. Jones, "The Toba Supervolcanic Eruption: Tephra-fall Deposits in India and Paleoanthropological Implications," in M. D. Petraglia and B. Allchin, eds., *The Evolution and History of Human Populations in South Asia*, New York: Springer/Kluwer Academic Publishers, 2010, p. 193.

92. Randi Haaland, "Crops and Culture: Dispersal of African Millets to the Indian Subcontinent and Its Cultural Consequences," *Dhaulagiri Journal of Sociology and Anthropology*, vol. 5, 2011, pp. 13–14.

93. Pandey and Tripathi, "The Settlement and Subsistence Pattern of the Early Farming Cultures of the Middle Ganga Plain," p. 78.

94. Purushottam Singh, "Origin of Agriculture in the Middle Ganga Plains," in Gopal, *History of Agriculture in India*, pp. 6–7.

95. Ibid., p. 8.

96. D. Q. Fuller, "Further Evidence on the Prehistory of Sesame," *Asian Agri-History*, vol. 7, no. 2, 2003, pp. 132–134.

97. D. P. Agrawala, *Ancient Metal Technology and Archaeology of South Asia: A Pan-Asian Perspective*, New Delhi: Aryna Books, 2000, p. 188.

98. Mortimer Wheeler, *Charsada, a Metropolis of the North-west Frontier, Being a Report on the Excavations of 1958*, London: Oxford University Press, 1962, pp. 33–34.

99. Kumkum Roy, *Historical Dictionary of Ancient India*, Lanham, MD: Scarecrow Press, 2009, p. 111.

100. Rakesh Tewari, "The Origins of Iron Working in India: New Evidence from the Central Ganga Plain and the Eastern Vindhyas," *Antiquity*, vol. 77, no. 297, September 2003, p. 541.

101. D. P. Agrawal suggested that the colonization of the Doab lands began during the same time as the rise of the PGW. See D. P. Agarwal, *The Copper Bronze Age in India: An Integrated Archaeological Study of the Copper Bronze Age in India in the Light of Chronological, Technological, and Ecological Factors, ca. 3000–500 B.C.*, Delhi: Munshiram Manoharlal, 1971, p. 244.

102. T. P. Verma, B. R. Mani, A. K. Singh, and R. Kumar, eds., *Pracyabodha: Indian Archeology and Tradition*, vol. 2, Delhi: B. R. Publishing, 2014, p. 223.

103. See Ratnagar, "Approaches to the Study of Ancient Technology," pp. 70–71.

104. D. D. Kosambi, *An Introduction to the Study of Indian History*, rev. 2nd ed., Mumbai: Popular Prakashan, 1975, pp. 23, 70–71.

105. R. S. Sharma, "Stages in Ancient Economy," reprinted in *Rethinking India's Past*, New Delhi: Oxford University Press, 2010, p. 147.
106. See R. S. Sharma, "Material Background to the Origins of Buddhism," in Mohit Sen and M. B. Rao, eds., *Das Kapital Centenary Volume*, New Delhi: People's Publishing, 1968, pp. 49–66. Sharma raised these questions relating to the material culture during the age of Buddhism in a paper, "Material Culture Down to the Age of the Buddha," presented at the Seminar on Social and Economic History of India, Indian Institute of Advanced Study, Simla, in 1966.
107. K. A. Chowdhury, *Ancient Agriculture and Forestry in North India*, New York: Asia Publishing, 1978, p. 69.
108. For a recent and admirably succinct view, see Thomas Trautmann, *India: Brief History of a Civilization*, New York: Oxford University Press, 2011, pp. 49–50.

Chapter 4. Rise of the Warring Kingdoms

1. F. Max Muller, *The Upanishads, Part 2*, Oxford: Clarendon Press, 1884, pp. 289–290.
2. A. L. Basham, *The Wonder That Was India*, New York: Grove Press, 1964, p. 246.
3. Lewis Mumford, *The City in History*, New York: Harcourt, Brace and World, 1961, p. 111.
4. D. D. Kosambi, *Ancient India: A History of Its Culture and Civilization*, New York: Pantheon Books [1965] 1966, p. 120.
5. A. Ghosh, *The City in Early Historical India*, Simla, India: Indian Institute of Advanced Study, 1973, p. 22.
6. Sharma, "Material Background on the Origins of Buddhism," pp. 59–66.
7. R. S. Sharma, *Material Culture and Social Formations in Ancient India*, New Delhi: Macmillan, 1983, pp. 90–110.
8. For a succinct restatement of these arguments see Sharma, "Stages in Ancient Economy," pp. 147–148, and *Aspects of Political Ideas and Institutions in Ancient India*, 4th rev. ed., New Delhi: Motilal Banarsidass, 1996, p. 198.
9. Problems with such sweeping arguments about the connection between iron, an agricultural revolution, and the rise of urbanization were first pointed out by Nihar Ranjan Ray, the eminent historian of ancient Bengal. See Nihar Ranjan Ray, "Technology and Social Change in Early Indian History: A Note Posing a Theoretical Question," *Puratattva*, vol. 8, 1976, pp. 132–138. Similar concerns were raised by Dilip K. Chakrabarti, "Iron and Urbanization: An Examination of the Indian Context," *Puratattva*, vol. 15, 1984–1985, pp. 68–74.
10. Sudarsan Seneviratne, "The Mauryan State," in Henri J. M. Claessen, Peter Skalnik, et al., eds., *The Early State*, The Hague: Mouton, 1978, p. 282.
11. Ester Boserup, *The Conditions of Agricultural Growth: The Economics of Agrarian Change under Population Pressure*, London: Allen and Unwin, 1965, pp. 6, 13.
12. Ibid., p. 65.

13. J. P. Sharma, *Republics in Ancient India, C. 1500 B.C.–C. 500 B.C.*, Leiden, The Netherlands: E. J. Brill, 1968, pp. 86–87.
14. Ibid., p. 87.
15. Ibid., p. 92.
16. Narayanchandra Bandyopadhyaya, "Types of Indian States," *Calcutta Review*, vol. 14, no. 2, 1925, p. 379.
17. Edward B. Cowell, et al., eds., *The Jataka: Or, Stories of the Buddha's Former Births*, vol. 5, reprint [1895], New Delhi: Asian Educational Services, 2000, p. 316.
18. See R. S. Sharma, "Urbanism in Early Historic India," in Indu Banga, ed., *The City in Indian History: Urban Demography, Society, and Politics*, Columbia, MO: South Asia Publications, 1991, pp. 9–18.
19. Bhikkhu Bodhi, *The Numeral Discourses of the Buddha: A Complete Translation of the Anguttara Nikaya*, Boston: Wisdom, 2012, p. 1010.
20. Ibid., pp. 1011–1012.
21. Kathleen Kuiper, *Ancient Greece: From the Archaic Period to the Death of Alexander the Great*, New York: Britannica Educational, 2010, pp. 22–23.
22. Thapar, *Early India*, pp. 121–123.
23. This story is told in the "Maha-Govinda Suttanta" section of the *Digha Nikaya*. See T. W. Rhys Davids and J. Estlin Carpenter, eds., *The Digha Niakaya*, vol. 2, London: Henry Frowde, Pali Text Society, 1903, p. 235.
24. Maurice Walshe, *The Long Discourses of the Buddha: A Translation of the Digha Nikaya*, Boston: Wisdom, [1987] 1995.
25. T. N. Adhikari, *Gopatha Brahmana, a Critical Study: An Analysis of the Topics and a Comprehensive Critique of the Brahmana Text of the Atharva-veda*, Calcutta: Sanskrit Pustak Bhandar, 1994, p. 16.
26. George Erdosy, "City States of North India and Pakistan at the Time of the Buddha," in *The Archeology of Early Historic South Asia: The Emergence of Cities and States*, Cambridge, UK: Cambridge University Press, 1995, p. 117.
27. The story has been reproduced from the *Ragovada Jataka*, discussed in T. W. Rhys Davids and Caroline Rhys Davids, eds., *Dialogues of the Buddha*, vol. 2, London: Henry Frowde, 1899, pp. 160–161.
28. T. W. Rhys Davids, *Dialogues of the Buddha: Translated from the Pali of the Digha Nikaya*, Delhi: Motilal Banarsidass, 2000, p. 59.
29. Ibid., p. 73.
30. John Holder, *Early Buddhist Discourses*, Indianapolis, IN: Hackett, 2006, p. 175.
31. Davids and Davids, *Dialogues of the Buddha*, vol. 2, p. 242.
32. Walshe, *Long Discourses of the Buddha*, p. 249.
33. Bhikkhu Nanamoli and Bikkhu Bodhi, eds., *The Middle Length Discourses of the Buddha: A Translation of the Majjhima Nikaya*, Boston: Wisdom, 1995, p. 22.
34. See Sharma, "Urbanism in Early Historic India," pp. 9–18.
35. Ibid., p. 11.

36. W. W. Tarn, *The Greeks in Bactria and India*, Cambridge, UK: Cambridge University Press, reprint [2nd ed. 1951] 1966, pp. 228–229.
37. Sharma, "Urbanism in Early Historic India," p. 14.
38. Cowell et al., *The Jataka*, p. 316.
39. The story is vividly narrated from the Jatakas in Wolfgang Schumann, *The Historical Buddha: The Times, Life, and Teachings of the Founder of Buddhism*, Delhi: Motilal Banarsidass, 2004, pp. 242–243.
40. Shohei Ichimura, *Buddhist Critical Spirituality: Prajña and Śunyata*, Delhi: Motilal Banarsidass, 2001, p. 383.
41. Erdosy, "City States of North India and Pakistan," p. 118.
42. See the informative Anonymous, "Rajagriha: Its History and Shrines," *Modern Review*, vol. 3, no. 3, 1908, p. 351.
43. Chakrabarti, *Archeological Geography of the Ganga Plain*, pp. 184–185.
44. Ibid., p. 185. See also R. K. Harding, "Rajagriha and Its Hinterland," PhD thesis, University of Cambridge, 2001.
45. James Burgess, "On the Identification of the Various Places in the Kingdom of Magadha Visited by Chih-Fah-Hian, AD 400–415," *Indian Antiquary*, vol. 1, 1872, p. 18.
46. S. D. Prasad, *District Census Handbook, Bihar*, Patna, India: Government of Bihar, 1965–1966, p. lxvi. See also F. R. Allchin, *The Archaeology of Early Historic South Asia: The Emergence of Cities and States*, Cambridge, UK: Cambridge University Press, 1995, p. 70.
47. Allchin, *Archaeology of Early Historic South Asia*, pp. 57–58. For a useful summary, see A. Ghosh, ed., *Indian Archaeology 1957–58: A Review*, New Delhi: Archaeological Survey of India, reprint [1985] 1993, pp. 48–49.
48. G. R. Sharma, *Excavations at Kausambi, 1949–50*, Delhi: Archaeological Survey of India, 1969, p. 24.
49. For a lucid description of the antiquity and continuity in the archaeology of Kaushambi, see Glucklich, *Strides of Vishnu*, pp. 48–49.
50. G. M. Bongard-Levin, *Ancient Indian History and Civilization*, Delhi: Ajanta, 1998, p. 37.
51. B. C. Law, *Life and Work of Buddhaghosha*, New Delhi: Asian Educational Services, 1997, p. 107.
52. Cowell et al., *The Jataka*, p. 281.
53. Allchin, *Archeology of Early Historic South Asia*, p. 108.
54. D. K. Chakrabarti, "Relating History to Land," in Patrick Olivelle, ed., *Between the Empires: Society in India 300 BCE to 400 CE*, New York: Oxford University Press, 1996, pp. 19–20.
55. B. B. Lal and K. N. Dikshit, "Sringaverapura: A Key Site in the Central Ganga Valley," *Puratattva*, vol. 8, 1978, pp. 1–8.

56. Chakrabarti, "Relating History to Land," p. 15.
57. George Erdosy, *Urbanisation in Early Historic India*, Oxford, UK: British Archeological Reports, 1988, p. 51.
58. Allchin, *Archeology of Early Historic South Asia*, p. 69. See also George Erdosy, "Settlement Archeology of the Kaushambi Region," *Man and Environment*, vol. 9, 1985, pp. 61–80.
59. Sharma, *Excavations at Kausambi*, pp. 83–84. See also O. P. L. Srivastava, "The Coins of Kaushambi-Jethamitra I," *Journal of the Numismatic Society of India*, vols. 62–63, 2001, pp. 54–55.
60. R. C. Dutt, *A History of Civilization in Ancient India, Based on Sanscrit Literature*, Calcutta: Thacker, Spink, 1889–1990, p. 279.
61. S. C. Gupta, "Rajgriha: Its History and Shrines," *Modern Review*, vol. 3, January 1908, p. 355.
62. Sherab Chödzin, *A Life of the Buddha*, Boston: Shambhala, 2009, pp. 89–90.
63. Gupta, "Rajgriha: Its History and Shrines," p. 354.
64. Allchin, *Archeology of Early Historic South Asia*, p. 114.
65. The *Vinaya Pitaka* gives this early definition of the term. See *The Book of the Discipline (Vinaya-Pitaka)*, vol. 3, *Sutta Vibhanga*, trans. I. B. Horner, Oxford, UK: Pali Text Society, 2004, p. 179.
66. William Stede and T. W. Rhys Davids, *The Pali-English Dictionary*, New Delhi: Asian Educational Services, 2004, p. 248.
67. S. Bhattacharya, "The Meaning and Significance of the Term *Gahapati* (During the Post-Mauryan and Pre-Gupta Period)," *Archiv Orientální*, vol. 44, 1976, p. 150.
68. See Uma Chaktavarti, *The Social Dimensions of Early Buddhism*, New Delhi: Munshiram Manoharlal, 1996, pp. 69–70; and Ranabir Chakrabarty, "The Kutumbikas of Early India," in V. K. Thakur and A. Aounshuman, eds., *Peasants in Indian History*, Patna, India: Janaki Prakashan, 1996, pp. 179–198.
69. Monica L. Smith, "The Archaeology of South Asian Cities," *Journal of Archeological Research*, no. 14, 2006, pp. 108–109.
70. Sharma, *Aspects of Political Ideas and Institutions in Ancient India*, p. 221.
71. The Sinhalese Buddhist sources have unreliable and conflicting dates, but they all suggest that the last of the Haryanka kings was Nagadasaka, who ruled for at least twenty years before he was deposed by the minister Shishunaga. See G. N. Prasad, *Chronology of North Indian Kings*, Delhi: Agam Kala Prakashan, 1990, pp. 75–76. Historians have long noted the irreconcilable disagreement between the Puranas and the Sinhala chronicles as to the dates of the Hariyanka and Shishunaga Dynasties. See Raychaudhuri, *Political History of Ancient India*, p. 198.
72. Charles Drekmeier, *Kingship and Community in Early India*, Stanford, CA: Stanford University Press, 1962, p. 165.

73. B. C. Law, *The Magadhas in Ancient India,* London: Royal Asiatic Society, 1946, p. 11.
74. Hemchandra Raychaudhuri, "India in the Age of the Nandas," in K. A. Nilakantha Shastri, ed., *Age of the Nandas and Mauryas,* Delhi: Motilal Banarsidass, 1988, pp. 9–10.
75. Radhakumud Mookerji, *Chandragupta Maurya and His Times,* 4th ed., Delhi: Motilal Banarsidass, 1966, p. 19.
76. Raychaudhuri, "India in the Age of the Nandas," p. 24.
77. Legends of the wealth of the Nandas reached far and wide and is recorded in Tamil Cankam poetry as wealth from "Patali" hidden in the "flood of the Ganges." See S. Krhishnaswami Aiyangar, *Beginnings of South Indian History,* Madras: Modern Printing Works, 1918, p. 89.
78. Kautilya, *The Arthashastra,* trans. L. N. Rangarajan, New Delhi: Penguin Books India, 1992, p. 4.
79. Mookerji, *Chandragupta Maurya and His Times,* p. 21.
80. Bratindranatha Mukhopadhyaya, *Ganga-Banga,* Kolkata: K. Mitra, 2004, p. 14.
81. Quintus Curtius Rufus, *The History of Alexander,* trans. John Yardley, reprint, London: Penguin Classics, 1984, p. 15.
82. In these accounts, multiple Greek names appear for the Nanda kings. Curtius Rufus calls Ugrasena "Aggrammes" and Diodorus calls his son Xandrames, whose Sanskrit equivalent is Candramas, and whose name is not clear from the extant Indian records. See Sharma, *Republics in Ancient India,* pp. 48–49.
83. Curtius Rufus, *History of Alexander,* p. 215.
84. A. B. Bosworth, *Alexander and the East: The Tragedy of Triumph,* reprint 2004, Oxford, UK: Oxford University Press, 1996, pp. 74–75.
85. Grant Parker, *The Making of Roman India,* Cambridge, UK: Cambridge University Press, 2008, pp. 36–37.
86. E. A. Schwanbeck and J. W. McCrindle, *Ancient India as Described by Megasthenes and Arrian,* London: Trubner, 1877.
87. Arrian, *Alexander the Great, the Anabasis and the Indica,* trans. Martin Hammond, Oxford, UK: Oxford University Press, 2013, p. 291.
88. Bosworth, *Alexander and the East,* pp. 186–200.
89. Waldemar Heckel, *The Conquests of Alexander the Great,* Cambridge, UK: Cambridge University Press, 2008, pp. 6–7.
90. Yuval Shahar, *Josephus Geographicus: The Classical Context of Geography in Josephus,* Tübingen: Mohr Siebeck, 2004, p. 27.
91. Strabo, *The Geography of Strabo,* vol. 1, London: W. Heinemann, 1917, p. 125.
92. Strabo, *The Geography of Strabo,* vol. 3, London: Henry G. Bohn, 1857, p. 96.
93. Arrian, *Alexander the Great,* p. 264.
94. Schwanbeck and McCrindle, *Ancient India as Described by Megasthenes and Arrian,* p. 34.

95. Parker, *The Making of Roman India*, p. 44.
96. Ibid., p. 64.
97. Raychaudhuri, *Political History of Ancient India*, pp. 209–210.
98. Plutarch, *Alexander*, ed. J. R. Hamilton, London: Bristol Classical Press, 1999, p. 175; see also Mookerji, *Chandragupta Maurya and His Times*, p. 26.
99. Tarn, *The Greeks in Bactria and India*, p. 100.
100. For an account of the state of the Greek empire and its disputed provinces after Alexander's death, see Diodorus Siculus, *Diodorus of Sicily: In Twelve Volumes, Books XVIII and XIX, 1–65*, vol. 9, trans. Russel M. Geer, Cambridge, MA: Harvard University Press, 1984, pp. 13, 19.
101. Justin, *Epitome of the Philippic History of Pompeius Trogus, Volume II, Books 13–15: The Successors to Alexander the Great*, trans. J. C. Yardley, Oxford, UK: Oxford University Press, 2011, p. 294.
102. Paul J. Kosmin, *The Land of the Elephant Kings: Space, Territory, and Ideology in the Seleucid Empire*, Cambridge, MA: Harvard University Press, 2014, p. 33.
103. Raychaudhuri, *Political History of Ancient India*, p. 247.
104. Ibid., p. 256.
105. Mookerji, *Chandragupta Maurya and His Times*, p. 51.
106. For a perceptive and nuanced discussion about the authorship of the *Arthashastra*, see K. A. Nilakantha Shastri, "Mauryan Polity," in Shastri, *Age of the Nandas and Mauryas*, pp. 172–173. See also Raychaudhuri, "India in the Age of the Nandas," p. 12.
107. Shastri, "Mauryan Polity," pp. 194–195.
108. Ibid., p. 190.
109. Kautilya, *Arthashastra*, p. 127.
110. Raychaudhuri, *Political History of Ancient India*, p. 259.
111. R. P. Kangle, "The Vyasanas According to Kautilya," *Indian Antiquary*, vol. 1, 1964, p. 146.
112. Kautilya, *Arthashastra*, pp. 111, 137.
113. Ibid., p. 76.
114. Hemchandra Raychaudhuri suggested that *grāmika*s during the Maurya period took over the older office of the Vedic *adhikṛta*s. See Raychaudhuri, "India in the Age of the Nandas," p. 22.
115. Shastri, "Mauryan Polity," p. 181.
116. J. W. McCrindle, ed., *Ancient India as Described by Megasthenes and Arrian*, Calcutta: Thacker, Spink, 1877, p. 33.
117. The word *ἀγρονόμοι* appears in Strabo's account of India based on Megasthenes's *Indika*. The expression, as in classical Greek, is used generally for wardens or magistrates of the countryside, and in this case specifically for the inspectors of drainage and irrigation. See Roy, *Historical Dictionary of Ancient India*, p. 8.
118. McCrindle, *Ancient India*, p. 86.

119. The most plausible interpretation was given by G. Bühler in "Asokas Rajukas oder Lajukas," *Zeitschrift der Deutschen Morgenländischen Gesellschaft*, vol., 47, 1893, p. 467. Bühler associated the word with the Pali word *rajjugāhaka*, or "holder of the rope." See also Shastri, "Mauryan Polity," p. 224.
120. U. N. Ghoshal, "Weights, Measures and Taxes," in Shastri, *Age of the Nandas and Mauryas*, p. 270.
121. Kautilya, *Arthashastra*, p. 68.
122. Ibid., p. 113.
123. Shastri, "Introduction," in *Age of the Nandas and the Mauryas*, p. 4.
124. Hemchandra Raychaudhuri, "Chandragupta and Bimbisara," in Shastri, *Age of the Nandas and Mauryas*, p. 167.
125. Shastri, "Asoka and His Successors," in *Age of the Nandas and the Mauryas*, p. 203.
126. Romila Thapar suggests that during this period, especially as far as the *Arthashastra* is considered, it is difficult to distinguish clearly between the king's own land as property and land belonging to the state because in practice the monarch and the state were virtually indistinguishable. See Romila Thapar, *Asoka and the Decline of the Maurya Empire*, 3rd ed., New Delhi: Oxford University Press, [1973] 1997, pp. 82–83.
127. Richard Salomon, *Indian Epigraphy: A Guide to the Study of Inscriptions in Sanskrit, Prakrit, and the other Indo-Aryan Languages*, New York: Oxford University Press, 1998, p. 138.
128. Pillar Edict V, reproduced in *The Edicts of Asoka*, ed. and trans. N. A. Nikam and Richard P. McKeon, Chicago: University of Chicago Press, 1959, pp. 55–56.
129. Raychaudhuri, "Chandragupta and Bimbisara," p. 171.
130. Ibid., p. 191.
131. John S. Strong, *Relics of the Buddha*, Delhi: Motilal Banarsidass, 2007, p. 26.

Chapter 5. Guardians of the Middle Country

1. Martin Lerner and Steven Kossak, *The Lotus Transcendent: Indian and Southeast Asian Art from the Samuel Eilenberg Collection*, New York: Metropolitan Museum of Art, 1991, p. 91.
2. Susan L. Huntington and John C. Huntington, *The Art of Ancient India: Buddhist, Hindu, Jain*, Delhi: Motilal Banarsidass, 2014, pp. 71–72.
3. Julia Shaw, *Buddhist Landscapes in Central India: Sanchi Hill and Archaeologies of Religious and Social Change, C. Third Century BC to Fifth Century AD*, London: British Association for South Asian Studies, 2007, p. 23; Sister Vajira and Francis Story, *Last Days of the Buddha: The Mahaparinibbana Sutta*, 2nd rev. ed., Kandy, Sri Lanka: Buddhist Publication Society, [1964] 1998, p. 18.
4. John Walters, "Stupa, Story, and Empire: Constructions of the Buddha Biography in Post-Asokan India," in Juliane Schober, ed., *Sacred Biography in the*

Buddhist Traditions of South and Southeast Asia, Delhi: Motilal Banarsidass, 2002, pp. 169–170.

5. Ibid., p. 174.
6. Vajira and Story, *Last Days of the Buddha*, p. 18.
7. Jean Philippe Vogel, *Indian Serpent-lore: Or, the Nagas in Hindu Legend and Art*, New Delhi: Asian Educational Services, 1926, pp. 117–118.
8. Alexander Peter Bell, *Didactic Narration: Jataka Iconography in Dunhuang with a Catalogue of Jataka Representations in China*, Münster, Germany: LIT, 2002, p. 17.
9. Ibid., pp. 21, 23.
10. Alexander Cunningham, *The Stûpa of Bharhut: A Buddhist Monument Ornamented with Numerous Sculptures Illustrative of Buddhist Legend and History in the Third Century B.C.*, London: W. H. Allen, 1879, p. 4.
11. S. C. Kala, *Terracottas in the Allahabad Museum*, New Delhi: Abhinav, 1980, p. xviii.
12. Ananda Kentish Coomaraswamy, *Yaksas: Part II*, Washington, DC: Freer Gallery of Art, Smithsonian Institution, 1931, p. 1.
13. Cunningham, *Stûpa of Bharhut*, p. 20.
14. Richard S. Cohen, "Shakyamuni: Buddhism's Founder in Ten Acts," in David Noel Freedman and Michael James McClymond, eds., *The Rivers of Paradise: Moses, Buddha, Confucius, Jesus, and Muhammad as Religious Founders*, Grand Rapids, MI: W. B. Eerdmans, 2001, pp. 184–185.
15. Richard S. Cohen, "Nāga, Yakṣiṇī, Buddha: Local Deities and Local Buddhism at Ajanta," *History of Religions*, vol. 37, no. 4, 1998, pp. 360–400.
16. Kala, *Terracottas in the Allahabad Museum*, p. 53.
17. Padmanabh Shreevarma Jaini, *Collected Papers on Jaina Studies*, Delhi: Motilal Banarsidass, 2000, pp. 272–273.
18. Coomaraswamy, *Yaksas: Part II*, p. 5.
19. Alice Getty, *The Gods of Northern Buddhism: Their History and Iconography*, New York: Dover, 1988, pp. 84–85.
20. Ananda Kentish Coomaraswamy, *Yaksòas: On the Worship and Iconography of Yaksòas*, Washington DC: Freer Gallery of Art, Smithsonian Institution, 1928, p. 9.
21. D. Wujastyk, *The Roots of Ayurveda: Selections from Sanskrit Medical Writings*, London: Penguin Books, 2003, p. 165.
22. Richard S. Cohen, *The Splendid Vision: Reading a Buddhist Sutra*, New York: Columbia University Press, 2012, p. 110; see also Cohen, "Nāga, Yakṣiṇī, Buddha," p. 382.
23. Coomaraswamy, *Yaksòas*, p. 18.
24. Ibid., p. 28.
25. Ibid., p. 80.
26. Arabinda Ghosh, *Remains of the Bharhut Stupa in the Indian Museum*, Calcutta: Indian Museum, 1978, p. 28.
27. Ibid., pp. 37–38.

28. Ibid., pp. 41, 44.
29. Vogel, *Indian Serpent-lore*, p. 95.
30. Ibid., p. 113.
31. Ibid., pp. 87–98.
32. Ibid., pp. 107–108.
33. Lowell W. Bloss, "The Buddha and the Naga: A Study in Buddhist Folk Religiosity," *History of Religions*, vol. 13, no. 1, 1973, p. 7.
34. E. H. Brewster, *The Life of Gotama the Buddha: Compiled Exclusively from the Pali Canon*, New York: Routledge, 2013, pp. 51–52.
35. T. W. Rhys Davids, *Buddhist India*, London: T. Fisher Unwin, 1903, pp. 220–221.
36. Jaini, *Collected Papers*, p. 274.
37. Shaw, *Buddhist Landscapes in Central India*, p. 187.
38. Riben Fais, "Birth of the Buddha in the Early Buddhist Art Schools," in Tiziana Pontillo and Maria Piera Candotti, eds., *Signless Signification in Ancient India and Beyond*, London: Anthem Press, 2014, pp. 201–241.
39. Ibid., pp. 344–345.
40. See the preface to R. Sarkar, ed., *Sriprayaga-Mahatmya*, Calcutta: Kebalram Chattopadhyay, 1910.
41. Brewster, *Life of Gotama the Buddha*, p. 49.
42. See Eugène Burnouf, *Introduction to the History of Indian Buddhism*, trans. Katia Buffetrille and Donald S. Lopez Jr., Chicago: University of Chicago Press, 2010, p. 369, fn. 265.
43. Brewster, *Life of Gotama the Buddha*, p. 50.
44. Ibid., p. 51.
45. Ibid., p. 52.
46. Ibid., p. 55.
47. Reginald A. Ray, *Buddhist Saints in India: A Study in Buddhist Values and Orientations*, New York: Oxford University Press, 1994, p. 295.
48. Burnouf, *Introduction to the History of Indian Buddhism*, pp. 369–370.
49. Thera Mahanama-Sthavira and Douglas Bullis, *Mahavamsa: The Great Chronicle of Sri Lanka*, Fremont, CA: Asian Humanities Press, 2012, p. 204.
50. H. S. S. Nissanka, *Maha Bodhi Tree in Anuradhapura, Sri Lanka: The Oldest Historical Tree in the World*, New Delhi: Vikas, 1994, pp. 34–35.
51. Victor W. Turner and Edith L. B. Turner, *Image and Pilgrimage in Christian Culture*, New York: Columbia University Press, 1978, p. 6.
52. Drekmeier, *Kingship and Community in Early India*, p. 113.
53. Shastri, "Asoka and His Successors," p. 230.
54. Hermann Oldenberg, *The Dipavamsa, an Ancient Buddhist Historical Record*, New Delhi: Asian Educational Services, 1982, pp. 8–9.
55. T. W. Rhys Davids, *Buddhism: Being a Sketch of the Life and Teachings of Gautama, the Buddha*, New York: E. and J. B. Young, 1894.

56. James P. McDermott,"The Kathāvatthu Niyāma Debates," *Journal of the International Association of Buddhist Studies*, vol. 12, no. 1, 1989, p. 139.
57. Story and Vajira, *Last Days of the Buddha*, p. 64.
58. Gregory Schopen, *Bones, Stones, and Buddhist Monks: Collected Papers on the Archaeology, Epigraphy, and Texts of Monastic Buddhism in India*, Honolulu: University of Hawaii Press, 1997, pp. 30, 100.
59. Ibid., p. 33.
60. Akira Hirakawa, *A History of Indian Buddhism: From Śākyamuni to Early Mahāyāna*, Delhi: Motilal Banarsidass, 1993, p. 134.
61. Bibhuti Baruah, *Buddhist Sects and Sectarianism*, New Delhi: Sarup, 2000, p. 69.
62. Carla Sinopoli and Thomas Trautmann, "In the Beginning was the Word: Excavating the Relations between History and Archeology in South Asia," in Norman Yoffee and Bradley L. Crowell, eds., *Excavating Asian History: Interdisciplinary Studies in Archaeology and History*, Tucson: University of Arizona Press, 2006, p. 212.
63. A. S. Amar, "Sacred Bodh Gaya: The Buddhaksetra of Gotama Buddha," in David Geary, Matthew R. Sayers, and Abhishek Singh Amar, eds., *Cross-Disciplinary Perspectives on a Contested Buddhist Site: Bodh Gaya Jataka*, London: Routledge, 2012, p. 30.
64. Shaw, *Buddhist Landscapes in Central India*, pp. 112–113.
65. Étienne Lamotte, *History of Indian Buddhism: From the Origins to the Saka Era*, Louvain-la-Neuve, France: Université Catholique de Louvain, Institut Orientaliste, 1988, p. 392. See also Romila Thapar, *Early India*, p. 210.
66. Vincent Smith, eminent Indologist and historian of ancient India, argued that Pushyamitra Shunga not only revived Brahmin-centered Hinduism but "indulged in a savage persecution of Buddhism." See Vincent Smith, *Early History of India*, 2nd rev. ed., Oxford, UK: Clarendon Press, 1907, p. 190. The archaeological evidence for Pushyamitra's destruction of Buddhist stupas and monasteries is scant. For debates about the so-called Brahminical revival advanced by Indian nationalist historians, see Daniel Michon, *Archaeology and Religion in Early Northwest India: History, Theory, Practice*, New York: Routledge, 2015, pp. 105–107.
67. Drekmeier, *Kingship and Community*, pp. 96–97.
68. Hirakawa, *History of Indian Buddhism*, p. 223.
69. Shaw, *Buddhist Landscapes in Central India*, 2007, p. 135.
70. Ibid., pp. 384–385.
71. Ibid., p. 137.
72. Richard F. Gombrich, *Theravada Buddhism: A Social History from Ancient Benares to Modern Colombo*, 2nd ed., New York: Routledge, 2006, pp. 53–54.
73. Lars Fogelin, *An Archaeological History of Indian Buddhism*, New York: Oxford University Press, p. 108.
74. J. E. Neelis, *Early Buddhist Transmission and Trade Networks: Mobility and Exchange Within and Beyond the Northwestern Borderlands of South Asia*, Leiden, The Netherlands: E. J. Brill, 2011, pp. 24–25.

75. Ibid., p. 56.
76. Hans T. Bakker, "Monuments to the Dead in Ancient North India," *Indo-Iran Journal*, no. 50, 2007, p. 17.
77. Ibid., pp. 40–42.
78. J. N. Tiwari, *Disposal of the Dead in the Mahabharata: A Study in the Funeral Customs in Ancient India*, Varanasi, India: Kishor Vidya Niketan, 1979, p. 24.
79. Bakker, "Monuments to the Dead," p. 15.
80. Julia Shaw, "Buddhist and Non-Buddhist Traditions in Ancient India," in Colin Renfrew, Michael J. Boyd, and Iain Morley, eds., *Death Rituals, Social Order and the Archeology of Immortality in the Ancient World*, Cambridge, UK: Cambridge University Press, 2016, p. 384.
81. On the long and unbroken tradition of megalithic burials dedicated in ashes and urns, see P. C. Pant, "The Megaliths of Jangal-Mahal, and the Vedic Tradition," in V. N. Mishra and Peter Bellwood, *Recent Advances in Indo-Pacific Prehistory*, Leiden, The Netherlands: E. J. Brill, 1985.
82. Fogelin, *Archaeological History of Indian Buddhism*, p. 185.
83. Kathryn R. Blackstone, *Women in the Footsteps of the Buddha: Struggle for Liberation in the Therigatha*. Surrey, London: Curzon Press, 1998, p. 48.
84. Susan Murcott, *First Buddhist Women: Poems and Stories of Awakening*, Berkeley, CA: Parallax Press, 2006. pp. 58–59.
85. Ibid., pp. 60–61.
86. B. C. Law, *Geography of Early Buddhism*, Varanasi, India: Bharatiya, 1973, p. xvii.
87. Patrick Olivelle, *The Dharmasutras: The Law Codes of Apastamba, Gautama, Baudhayana, and Vasistha*, Oxford, UK: Oxford University Press, 1999, p. 134.
88. Ibid., p. xxxiii.
89. Law, *Geography of Early Buddhism*, p. 3.
90. B. N. Chaudhury, *Buddhist Centers in Ancient India*, Calcutta: Sanskrit College, 1969, pp. 12–13.
91. Ibid., pp. 8–9.
92. Neelis, *Early Buddhist Transmission*, p. 10.
93. Moti Chandra, *Kashi Ka Itihas*, Bombay: Hindi Grantha-Ratnakar, 1962, p. 16.
94. H. Saddhatissa, *The Sutta-Nipata: A New Translation from the Pali Canon*, New York: Routledge, 2013, p. 5.
95. Chaudhury, *Buddhist Centers*, p. 85.
96. Neelis, *Early Buddhist Transmission*, p. 10.
97. Erik Zürcher and J. A. Silk, *Buddhism in China: Collected Papers of Erik Zürcher*, Leiden, The Netherlands: E. J. Brill, 2014, p. 545.
98. Dilip Kumar Chakrabarti, *The Archaeology of Ancient Indian Cities*, Delhi: Oxford University Press, 1995, pp. 200–215.
99. For a history of Buddhist monastic orders and their intimate ties to trade, patronage, and urban centers, see James Heitzman, "The Urban Context of Early

Buddhist Monuments in South Asia," in Jason Hawkes and Akira Shimada, eds., *Buddhist Stupas in South Asia: Recent Archeological, Art-historical and Historical Perspectives*, Delhi: Oxford University Press, 2009, pp. 192–215.

100. Kevin Trainor, *Relics, Ritual, and Representation in Buddhism: Rematerializing the Sri Lankan Theravada Tradition*, Cambridge, UK: Cambridge University Press, 1997, pp. 129–130.
101. Christopher Y. Tilley, *Interpreting Landscapes: Geologies, Topographies, Identities*, Walnut Creek, CA: Left Coast Press, 2010, p. 27.
102. Denis E. Cosgrove, *Social Formation and Symbolic Landscape*, Madison: University of Wisconsin Press, 1998, pp. 33–34.

Chapter 6. The Goddess of Fortune

1. John F. Fleet, *Corpus Inscriptionum Indicarum*, vol. 3, Calcutta: Government of India, 1888, pp. 53–54.
2. Ibid., p. 54.
3. Ibid., p. 125.
4. See Ramakrishna Gopal Bhandarkar, "On the Date of Patanjali and the King in Whose Reign He Lived," in *Indian Antiquary*, October 4, 1872, pp. 299–300.
5. Michael Witzel, "Brahmanical Reactions to Foreign Influences and to Social and Religious Change," in Patrick Olivelle, ed., *Between the Empires: Society in India 300 BCE to 400 CE*, New York: Oxford University Press, 2006, pp. 478–479.
6. See Raychaudhuri, *Political History of Ancient India*, pp. 418–419; and Radhakumud Mookerji, *The Gupta Empire*, Bombay: Hind Kitabs, 1947, pp. 3–4.
7. Mookerji, *Gupta Empire*, p. 8.
8. Sonya R. Quintanilla, *History of Early Stone Sculpture at Mathura, ca. 150 BCE–100 CE*, Leiden, The Netherlands: E. J. Brill, 2007, p. 257.
9. Dineschandra Sircar, *Studies in the Religious Life of Ancient and Medieval India*, Delhi: Motilal Banarsidass, 1971, p. 134. See also D. K. Chakrabarti, "Post-Mauryan States of Mainland South Asia (c. BC 185–AD 320)," in Allchin, *Archaeology of Early Historic South Asia*, p. 295.
10. Umakant Premamand Shah, *Jaina-Rupa-Mandana*, New Delhi: Abhinav, 1987, p. 7.
11. Alexander Cunningham, *Report of Tours in the Gangetic Provinces from Badaon to Bihar in 1875–76 and 1877–78*, Calcutta: Archaeological Survey of India, 1880, pp. 101–102.
12. My discussion here is based on J. H. Marshall's original report. See J. H. Marshall, "Archaeological Exploration in India, 1909–10," *Journal of the Royal Asiatic Society*, vol. 43, no. 1, January 1911, pp. 127–158. See also Charles Higham, *Encyclopedia of Ancient Asian Civilizations*, New York: Facts on File, 2004, pp. 49–51; and Lahiri, "Archeology and Some Aspects of Social History of Early India," p. 14.
13. J. H. Marshall, "Archaeological Exploration in India," p. 129.

14. J. H. Marshall, "Excavations at Bhita," *Annual Report of the Archaeological Survey of India, 1911–12*, Calcutta: Superintendent of Government Printing, 1912, p. 48.
15. F. R. Allchin, "Mauryan Architecture and Art," in *Archaeology of Early Historic South Asia*, p. 231.
16. Ibid., p. 236.
17. F. Kielhorn, ed., *The Vyakarana-Mahabhasya of Patanjali*, Bombay: Government Central Book Depot, 1880, p. 380.
18. Chakrabarti, *Archaeology of Ancient Indian Cities*, p. 189.
19. Alexander Cunningham, *The Ancient Geography of India*, vol. 1, London: Trubner, 1871, pp. 360–362.
20. Ibid., p. 371.
21. Hirananda Shastri, "Excavations at Sankisa," *Journal of the United Provinces Historical Society*, vol. 3, pt. 1, May 1922, p. 113.
22. D. Schlingloff, *Fortified Cities of Ancient India: A Comparative Study*, London: Anthem Press, 2014, pp. 78–79.
23. Chakrabarti, *Archaeology of Ancient Indian Cities*, p. 196.
24. Raychaudhuri, *Political History of Ancient India*, p. 463.
25. Schlingloff, *Fortified Cities of Ancient India*, p. 32.
26. Ibid., p. 26.
27. P. A. Eltsov, *From Harappa to Hastinapura: A Study of the Earliest South Asian City and Civilization*, Boston: E. J. Brill, 2008, p. 91.
28. McCrindle, *Ancient India*, p. 66.
29. F. R. Allchin, "The Mauryan State and Empire," in *Archaeology of Early Historic South Asia*, pp. 202–203.
30. Upinder Singh, *Political Violence in Ancient India*, Cambridge, MA: Harvard University Press, 2017, p. 253. Bahasatimitra or Brihaspatimitra may have been none other than the main ruler of the Shunga Dynasty, Pushyamitra; see K. P. Jayaswal and R. D. Banerji, "The Hatigumpha Inscription of Kharavela," *Epigraphia Indica, Vol. XX, 1929–30* [1933] reprint, Delhi: Archaeological Survey of India, 1983, pp. 75–76.
31. *Epigraphia Indica, Vol. XX*, pp. 86–89.
32. Hektar Alahokon, *The Later Mauryas*, Delhi: Munshiram Manoharlal, 1980, p. 117. See also Raychaudhuri, *Political History of Ancient India*, p. 371.
33. Raychaudhuri, *Political History of Ancient India*, p. 469; see also *The Vayu Purana: Part 1*, trans. G. V. Tagare, Delhi: Motilal Banarsidass, 1987, p. lix. English translation is by the author.
34. Mookerji, *Gupta Empire*, p. 12.
35. R. C. Majumdar and A. S. Altekar, eds., *The Vakataka-Gupta Age*, Delhi: Motilal Banarsidass, 1986, p. 118.
36. R. D. Banerji, *The Age of Imperial Guptas*, Benares: Benares Hindu University, 1933, p. 4.
37. Fleet, *Corpus Inscriptionum*, vol. 3, p. 13.

38. Ibid., p. 340.

39. Ibid., p. 9.

40. For an overview, see Stephen C. Berkwitz, *South Asian Buddhism: A Survey*, London; New York: Routledge, 2010, pp. 28–30.

41. Susan Huntington was among the first scholars to question the widely accepted theory that anthropomorphic images of the Buddha were largely shunned in South Asia in favor or indirect symbols. See S. L. Huntington, "Early Buddhist Art and the Theory of Aniconism," *Art Journal*, vol. 49, no. 4, Winter 1990, pp. 401–402. Other scholars such as Klemens Karlsson have suggested that this was not so much due to doctrinal strictures but a generally de facto norm of everyday visual culture of popular Buddhism as it developed in India. See K. Karlsson, "The Formation of Early Buddhist Visual Culture," *Material Religion*, vol. 2, no. 1, pp. 68–95.

42. Some of the earliest images of the Buddha were directly inspired by the form in which *yakṣas* had been carved in stone. On this, see Pramod Chandra, *The Sculpture of India, 3000 B.C.–1300 A.D.*, Washington, DC: National Gallery of Art, 1985. Robert DeCaroli has pointed out recently that rather than the Mahayana schools initiating the practice of Buddha being venerated in his human form, it might have been the rising popularity of the Buddha image itself that might have inspired certain aspects of the early Mahayana doctrines in India. See Robert DeCaroli, *Image Problems: The Origin and Development of the Buddha's Image in Early South Asia*, Seattle: University of Washington Press, 2015, pp. 7–8.

43. Yuvraj Krishan, *The Buddha Image: Its Origin and Development*, New Delhi: Munshiram Manoharlal, 1996, pp. 42–43.

44. A. K. Coomaraswamy, "The Origin of the Buddha Image," *Art Bulletin*, vol. 9, no. 4, 1927, pp. 317–318.

45. A. K. Coomaraswamy, *History of Indian and Indonesian Art*, New York: Dover, 1985, p. 71.

46. Ibid., p. 46.

47. DeCaroli, *Image Problems*, p. 77.

48. Stella Kramrisch, *Exploring India's Sacred Art*, New Delhi: Indira Gandhi National Centre for the Arts, 1994, pp. 182–184. See also Pratapaditya Pal, *Indian Sculpture: A Catalogue of the Los Angeles County Museum of Art Collection*, Los Angeles: Los Angeles County Museum of Art, 1986, p. 212.

49. Frederick Asher makes the important point that these early Gupta-era Buddhas of the eastern Ganges valley were influenced by Kushan-era workshops of the Mathura region, but not necessarily the work of the Mathura School. See F. Asher, *The Art of Eastern India, 300–800*, Minneapolis: University of Minnesota Press, 1980, pp. 19–20.

50. Such nimbi around the Buddha head are found in the third- and fourth-century Kushan-era murals from Belahan, Central Asia.

51. Majumdar and Altekar, *Vakataka-Gupta Age*, p. 298.

52. See Osmund Bopearachchi and Wilfried Pieper, eds., *Ancient Indian Coins*, Turnhout, Belgium: Brepols, 1998, pp. 52–53.
53. Alexandra van der Geer, *Animals in Stone*, Leiden, The Netherlands: E. J. Brill, pp. 201–202.
54. A. K. Coomaraswamy, *Christian and Oriental Philosophy of Art*, New York: Dover, 1990, pp. 47–48.
55. Banerji, *Age of the Imperial Guptas*, p. 72. See also the *Annual Report of the Archeological Survey, Bengal Circle, for the Year Ending with 1904*, Calcutta: Bengal Secretariat Press, 1904, pp. 18–19.
56. Chandra, *Sculpture of India*, p. 54.
57. For a detailed description of this iconography, see Debala Mitra, "Varaha Cave at Udayagiri—An Iconographic Study," *Journal of the Asiatic Society of Bengal*, vol. 5, 1963, pp. 99–103. See also J. C. Harle, *Gupta Sculpture: Indian Sculpture of the Fourth to the Sixth Centuries AD*, Oxford, UK: Clarendon Press, 1974, pp. 38–42.
58. This point is made by H. V. Stietencron in his essay "Political Aspects of Indian Religious Art," in *Approaches to Iconology*, Leiden, The Netherlands: E. J. Brill, 1986, pp. 19–20.
59. D. C. Sircar, *Studies in Indian Coins*, Delhi: Motilal Banarsidass, 2008, p. 3.
60. Mookerjee, *Gupta Empire*, p. 31.
61. Ibid., p. 35.
62. Banerji, *Age of the Imperial Guptas*, p. 215.
63. Chhanda Mukhopadhyay, "Goddess Ganga on Gupta Coins," *Journal of the Numismatic Society of India*, vol. 44, 1982, pp. 146–148. See also B. N. Mukherjee, *Numismatic Art of India*, New Delhi: Indira Gandhi National Centre for the Arts, 2007, p. 84.
64. See R. N. Saletore, *Early Indian Economic History*, London: Curzon Press, 1973, pp. 334–337.
65. Francine Tissot, ed., *Catalogue of the National Museum of Afghanistan, 1931–1985*, Paris: UNESCO, 2006, pp. 135–136.
66. John M. Rosenfield, *The Dynastic Art of the Kushans*, Berkeley: University of California Press, 1967, p. 51.
67. G. A. Pugachenkova, S. R. Dar, R. C. Sharma, and M. A. Joyenda, "Kushan Art," in Janos Hermatta et al., eds., *History of Civilizations of Central Asia, Vol. 2: The Development of Sedentary and Nomadic Civilizations, 700 B.C. to A.D. 250*, Delhi: Motilal Banarsidass, 1999, p. 328.
68. Calambur Sivaramamurti, *Approach to Nature in Indian Art and Thought*, New Delhi: Kanak, 1980, p. 35.
69. Catherine Glynn, "Some Reflections on the Origins of the Type of Ganga Image," *Journal of the Indian Society of Oriental Art*, vol. 5, 1972, p. 23. For a comparison of these two sites, see C. Sivaramamurti, *The Art of India*, New York: Harry Abrams, 1977, p. 537.

70. Mookerjee, *Gupta Empire*, pp. 145–146.
71. Odette Viennot, *Les Divinités Fluviales Gaṅgā et Yamunā aux Portes des Sanctuaires de l'Inde: Essai d'Évolution d'un Thème Décoratif*, Paris: Presses Universitaires de France, 1964, pp. 9–10.
72. Harle, *Gupta Sculpture*, p. 117.
73. Madho Sarup Vats, *The Gupta Temple at Deoghar*, New Delhi: Archaeological Survey of India, 1952, p. 31.
74. J. F. Blakiston, ed., *Annual Report of the Archeological Survey of India, 1924–25*, Calcutta: Government of India, 1927, pp. 98–99.
75. Walter M. Spink, *Ajanta: History and Development, Volume 5*, Leiden, The Netherlands: E. J. Brill, 2005, p. 127.
76. Stella Kramrisch, *The Hindu Temple: Volume 2*, Delhi: Motilal Banarsidass, 1996, 315–316.
77. The inscription has the following phrase: "*bhagīrathyāmalajala mūrdhābhiṣikta.*" See Fleet, *Corpus Inscriptionum*, vol. 2, pp. 237, 245.
78. Raychaudhuri, *Political History of Ancient India*, p. 149.
79. Bhasa, *Abhisekanatakam: A Play in Six Acts*, Mysore, India: Samskrita Sahitya Sadana, 1960.
80. Ibid., pp. 90–94.
81. Kalidasa, *The Raghuvamsa*, ed. Kasinatha Panduranga Paraba and Srinivasa Venkatarama Sarma, Bombay: Nirnaya-Sagar Press, 1898, p. 390.
82. Ibid., p. 259; see also the comments of Mallinatha discussed by C. R. Devadhar in *Works of Kalidasa: Vol. 2*, Delhi: Motilal Banarsidass, 1993, p. 550.
83. Ibid., p. 349.
84. J. C. Heesterman, *The Inner Conflict of Tradition: Essays in Indian Ritual, Kingship and Society*, Chicago: University of Chicago Press, 1985, pp. 148–149.
85. On lordship and cosmology in Puranic conceptions of medieval Indian kingship, see Daud Ali, "Cosmos, Realm and Property in Early Medieval South Asia," in Martha A. Selby and Indira V. Peterson, eds., *Tamil Geographies: Cultural Constructions of Space and Place in South India*, Albany: State University of New York Press, 2008, pp. 123–124.
86. *Si-yu-ki*, vol. 1, p. xxx.
87. Ibid., p. lvi.
88. It has been suggested that by the phrase "confluence of five rivers," Faxian meant the combined waters of the Ganges, Yamuna, Ghogra, Rapti, and Gandaki. See W. Hoey, "The Five Rivers of the Buddhists," *Journal of the Royal Asiatic Society of Great Britain and Ireland*, 1907, pp. 43–44.
89. Ibid., p. lvii.
90. Raychaudhuri, *Political History of Ancient India*, p. 495.
91. Ibid., p. 474.
92. Fleet, *Corpus Inscriptionum*, vol. 3, p. 206.

93. Ibid.
94. Hans Bakker, *World of the Skandapurana: Northern India in the Sixth and Seventh Centuries*, Leiden, The Netherlands: E. J. Brill, 2014, p. 60; Raychaudhuri, *Political History of Ancient India*, pp. 527–528.
95. D. Devahuti, *Harsha: A Political Study*, Delhi: Oxford University Press, 1998, p. 79.
96. *The Harsa-Carita of Bana*, trans. E. B. Cowell and F. W. Thomas, London: Royal Asiatic Society, 1897, p. 187.
97. For a succinct summary, see C. V. Vaidya, "Harsha and his Times," *Journal of the Asiatic Society of Bombay*, vol. 24, 1914–1917, pp. 236–276.
98. *Harsa-Carita*, p. 195. See also *Si-yu-ki*, vol. 1, p. 213; and Radhakumud Mookerji, *Harsha*, London: Oxford University Press, 1926, p. 29.
99. Mookerji, *Harsha*, pp. 33–34.
100. *Harsa-Carita*, p. 200; *Si-yu-ki*, vol. 1, p. 215.
101. *Si-yu-ki*, vol. 1, p. 218.
102. Ibid., pp. 76–78.
103. Mookerji, *Harsha*, p. 80; see also *Si-yu-ki*, vol. 1, p. 233.
104. Ibid., p. 81.
105. Ibid., p. 87.
106. *Harsa-Carita*, p. 57.
107. Ibid., pp. 58, 60.
108. Ibid., p. 59.
109. Ibid., pp. 80–81.
110. Devahuti, *Harsha*, p. 45.
111. *Harsa-Carita*, p. 185.
112. Xuanzang, *Si-yu-ki, Buddhist Records of the Western World*, translated by Samuel Beal, vol. 2, London: Trubner, 1884, p. 42.
113. Ibid., p. 118.
114. *Si-yu-ki*, vol. 2, pp. 90–91.
115. For a description of these coins, see B. P. Sinha, *Dynastic History of Magadha*, New Delhi: Abhinav, 1977, pp. 133–134.
116. Ibid., p. 146; Mookerji, *Harsha*, p. 42.
117. *Harsa-Carita*, p. 153.
118. *Si-yu-ki*, vol. 1, p. 214.
119. *Harsa-Carita*, p. 124.
120. Ibid., p. 168.

Chapter 7. Crucible of Empires

1. *Si-yu-ki*, vol. 1, p. 188.
2. The Tamilian author Kalki (R. A. Krishnamurti) made the siege of Vatapi and the exploits of the commander of the Pallava army Paranjothi the plot of his famous historical novel *Civakamiyin Capatam* [The Vow of Sivagami].

3. R. Gopalan, *History of the Pallavas of Kanchi*, Madras: University of Madras, 1928, p. 104.
4. See "Gadval Copper Plate Inscription of Vikramaditya I," *Epigraphia Indica*, vol. 10, p. 105.
5. D. P. Dikshit, *Political History of the Chalukyas of Badami*, Delhi: Abhinav, 1980, pp. 163, 165.
6. George Mitchell, *Pattadakal*, Delhi: Oxford University Press, 2001, p. 60.
7. C. Sivaramamurti, "The Story of Ganga and Amrita at Pattadakal," *Oriental Art*, vol. 3, 1957, pp. 20–24.
8. Cathleen Cummings, *Decoding a Hindu Temple: Royalty and Religion in the Iconographic Program of the Virupaksha Temple, Pattadakal*, Woodland Hills, CA: South Asia Studies Association, 2014, pp. 33, 38.
9. Ibid., pp. 74–75, 81, 94; see also Michael Lockwood, *Pallava Art*, Madras: Tambaram Research Associates, 2001, p. 68.
10. Padma Kaimal, "Playful Ambiguity and Political Authority in the Large Relief at Mamallapuram," *Ars Orientalis*, vol. 24, 1994, p. 8.
11. Gabriel Jouveau-Dubreuil and V. S. Swaminadha Dikshitar, *The Pallavas*, New Delhi: Asian Educational Services, 1995, p. 37.
12. E. Hultzsch, ed., *South Indian Inscriptions Vol. II: Part II*, Madras: Superintendent, Government Press, 1892, pp. 347–348, 355. I have translated this line slightly differently.
13. Sivaramamurti, *Art of India*, pp. 211–212.
14. K. R. Srinivasan, *Cave-Temples of the Pallavas*, New Delhi: Archaeological Survey of India, 1993, p. 2.
15. Ibid., p. 5.
16. Marilyn Hirsh, "Mahendravarman I Pallava: Artist and Patron of Māmallapuram," *Artibus Asiae*, vol. 48, no. 1/2, 1987, p. 125. See also Srinivasan, *Cave-Temples of the Pallavas*, p. 79.
17. Ibid., p. 209.
18. *Buddhist Records*, vol. 2, p. 206.
19. Cunningham, *The Ancient Geography of India*, vol. 1, pp. 380–381.
20. I am indebted to Ronald Inden for his thoughts on this idealized conception of territorial authority and its attendant geography centered on an archetypal fluvial, mountainous terrain that is evident in the manner in which the Chalukyas, Rashtrakutas, and Cholas sought to organize and represent their empires. See R. B. Inden, *Imagining India* [1990], Oxford, UK: Blackwell, 1994, pp. 256–257.
21. R. S. Tripathi, *History of Kanauj: To the Moslem Conquest*, Delhi: Motilal Banarsidass, 1959, p. 189.
22. Ibid., p. 193.
23. Ibid., p. 198.
24. The Khalimpur copper plate inscription from the thirty-second regnal year of King Dharmapala states that Gopala, the founder of the Pala line, nominated by

the people themselves, put an end to the "practice of the fishes." See R. C. Majumdar, ed., *The History of Bengal: Vol. 1*, Dacca: University of Dacca, 1943, p. 97.

25. Dinesh Chandra Sen, *Brihat Banga*, vol. 2, Calcutta: Dey's Publishing, 1993, p. 467.
26. R. C. Majumdar, *The Ramacaritam of Sandhyakaranandin*, Rajshahi, Bangladesh: Varendra Research Museum, 1939, p. 5.
27. Pandit Bisheshwar Nath, *History of the Rashtrakutas*, Jaipur: Jaipur Publication Scheme, 1997, p. 4.
28. R. C. Majumdar, *Age of Imperial Kanauj*, Bombay: Bharatiya Vidya Bhavan, 1955, p. 2.
29. A. S. Altekar, *The Rashtrakutas and Their Times*, Pune: Oriental Book Agency, 1934, p. 46.
30. Majumdar, *Ancient India*, p. 283.
31. Fleet, *Corpus Inscriptionum*, vol. 3, p. 14.
32. Ibid., p. 31. See the Sanchi stone inscription of 412–413 C.E.
33. *Harsa-Carita*, p. 225.
34. Steven H. Rutledge, *Ancient Rome as a Museum: Power, Identity, and the Culture of Collecting*, Oxford, UK: Oxford University Press, 2011, pp. 124–125.
35. Ibid., p. 131.
36. Margaret Melanie Miles, *Art as Plunder: The Ancient Origins of Debate about Cultural Property*, New York: Cambridge University Press, 2008, pp. 1–2.
37. *Epigraphia Indica*, vol. 18, p. 238. See also Vibhuti Bhushan Mishra, *The Gurjara Pratiharas and Their Times*, Delhi: S. Chand, 1966, p. 18; and Altekar, *Rashtrakutas*, p. 41.
38. Altekar, *Rashtrakutas*, pp. 53, 57–58. On the *biruda* of Kalivallabha, see J. F. Fleet, "Nilgund Inscription of the Time of Amoghavarsha I," in *Corpus Inscriptonum Indicarum*, vol. 6, p. 105. See also Annette Schmiedchen, *Herrschergenealogie und Religiöses Patronat die Inschriftenkultur der Rāṣṭrakūṭas, Śilāhāras und Yādavas*, Leiden, The Netherlands: E. J. Brill, 2014, p. 76.
39. See "The Baroda Copper Plate Inscription," *Indian Antiquary*, vol. 12, p. 158; and *Epigraphia Indica*, vol. 6, pp. 243, 248.
40. "*Śaradindupadadhavalam-chatradvayaṃ.*" See the Radhapur copper plate inscription of Govinda III in *Epigraphia Indica*, vol. 6, p. 197.
41. This is my own liberal paraphrase of the passage: "*gaṅgāyamunormmadhye-lakṣ mīlīlāravindāni sveta chatrāni.*" See the Sanjan Copper Plate of Amoghavarsha I, *Epigraphia Indica*, vol. 18, p. 244.
42. Mahalingam, *South Indian Polity*, Madras: University of Madras, 1967, pp. 86–87.
43. *Epigraphia Indica*, vol. 6, p. 34.
44. P. L. Gupta, "Nesarika Grant of Govinda III," in D. C. Sircar, ed., *Epigraphia Indica*, vol. 34, Delhi: Government of India, 1963, pp. 125, 131.

45. D. C. Sircar, "Note on the Nesarika Grant of Govinda III," in *Epigraphia Indica*, vol. 34, p. 137.
46. For an analysis of the extended symbolism of the *pālidhvaja*, see Inden, *Imagining India*, p. 251.
47. Mahalingam, *South Indian Polity*, pp. 84, 87.
48. Nath, *History of the Rashtrakutas*, p. 62.
49. The passage in question is as follows: *"gāṅgaugha-santati-nirodha-vivṛddhakīrtih."* See the Surat plates of Karakkaraja Suvarnavarsha, *Epigraphia Indica*, vol. 21, p. 137.
50. "Baroda Copper Plate Inscription," pp. 159, 163.
51. Susan L. Huntington, *The "Pala-Sena" Schools of Sculpture*, Leiden, The Netherlands: E. J. Brill, 1984, p. 186.
52. Tripathi, *History of Kanauj*, p. 233.
53. *Epigraphia Indica*, vol. 18, p. 240.
54. J. F. Fleet, "Sanskrit and Old Canarese Inscriptions," *Indian Antiquary*, vol. 12, 1883, p. 248.
55. Geri H. Malandra, *Unfolding a Maṇḍala: The Buddhist Cave Temples at Ellora*, Albany: State University of New York Press, 1993, p. 9.
56. S. K. Dikshit, "Ellora Plates of Dantidurga," *Epigraphia Indica*, vol. 25, 1940, pp. 29–30.
57. See Ramchandra Gopal Bhandarkar's discussion of the inscription in "The Rashtrakuta King Krishnaraja and Elapura," *Indian Antiquary*, vol. 12, 1883, p. 228.
58. Ramkrishna Gopal Bhandarkar, *Early History of the Dekkan Down to the Mahomedan Conquest*, Bombay: Government Central Press, 1884, p. 48.
59. For a detailed description of the caves and sculptures of Ellora, see Sivaramamurti, *Art of India*, pp. 475–481.
60. Inden, *Imagining India*, p. 258.
61. K. A. Nilakantha Sastri, *The Colas*, 2nd rev. ed., Madras: University of Madras, 1955, p. 176.
62. D. B. Diskalkar, "An Odd Plate of Paramara Siyaka of Vikrama Samvat 1026," *Epigraphia Indica*, vol. 19, 1927–1928, p. 178. See also B. N. Reu, *History of the Rashtrakutas*, Jodhpur, India: Jodhpur Archeological Department, 1933, pp. 89–90.
63. K. G. Krishnan, *Karandai Tamil Sangam Plates of Rajendrachola I*, New Delhi: Archaeological Survey of India, 1984, pp. 9, 17.
64. *Culavamsa, Being the Most Recent Part of the Mahavamsa: Part I*, trans. William Geiger, London: Pali Text Society, 1929, p. 140.
65. Ibid., p. 141.
66. E. Hultzsch, "Inscriptions in the Pasupatisvara Temple at Karuvur," in E. Hultzsch, ed., *South Indian Inscriptions: Volume 3*, Madras: Superintendent, Government Press, 1899, pp. 32–33.

67. Ibid., p. 39.
68. On the phrase "*taḷaikoṇḍa,*" see Sastri, *Colas,* p. 143.
69. *Epigraphy,* July 2, 1906, Report of the Archaeological Survey of India, Southern Circle, p. 68.
70. Ramanujan, "Introduction," in *Speaking of Shiva,* p. 21.
71. A. K. Ramanujan, *Poems of Love and War: From the Eight Anthologies and the Ten Long Poems of Classical Tamil,* New York: Columbia University Press, 1985, p. 289.
72. Ibid., p. 118.
73. Ibid., p. 115.
74. Ibid., p. 113.
75. Sastri, *Colas,* p. 184.
76. David Dean Shulman, *The King and the Clown in South Indian Myth and Poetry,* Princeton, NJ: Princeton University Press, 1985, p. 401.
77. Sastri, *Colas,* p. 207.
78. "Tirukkalar Plate of Rajendra Chola I," in E. Hultzsch and H. Krishna Shastri, eds., *South Indian Inscriptions: Volume III, Miscellaneous Inscriptions from the Tamil Country,* Madras: Government Press, 1929, p. 469.
79. "The Tiruvalangadu Plates," in H. Krishna Sastri, *South-Indian Inscriptions, Vol. 3,* Madras: Superintendent, Government Press, 1920, p. 424.
80. Sastri, *South-Indian Inscriptions, Vol. 3,* pp. 424–425.
81. Sastri alludes to the classical literary conventions for royal eulogy known as *praśasti.* See Sastri, *Colas,* p. 207.
82. P. V. Jagadisa Ayyar, *South Indian Shrines,* New Delhi: Asian Educational Services, 1982, pp. 291–292.
83. S. R. Balasubrahmanyam, *Middle Chola Temples: Rajaraja I to Kulottunga I,* Faridabad, India: Thomson Press, 1975, p. 254; Sastri, *Colas,* pp. 209–210.
84. Richard Davis, "Indian Art Objects as Loot," in Sunil Kumar, ed., *Demolishing Myths or Mosques and Temples? Readings on History and Temple Destruction in Medieval India,* Gurgaon, India: Three Essays Collective, 2008, pp. 54–55. See also Balasubrahmanyam, *Middle Chola Temples,* p. 249.
85. Ali, "Cosmos, Realm and Property," pp. 127–128.
86. Sastri, *South-Indian Inscriptions, Vol. 3,* pp. 416–417.
87. Padma Kaimal, "Shiva Nataraja: Shifting Meanings of an Icon," *Art Bulletin,* vol. 81, no. 3, 1999, p. 391.
88. Ibid., p. 393.
89. Chandra, *Kashi Ka Itihas,* p. 114.
90. Ibid., p. 112. We have evidence of the Ashtamahasthanshailagandhakuti Vihara established by Sthirapala, the Dharmachakramahasthana Vihara built by Queen Kumaradevi, and the Saddharmapravartanachakra Vihara.
91. Al-Mas'ūdī, *El-Mas'udi's Historical Encyclopaedia, Entitled "Meadows of Gold and Mines of Gems,"* London: Oriental Translation Fund of Great Britain and Ireland, 1841, pp. 193–194; Tripathi, *History of Kanauj,* pp. 268–269.

92. Al-Utbi, *The Kitab-i-Yamini*, trans. James Reynolds, London: W. H. Allen, 1858, p. 450.
93. Ibid., pp. 454–455.
94. On the repurposed resources from his Indian raids, see Richard Eaton, "Temple Desecration in Pre-modern India," *Frontline*, December 22, 2000, p. 63.
95. "Tarikh-us Sabuktigin of Baihaki," in H. M. Elliot and John Dowson, *The History of India as Told by Its Own Historians*, vol. 2, London: Trubner, 1869, pp. 123–124.
96. Chandra, *Kashi Ka Itihas*, p. 245.
97. Ibid., p. 146.
98. "Kamauli Plates of the Kings of Kanauj," *Epigraphia Indica*, vol. 4, pp. 28, 101, 104.
99. Hemchandra's *Kumarapala Carita* speaks of the *niarāmayaniśaṅkah* (cured and the free); the *santuṣṭah* (contented); and the *paramāyuṣah* (ageless), oblivious of the dictates of time (*kāla*), cited in Chandra, *Kashi Ka Itihas*, p. 137.
100. Ibid., pp. 121–122.
101. Ibid., p. 123.
102. J. G. Buhler, "On the Age of the Naishada-Charita of Sriharsha," *Journal of the Bombay Branch of the Royal Asiatic Society*, vol. 28, 1871–1872, p. 32.
103. "Deopara Inscription of Vijayasena," *Epigraphia Indica*, vol. 1, pp. 307, 309.
104. G. Bühler, "The Udepur Prasasti of the Kings of Malwa," *Epigraphia Indica*, vol. 1, p. 234.
105. F. Kielhorn, "Badaun Stone Inscription of Lakhanapala," *Epigraphia Indica*, vol. 1, p. 64.
106. Alka Patel, "Architectural Cultures and Empire: The Ghurids in Northern India (ca. 1192–1210)," *Bulletin of the Asia Institute*, New Series, vol. 21, 2007, pp. 38, 45, 47.
107. Problems with the historical veracity and the various dénouements of the Prithviraj story in Rajput and other chronicles have been dealt with at length in Norbert Peabody, *Hindu Kingship and Polity in Precolonial India*, New York: Cambridge University Press, 2003, pp. 19–20; and Cynthia Talbot, *The Last Hindu Emperor: Prithviraj Chauhan and the Indian Past, 1200–2000*, Cambridge, UK: Cambridge University Press, 2016, pp. 2–4, 62–66.
108. *"Tarikh-i Mubarakshahi,"* in Elliot and Dowson, *History of India*, vol. 4, p. 11.
109. "Jayacandra Prabandha," in the *Puratana Prabandha Sangraha*, ed. Jinavijaya Muni, Calcutta: Adhishthata Sindhi Jaina Gyanpith, 1936, p. 90.
110. James Tod, *Annals and Antiquities of Rajasthan*, vol. 1, London: Routledge, 1914, p. 210.
111. Hasan Nizami, *"Taj ul Masir,"* in Elliot and Dowson, *History of India*, vol. 2, p. 221.
112. Ibid., p. 248.
113. Mahomed Kasim Ferishta, *Rise of the Mahomedan Power in India*, vol. 1, Calcutta: Asiatic Society, 1829, p. 108.
114. See *The Tabaqat-i Nasiri of Minhaj Siraj Juzjani*, ed. Abdul Haiy Habibi Afghani, Lahore, India: University of the Punjab, 1954, p. 478.

115. See G. R. Smith's discussion of the conventions and stereotypes dictating the account of battles such as A-Qadisiyyah and Nihawand in Al-Tabari's account in the "Translator's Foreword," Tabari, *The Conquest of Iran*, Albany: State University of New York Press, 1994, p. xv.
116. F. P. Lock, *The Rhetoric of Numbers in Gibbon's History*, Newark: University of Delaware Press, 2012, p. 22.
117. James Fergusson, *A History of Architecture in All Countries, from the Earliest Times to the Present Day*, London: J. Murray, 1862, p. 648.
118. Ibid.
119. Ibid., p. 647.
120. Maria Fabricius Hansen, *The Eloquence of Appropriation: Prolegomena to an Understanding of Spolia in Early Christian Rome*, Rome: L'Erma di Bretschneider, 2003, p. 263.
121. Finbarr B. Flood, *Objects of Translation: Material Culture and Medieval "Hindu-Muslim" Encounter*, Princeton, NJ: Princeton University Press, 2009, pp. 36–37, 155–157.
122. Richard B. Eaton, "Temple Desecration in Premodern India," in Kumar, *Demolishing Myths*, p. 99.
123. Ferishta, *Rise of the Mahomedan Power*, p. 119.
124. Yahiya Bin Ahmad Sirhindi, *The Tarikh-i Mubarakshahi*, Baroda, India: Oriental Jadunath Institute, 1932, p. 20.
125. J. A. Page, *An Historical Memoir on the Qutb, Delhi*, New Delhi: Lakshmi Book Store, 1970, p. 29.
126. Sunil Kumar, "Qutb and Modern Memory," in Suvir Kaul, ed., *The Partitions of Memory: The Afterlife of the Division of India*, Delhi: Permanent Black, 2001, p. 175.
127. *Epigraphia Indica*, vol. 1, p. 62.
128. A. B. M. Habibullah, *The Foundation of Muslim Rule in India: A History of the Establishment and Progress of the Turkish Sultanate of Delhi, 1206–1290 A.D.*, 3rd rev. ed. [1961], Allahabad, India: Central Book Depot, 1976, p. 58.

Chapter 8. The Making of the Agrarian Heartland

1. *The Matsya Puranam*, eds. A Taluqdar of Oudh and Srisa Chandra Vasu, New York: AMS Press, 1974, p. 288.
2. Dave, *Birds in Sanskrit Literature*, pp. 423–425.
3. For a detailed discussion, see R. C. Hazra, *Studies in the Puranic Records on Hindu Rites and Customs*, Delhi: Motilal Banarsidass, 1987, pp. 45–46.
4. Bana, *Princess Kadambari*, trans. David Smith, New York: New York University Press, 2009, pp. 39–40, 47.
5. Ibid., p. 80.
6. Ibid., pp. 70–71.

7. *Harsa-Carita*, p. 190.
8. Ibid., p. 194.
9. Ibid., p. 225.
10. Ibid., pp. 226–228.
11. Wheeler M. Thackston, trans., *The Baburnama: Memoirs of Babur, Prince and Emperor*, New York: Modern Library, 2002, pp. 338–342.
12. Ibid., pp. 335–338.
13. Ibid., p. 343.
14. Sunil Kumar, "When Slaves Were Nobles: The Shamsi Bandagan in the Early Delhi Sultanate," *Studies in History*, vol. 10, no. 3, 1994, p. 39.
15. The word derives from the Arabic *qiṭ'ā*—section, unit, apportionment. See also Sunil Kumar, *Emergence of the Delhi Sultanate, 1192–1286*, Delhi: Permanent Black, 2007, p. 298.
16. For the balance between plunder and taxation, see Habibullah, *The Foundation of Muslim Rule in India*, p. 211.
17. Sunil Kumar, "Courts, Capitals and Kingship: Delhi and its Sultans in the Thirteenth and Fourteenth Centuries CE," in Albrecht Fuess and Jan-Peter Hartung, eds., *Court Cultures in the Muslim World: Seventh to Nineteenth Centuries*, New York: Routledge, 2011, p. 141.
18. In fact Alauddin Khalji built a new garrison city north of Delhi in Siri. Peter Jackson, "Delhi: The Problem of a Vast Military Encampment," in R. E. Frykenberg, ed., *Delhi through the Ages: Essays in Urban History, Culture and Society*, Delhi: Oxford University Press, 1986, pp. 20–21.
19. Ibid., pp. 19–20.
20. Habibullah, *Foundation of Muslim Rule*, pp. 107, 211.
21. W. H. Moreland, *The Agrarian System of Moslem India*, Allahabad: Central Book Depot, 1929, p. 23.
22. Irfan Habib, "Agrarian Economy," in Tapan Raychaudhuri and Irfan Habib, eds., *The Cambridge Economic History of India, Volume I: c. 1200–c. 1750*, Cambridge, UK: Cambridge University Press, 1982, p. 54.
23. Habibullah, *Foundation of Muslim Rule*, pp. 209, 218.
24. Ibid., p. 212.
25. Ibid., pp. 220–221.
26. Elliot and Dowson, *History of India, Vol. 3*, pp. 184–185.
27. Ibid., pp. 55–56.
28. S. Nurul Hasan, "Zamindars under the Mughals," in R. E. Frykenberg, ed., *Land Control and Social Structure in Indian History*, Madison: University of Wisconsin Press, 1969, p. 21; Habib, "Agrarian Economy," p. 56.
29. Mohammad Habib, "Introduction to Elliot and Dowson's History of India: Vol. II," in Khaliq Ahmed Nizami, ed., *Elliot and Dowson's History of India as Told by Its Own Historians (Vol. II)*, Aligarh, India: Cosmopolitan, 1936, pp. 69–71.

30. E. Dennison Ross, "The Genealogies of Fakhr-ud-Din," in T. W. Arnold and R. A. Nicholson, eds., *A Volume of Oriental Studies*, Cambridge, UK: Cambridge University Press, 1922, p. 398.
31. Ibid., p. 299.
32. Mohammad Habib and Afsar Khan, *The Political Theory of the Delhi Sultanate Including a Translation of Ziauddin Barani's Fatawa-i Jahandari, Circa 1358–9 A.D.*, Allahabad: Kitab Mahal, 1958, pp. 86, 92.
33. Ibid., pp. 47–48.
34. Ibid., p. 92.
35. Chandra, *Kashi Ka Itihas*, p. 190.
36. Jinaprabha Suri, *Tirthakalpah: A Treatise on the Sacred Places of the Jainas*, ed. D. R. Bhandarkar and Kedarnath Sahityabhusana, Calcutta: Asiatic Society of Bengal, 1942, pp. 243–244.
37. Ibid., p. 251.
38. Chandra, *Kashi Ka Itihas*, p. 193.
39. Habibullah, *Foundation of Muslim Rule*, p. 121.
40. A. Ghosh, J. Gupta, and H. Chattopadhyay, *Bikarampura Ramapalera Itihasa*, Kolkata: Dey's Publishing, 2004, p. 254.
41. Habibullah, *Foundation of Muslim Rule*, pp. 125–126.
42. Kumar, *Emergence of the Delhi Sultanate*, p. 168.
43. Sayid-Ahmad Khan, ed., *The Tarikh-i Ferozshahi of Zia-al-Din Barni*, Calcutta: W. N. Lees, 1862, p. 58.
44. Kumar, *Emergence of the Delhi Sultanate*, pp. 333–335.
45. Ibid., pp. 57, 59.
46. Ibid.
47. Abdul Malik Isami, *Futuh us-Salatin*, ed. A. S. Usha, Madras: Madras University Press, 1948, pp. 324–325.
48. Habib, "Introduction," pp. 69–70.
49. R. H. Phillimore, *Historical Records of the Survey of India: Volume 1, The Eighteenth Century*, Dehra Dun: Office of the Geodetic Branch, Survey of India, 1945, p. 23.
50. Ibid., p. 26.
51. Ibid., p. 29.
52. Edward Dowdeswell Lockwood, *Natural History, Sport, and Travel*, London: W. H. Allen, 1878, p. 2.
53. E. P. Stebbing, *The Forests of India*, vol. 1, London: John Lane, 1922, pp. 35–36.
54. Dietrich Brandis, "An Enumeration of the Dipterocarpaceae," *Journal of the Linnean Society*, vol. 31, 1895–97, pp. 6–7.
55. For a succinct summary, see Raychaudhuri and Habib, *Cambridge Economic History*, pp. 4–5.
56. Neel Amin, "Those Who Wandered Were Lost: Settling the Nomadic Banjaras of British India," PhD diss., University of California, Davis, 2016, pp. 92–139.

57. Elliot and Dowson, *History of India*, vol. 3, p. 158.
58. B. K. Sarkar, *Inland Transport and Communication in Medieval India*, Calcutta: Calcutta University Press, 1925, p. 48.
59. Thomas Roe, *The Embassy of Sir Thomas Roe to the Court of the Great Mogul, 1615–1619, as Narrated in His Journal and Correspondence*, London: Hakluyt Society, 1899, p. 88.
60. Jean Baptiste Tavernier, *Travels in India*, vol. 1, trans. V. Ball, London: Macmillan, 1889, p. 40.
61. Tim Dyson, "India's Population: The Past," in Tim Dyson, Robert Cassen, and Leela Visaria, eds., *Twenty-First Century India: Population, Economy, Human Development, and the Environment*, New Delhi: Oxford University Press, 2005, p. 17. See also John F. Richards, *The Mughal Empire*, Cambridge, UK: Cambridge University Press, 1993, p. 190.
62. Irfan Habib, "Population," in Raychaudhuri and Habib, *The Cambridge Economic History of India, Volume I*, pp. 166–167; see also Shireen Moosvi, "Production, Consumption and Population in Akbar's Time," in *Indian Economic and Social History Review*, vol. 10, no. 2, 1973, pp. 194–195.
63. Kishori Saran Lal, *Growth of Muslim Population in Medieval India, A.D. 1000–1800*, Delhi: Research Publications in Social Sciences, 1973, pp. 43–44.
64. Khan, *Tarikh-i Ferozshahi*, p. 147.
65. K. A. Nizami, *Supplement to Elliot and Dowson's History of India: Vol. III, the Khaljis and the Tughluqs*, Delhi: Idarah-i Adabiyat-i Delli, 1981, p. 21.
66. Isami, *Futuh us-Salatin*, p. 314.
67. Elliot and Dowson, *History of India*, vol. 3, p. 193.
68. Ibid., p. 196.
69. Ibid., p. 197.
70. For a comprehensive discussion of Aluddin's market reforms, see Satish Chandra, *Medieval India: From Sultanat to the Mughals: Part One*, New Delhi: Har-Anand, 2004, pp. 81–86.
71. Isami, *Futuh us-Salatin*, pp. 314–315.
72. Irfan Habib, "The Price Regulations of Alauddin Khalji—A Defence of Zia Barani," *Indian Economic and Social History Review*, vol. 21, no. 4, 1984, pp. 393–414.
73. Cited by Irfan Habib, "The Economic History of Medieval India: A Survey," in *Essays in Indian History: Towards a Marxist Perception*, London: Anthem Press, 2002, p. 385. See also W. Ivanow, "Letters of Mahru," *Journal of the Royal Asiatic Society of Great Britain and Ireland*, no. 4, October 1922, pp. 579–580.
74. Khan, *Tarikh-i Ferozshahi*, pp. 473–474.
75. Thackston, *Baburnama*, p. 335.
76. See the *Malfuzat-i Timuri* in Elliot and Dowson, *History of India*, vol. 3, p. 429.
77. Thackston, *Baburnama*, p. 334.
78. *Tarikh-i Ferozshahi*, pp. 568–570.

79. Elliot and Dowson, *History of India*, vol. 3, p. 302.
80. Henry Piddington, *A Letter to the Most Noble James Andrew, Marquis of Dalhousie, On the Storm-Waves of the Cyclones in the Bay of Bengal and Their Effects in the Sunderbunds*, Calcutta: Baptist Mission Press, 1853, pp. 5–6.
81. R. C. Majumdar, *History of Ancient Bengal*, Calcutta: G. Bharadwaj, 1971, pp. 437–438. See also N. K. Sengupta, *Land of Two Rivers: A History of Bengal from the Mahabharata to Mujib*, New Delhi: Penguin Books India, 2011, p. 48.
82. R. C. Majumdar et al., eds., *The Ramacaritam of Sandhyakaranandin*, Rajshahi, Bangladesh: Varendra Research Museum, 1939, pp. 19–20.
83. Chattopadhyaya, *Studying Early India*, p. 59.
84. Benoy Ghosh, *Pascimbangera Samskrti*, Calcutta: Pustaka Prakasa, 1957, pp. 496–497.
85. Pranabendra Nath Ghosh, ed., *Ibn Batutah's Account of Bengal*, Calcutta: Prajna, 1978, pp. 13–17.
86. Sukumar Sen, *Manasa-Vijaya: A 15th Century Bengali Text*, Calcutta: Asiatic Society, 1953, pp. 142–143.
87. Atul Chandra Roy, *History of Bengal: Turko-Afghan Period*, New Delhi: Kalyani, 1986, p. 110.
88. Sudipta Sen, "Betwixt Hindus and Muslims: The Many Lives of Zafar Khan, the Ghazi of Tribeni," *Asian Ethnology*, vol. 76, no. 2, 2017, p. 13.
89. Ghosh, *Pascimbangera Samskrti*, p. 490. English translation is by the author.
90. Richard Eaton, *The Rise of Islam and the Bengal Frontier, 1204–1760*, Berkeley: University of California Press, 1993, p. 77.
91. Abdur Rahim, ed., *Gaji Kalu o Campabati Kanyara Punthi*, Dhaka, Bangladesh: Hamidiya Library, 1961, p. 16.
92. Ibid., p. 77.
93. Ibid., p. 78. English translation is by the author.
94. Kanangopal Bagchi, *The Ganges Delta*, Calcutta: Calcutta University Press, 1944, pp. 32, 44–48. See also S. R. Khan and M. Badrul Islam, "Holocene Stratigraphy of the Lower Ganges-Brahmaputra River Delta in Bangladesh," *Frontiers of Earth Science in China*, vol. 2, no. 4, 2008, pp. 393–399.
95. Eaton, *Rise of Islam*, pp. xxiii–xxiv, 206–208.
96. Richard Eaton, "Forest Clearing and the Growth of Islam in Bengal," in *Islam in South Asia in Practice*, ed. Barbara D. Metcalf, Princeton, NJ: Princeton University Press, 2009, pp. 375–377.
97. R. Eaton, "Three Overlapping Frontiers in Early Modern Bengal: Religious, Agrarian, Imperial," in Bradley J. Parker, ed., *Untaming the Frontier in Anthropology, Archaeology, and History*, Tucson: University of Arizona Press, 2005, pp. 63–64.
98. Lal, *Twilight*, p. 44.
99. Salma Ahmed Farooqui, *Comprehensive History of Medieval India: From Twelfth to the Mid-Eighteenth Century*, Delhi: Pearson, 2011, p. 90.

100. Lal, *Twilight*, pp. 73–74.
101. Ibid., pp. 83, 104.
102. Ibid., p. 147; for a description of the depredations of the Sharqi army, see Hidayet Husain's genealogical sketch in Mulla Abdul Baqi Nahavandi's biography of Abdur-Rahim Khan-i-Khanan, in M. H. Husain, ed., *Ma 'asir-i Rahimi*, Calcutta: Baptist Mission Press, pp. 105–106.
103. Moreland, *Agrarian System*, p. 68.
104. Ibid., p. 194.
105. Ibid., p. 192.
106. D. N. Lorenzen, *Kabir Legends and Ananta-das's Kabir Parachai*, Albany: State University of New York Press, 1991, pp. 110–113.
107. Ibid., p. 17.
108. Thackston, *Baburnama*, pp. 334–335.
109. Moreland, *Agrarian System*, p. 49.
110. Ibid., p. 51.
111. Ibid., p. 79.
112. Abbas Khan Sarwani, "Tarikh-i Sher Shahi," in John Dowson, ed., *The History of India as Told by Its Own Historians, the Muhammadan Period: The Posthumous Papers of the Late Sir H. M. Elliot* [1872], Calcutta: S. Gupta, 1971, pp. 23–24. During the accession of Ibrahim Lodi, *jāgir*s seem to have been given as gifts along with robes of honor to some Afghan nobles, which would indicate that the term was in currency long before the adoption of the institution by Sher Shah. See Husain, *Ma 'asir-i Rahimi*, p. 480.
113. Moreland, *Agrarian System*, p. 70.
114. Kalikaranjan Qanungo, *Sher Shah: A Critical Study Based on Original Sources*, Calcutta: Kar, Majumder, 1921, pp. 353–354.
115. Abu-'l-Fadl Ibn-Mubarak, *The Ain-i Akbari: Vol. 2*, trans. H. S. Jarrett, Calcutta: Asiatic Society of Bengal, 1891.
116. Moreland, *Agrarian System*, p. 78.
117. K. R. Das, *Raja Todar Mal*, Calcutta: Saraswat Library, 1979, p. 150.
118. W. H. Moreland and A. Yusuf Ali, "Akbar's Land-Revenue System as Described in the 'Ain-i-Akbari,'" *Journal of the Royal Asiatic Society of Great Britain and Ireland*, January 1918, pp. 9, 11–12.
119. Das, *Raja Todar Mal*, pp. 191–192.
120. See Irfan Habib, *The Agrarian System of Mughal India, 1556–1707*, New Delhi: Oxford University Press, 1999, pp. 215–217.
121. Shibratan Mitra, ed., *Gopicandra*, Kolkata: Kaliprasanna Nath, 1919, p. 2.
122. Al-Badaoni, *A History of India: Muntakhabu-t-Tawarikh*, vol. 2, Delhi: Atlantic, 1990, p. 192.
123. J. N. Sarkar, *Studies in Mughal India*, Calcutta: M. C. Sarkar, 1919, pp. 172–174, 177.

124. Irfan Habib, "The Social Distribution of Landed Property in Pre-British India: A Historical Survey," in *Essays in Indian History: Towards a Marxist Perspective*, New Delhi: Tulika, 1995, pp. 98–99. See also Richards, *The Mughal Empire*, p. 143.

125. John F. Richards, *The Unending Frontier: An Environmental History of the Early Modern World*, Berkeley: University of California Press, 2003, p. 5.

126. Thackston, *Baburnama*, p. 439.

127. Sarwani, "Tarikh-i Sher Shahi," p. 42.

128. William Erskine, *History of India under the First Two Sovereigns of the House of Taimur, Babur and Humayun*, vol. 2, London: Longman, Brown, Green and Longmans: 1854, p. 131.

129. Sarwani, "Tarikh-i Sher Shahi," p. 75.

130. Ibid., pp. 72–73, 82–83.

131. Husain, *Ma 'asir-i Rahimi*, pp. 626–627.

132. Qanungo, *Sher Shah*, p. 185.

133. Ibid., p. 194.

134. Sarwani, "Tarikh-i Sher Shahi," p. 91.

135. Qanungo, *Sher Shah*, pp. 215–216.

136. Ibid., p. 219.

137. Sarwani, "Tarikh-i Sher Shahi," p. 100.

138. Husain, *Ma 'asir-i Rahimi*, p. 630.

139. J. D. Hooker, *Observations Made When Following the Grand Trunk Road across the Hills of Upper Bengal*, Calcutta: J. Thomas, 1848, p. 33.

140. Thackston, *Baburnama*, p. 437.

141. Ibid., p. 440.

142. Ibid., p. 457.

143. G. P. Tate, *The Kingdom of Afghanistan: A Historical Sketch*, Bombay: Bennett Coleman, 1911, p. 22.

144. See Mukerji, "Allahabad," p. 652; S. N. Sinha, *Subah of Allahabad under the Great Mughals, 1580–1707*, New Delhi: Jamia Millia Islamia, 1974, p. 22; and F. W. Porter, *Final Settlement Report of the Allahabad District*, Allahabad, India: North-western Provinces and Oudh Government Press, 1878, p. 2.

145. Abd al-Qadir Badauni, *Muntakhabut-Tawarikh*, vol. 2, ed. George S. A. Ranking, W. H. Lowe, and T. Wolsely Haig, Calcutta: Asiatic Society, 1884, p. 179.

146. *Si-yu-ki*, vol. 1, p. 232.

147. *Matsya Puranam*, p. 278.

148. Mukerji, "Allahabad," p. 634. See also the "Report on Monuments in the North-Western Provinces and Oudh Together with a Note on the Works Undertaken," *Report of the Curator of Ancient Monuments in India*, no. 1, 1882.

149. See H. Beveridge, "Garden of Climes (Hadiq-al-Aqalim)," *Imperial and Asiatic Quarterly Review*, 3rd series, vol. 9, no. 17–18, January–April 1900, p. 158.

150. Beveridge deciphered the original term *vratahārī* from oral and written sources. See H. Beveridge, "Note about Mukund Brahmachari," *Imperial and Asiatic Quarterly Review*, 3rd series, vol. 9, no. 17–18, January–April 1900.
151. Beveridge, "Garden of Climes," p. 161.
152. Muhammad Husain Azad, *Darbar-i Akbari*, Lahore, Pakistan: Azad Book Depot, 1910, p. 84.
153. Badauni, *Muntakhabut-Tawarikh*, p. 265.
154. See Beveridge, "Note about Mukund Brahmachari."
155. Jadu Nath Sarkar, *History of Aurangzib: Based on Original Sources, Vol. III, Northern India 1658–1681*, London: Longmans, Green, 1920, p. 301.
156. See Aurangzeb's Banaras *farmān* to Abul Hassan, February 28, 1659. The full text is cited in Sarkar, *History of Aurangzib*, p. 319.
157. Jadunath Sarkar and Raghubir Sinh, *Shivaji's Visit to Aurangzib at Agra; Rajasthani Records: A Collection of Contemporary Rajasthani Letters from the Jaipur State Archives*, Calcutta: Department of History, University of Calcutta, 1963, p. 36. Also Jadunath Sarkar, *Shivaji and His Times*, 2nd ed., London: Longmans, Green, 1920.
158. John Murdoch, *Kasi, or Benares: The Holy City of the Hindus*, London: Christian Literature Society for India, 1897, p. 26.

Chapter 9. The Ganges in the Age of Empire

1. James Rennell, "An Account of the Ganges and Burrampooter Rivers," *Philosophical Transactions of the Royal Society of London*, vol. xv, 1781–1785, p. 41.
2. Charles Lyell, *Principles of Geology*, vol. 1, Chicago: University of Chicago Press, 1990, p. 242.
3. Rennell, "An Account of the Ganges," p. 241.
4. Niharranjan Ray, *Bangalira Itihasa: Adiparba*, Calcutta: Dey's Publishing, [1949] 2008, pp. 85–86.
5. Deepak Kumar, Vinita Damodaran, and Rohan D'Souza, *The British Empire and the Natural World: Environmental Encounters in South Asia*, New Delhi: Oxford University Press, 2011, p. 1. For an overview of the literature, see Corey Ross, *Ecology and Power in the Age of Empire: Europe and the Transformation of the Tropical World*, Oxford, UK: Oxford University Press, 2017, p. 5.
6. *Kabikankana Candi: Dvitiya Bhaga*, ed. D. Sen, C. Bandyopadhyay, and H. Basu, Kolkata: Bangiya Sahitya Parishad, 1926, p. 669.
7. Charles Frederick Danvers, *The Portuguese in India: Being a History of the Rise and Decline of Their Eastern Empire*, New Delhi: Asian Educational Services, [1894] 1992, vol. 1, p. 422.
8. Jamini Mohan Ghosh, *Magh Raiders in Bengal*, Calcutta: Bookland, 1960, pp. 17–18.

9. Tamonash Das Gupta, *Pracina Bangala Sahityera Katha*, Calcutta: University of Calcutta, 1948, p. 96.
10. Ray, *Bangalira Itihasa*, vol. 1, pp. 75–76.
11. João de Barros, *Da Asia de João de Barros e de Diogo de Couto*, Lisboa: Na Regia Officina Typografica, 1778, vol. 2, p. 306.
12. Pedro Barreto de Resende and António Bocarro, *Livro das Plantas de Todas as Fortalezas, Cidades e Povoações do Estado da India Oriental*, ed. Francisco Paulo Mendez Da Luz, Lisbon: Centro de Estudos Historicos Ultrarinos Lisboan, 1960, pp. 91–92.
13. Ray, *Bangalira Itihasa*, vol. 1, pp. 85–86.
14. Father Pierre du Jarric, "A Missionary Tour in Bengal: Translated from the French and Annotated by the Reverend A. Saulière," *Bengal Past and Present*, vol. 14, no. 2, 1907, p. 149.
15. Ibid., pp. 3, 7.
16. Giovanni Pietro Maffei, Martinus Nutius, and John Hay et al., *De Rebus Iaponicis, Indicis et Peruanis Epistolae Recentiores*, Antverpiæ: Ex Officina Martini Nutij, 1605, p. 846.
17. Duarte Pacheco Pereira and Raphael Eduardo de Azevedo Bato, *Esmeraldo de Situ Orbis*, Lisboa: Imprensa Nacional, 1892, pp. 10, 15.
18. Samuel Purchas and Richard Hakluyt, *Hakluytus Posthumus, or Purchas his Pilgrimes*, vol. 10, Glasgow: J. Maclehose, 1905, pp. 113–114.
19. Ibid., p. 182.
20. Piddington, *A Letter to the Most Noble James Andrew*, p. 8.
21. Sanjay Subrahmanyam, "Written on Water: Designs and Dynamics in the Portuguese *Estado da India*," in Susan E. Alcock, Terence N. D'Altroy, Kathleen D. Morrison, and Carla M. Sinopoli, eds., *Empires: Perspectives from Archaeology and History*, New York: Cambridge University Press, 2001, pp. 45–47. See also Sanjay Subrahmanyam, *The Portuguese Empire in Asia: A Political and Economic History*, 2nd ed., Chichester, UK: John Wiley, 2012, pp. 233–234.
22. Das Gupta, *Pracina Bangala Sahityera Katha*, pp. 27–29.
23. Ibid., pp. 28–29.
24. J. J. A. Campos, *History of the Portuguese in Bengal*, Patna, India: Janaki Prakashan, 1979, p. 28.
25. Donald F. Lach, *Asia in the Making of Europe: Volume II*, Chicago: University of Chicago Press, 1970, p. 11.
26. K. M. Mathew, *History of the Portuguese Navigation in India, 1497–1600*, New Delhi: Mittal, 1988, p. 139.
27. The word *bandéi*, which means a foreign settlement or ward, most likely originated in India.
28. Resende and Bocarro, *Livro das Plantas*, pp. 7–9.
29. Stefan Halikowski Smith, *Creolization and Diaspora in the Portuguese Indies: The Social World of Ayutthaya, 1640–1720*, Leiden, The Netherlands; Boston. MA: E. J. Brill, 2011, pp. 16–17.

30. du Jarric, "A Missionary Tour in Bengal," p. 153.
31. Ibid., pp. 155–156.
32. Duarte Barbosa and Mansel Longworth Dames, *The Book of Duarte Barbosa: An Account of the Countries Bordering on the Indian Ocean and their Inhabitants*, New Delhi: Asian Educational Services, 2002, p. 147.
33. Sebastião Manrique, *Itinerário de Sebastião Manrique*, ed. Luís Silveira, Lisbon: Agência-Geral das Colonias, 1946, p. 45.
34. See excerpt from Kafi Kahn's account in D. G. Crawford, *A Brief History of the Hughli District*, Calcutta: Bengal Secretariat Press, 1902, pp. 7–8.
35. The reports of Portuguese firepower may have been exaggerated to provide further reasons for the Mughal siege; see Campos, *History of the Portuguese in Bengal*, p. 229.
36. Rev. H. Hosten, "Three Days at the Bandel Convent in 1920," *Bengal Past and Present*, vol. 26, no. 1, 1923, p. 71.
37. Abdul Hamid Lahori, "Badshah-Nama," in Elliot, *History of India*, vol. 7, pp. 31–32.
38. Christian J. Koot, *Empire at the Periphery: British Colonists, Anglo-Dutch Trade, and the Development of the British Atlantic, 1621–1713*, New York: New York University Press, 2011, pp. 105–106.
39. Henry Yule, ed., *The Diary of William Hedges*, London: Hakluyt Society, 1888, p. 233.
40. Bernier, *Travels in the Mogul Empire*, p. 439.
41. Streynsham Master, *Diaries, 1675–1680*, vol. 2, London: J. Murray, 1911, p. 83.
42. Bernier, *Travels in the Mogul Empire*, p. 365.
43. Ibid., p. 310.
44. Alexander Hamilton, *A New Account of the East Indies*, London: C. Hitch and A. Millar, 1746, p. 20.
45. Kalikinkar Dutta, *The Dutch in Bengal and Bihar, 1740–1825 A.D.*, Delhi: Motilal Banarsidass, 1968, pp. 1–2.
46. John Henry Grose and John Carmichael, *A Voyage to the East Indies*, London: S. Hooper, 1772, p. 306.
47. Bharatchandra, *Bharatacandrera Granthavali*, Kolkata: Basumati Sahitya Mandir, 1997, p. 13.
48. Master, *Diaries*, vol. 2, p. 68.
49. Bernier, *Travels in the Mogul Empire*, p. 439.
50. Ibid.
51. Grose and Carmichael, *Voyage to the East Indies*, p. 134.
52. Sudipta Sen, "Conquest of Marketplaces: Exchange, Authority and Conflict in Early Colonial North India," PhD diss., University of Chicago, 1994, p. 88.
53. Munshi Salimullah, *Tarikh-i-Bangala*, ed. S. M. Imamuddin, Dhaka: Asiatic Society of Bangladesh, 1979, p. 62.
54. C. R. Wilson, *The Early Annals of the English in Bengal*, London: W. Thacker, 1895, p. 142.

55. Alexander Hamilton, "Treats of the Towns, Cities, Country and Customs of Bengal, Particularly of Those Near the Famous Ganges," *Calcutta Journal of Medicine*, vol. 24, May 1906.
56. R. Kipling, "A Tale of Two Cities," in *Works of Rudyard Kipling*, New York: National Library, 1909, p. 294.
57. Adam Smith, *An Inquiry into the Nature and Causes of the Wealth of Nations*, Edinburgh: Thomas Nelson, 1843, p. 30.
58. Nick Robins, *The Corporation That Changed the World*, London: Pluto Press, 2006, p. 31.
59. Peter J. Marshall, *East India Fortunes: The British in Bengal in the Eighteenth Century*, Oxford, UK: Oxford University Press, 1976, p. 109.
60. Sen, "Conquest of Marketplaces," p. 110.
61. Sudipta Sen, *Empire of Free Trade*, Philadelphia: University of Pennsylvania Press, 1998, pp. 82–83.
62. Ibid., p. 115.
63. Jadunath Sarkar and R. C. Dutt, eds., *The History of Bengal*, vol. 2, Dacca: Dacca University, 1943–1948, p. 452.
64. John Zephaniah Holwell, *India Tracts*, 2nd ed., London: T. Becket and P. A. de Hondt, 1764, p. 255. Holwell's original count has been questioned and revised by subsequent commentators and historians. It is most likely that no more than sixty-four people were confined, of whom between eighteen and forty-three perished. See Brijen Kishore Gupta, *Sirajuddaullah and the East India Company, 1756–1757: Background to the Foundation of British Power in India*, Leiden, The Netherlands: E. J. Brill, 1966, pp. 77–78.
65. Karl Marx, *Notes on Indian History*, Moscow: Foreign Languages Publishing, 1947, p. 68.
66. John Malcolm, *The Life of Robert, Lord Clive*, London: J. Murray, 1836, p. 143.
67. David Cressy, *Saltpeter: The Mother of Gunpowder*, Oxford, UK: Oxford University Press, 2013, p. 147.
68. See the consultations of 3rd January, 1749 in James Long, *Selections from Unpublished Records of Government for the Years 1748–1767*, Calcutta: Superintendent of Government Printing, 1869, pp. 15–16.
69. G. B. Malleson, *The Decisive Battles of India: From 1746 to 1849*, London: W. H. Allen, 1883, pp. 112–113.
70. Arthur Broome, *History of the Rise and Progress of the Bengal Army*, vol. 1, Calcutta: W. Thacker, 1850, p. 270.
71. Sen, "Conquest of Marketplaces," p. 115.
72. Ibid., p. 116.
73. H. Vansittart, *A Narrative of the Transactions in Bengal, 1760–1764*, ed. A. C. Banerjee and B. K. Ghosh, Calcutta: K. P. Bagchi, [1766] 1976, pp. 148–149.

74. Ibid., pp. 192–193.
75. Ibid., pp. 182–183.
76. George Forrest, *The Life of Lord Clive*, vol. 2, London: Cassell, 1918.
77. Abdul Majed Khan, *The Transition in Bengal, 1756–1775: A Study of Seiyid Muhammad Reza Khan*, Cambridge, UK: Cambridge University Press, 2008, p. 100.
78. William Wilson Hunter, *Annals of Rural Bengal*, vol. 1, London: Smith, Elder, 1871, pp. 21–22.
79. Ibid., p. 25.
80. Ibid., p. 26.
81. Ibid.
82. Ghulam Husain Khan Tabatabai, *Seir Mutaqherin: Or Review of Modern Times*, vol. 3, Calcutta: R. Cambray, 1902, p. 26.
83. Letter to the Court of Directors dated November 3, 1772, cited in W. W. Hunter, *Bengal Ms. Records: A Selected List of 14,136 Letters in the Board of Revenue, Calcutta, 1782–1807*, London: W. H. Allen, 1894, p. 51.
84. Hunter, *Annals*, p. 19.
85. Edmund Burke, *The Works of the Right Honorable Edmund Burke*, vol. 7, London: G. Bell, 1889, p. 254.
86. Guillaume Thomas François Raynal, *A History of the Two Indies: A Translated Selection of Writings from Raynal's Histoire Philosophique et Politique des Etablissements des Europeens dans les des Deux Indes*, Aldershot, UK: Ashgate, 2006. A six-month store of grain was procured from the districts of Dhaka and Bakherganj for the troops in Bihar. See the *Report of the Indian Famine Commission, 1880–1885*, New Delhi: Agricole, [1898] 1989, p. 1.
87. Ranajit Guha, *A Rule of Property for Bengal: An Essay on the Idea of Permanent Settlement*, Durham, NC: Duke University Press, [1963] 1996, pp. 47–49.
88. Khan, *Transition in Bengal*, pp. 216, 219.
89. Ibid., pp. 304–306.
90. Extract of a Letter from Mr. Becher to the President, dated May 24, 1769, cited in *Reports from Committees of the House of Commons*, London: n. p., 1803, p. 931.
91. R. B. Ramsbotham, *Studies in the Land Revenue History of Bengal 1769–1787*, London: Oxford University Press, 1926, p. 16.
92. Ibid.
93. Burke, *Works*, p. 244.
94. B. B. Misra, *The Central Administration of the East India Company, 1773–1784*, Bombay: Oxford University Press, 1959, pp. 114–115.
95. Sudipta Sen, "Liberal Governance and Illiberal Trade: The Political Economy of 'Responsible Government' in Early British India," in Kathleen Wilson, ed., *The New Imperial History: Culture, Identity and Modernity 1660–1836*, Cambridge, UK: Cambridge University Press, 2004, pp. 143–144.

96. *The Fifth Report from the Select Committee of the House of Commons on the Affairs of the East India Company*, ed. Walter K. Firminger, Calcutta: R. Cambray, p. cccxi.
97. Sen, "Liberal Governance and Illiberal Trade," p. 148.
98. Husain, *Seir Mutaqherin*, pp. 55–56.
99. Sen, *Empire of Free Trade*, p. 98.
100. *A Brief Sketch of the Services of Sir G. H. Barlow, Governor of Madras*, London: E. Blackader, 1811, pp. 9–10.
101. *Barlow's Report on the Trade and Coinage of Benares*, cited in Sen, *Empire of Free Trade*, p. 103.
102. For a comprehensive history of the operation of the Permanent Settlement in Bengal, see N. K. Sinha, *Economic History of Bengal: From Plassey to the Permanent Settlement*, vol. 2, Calcutta: Firma K. L. Mukhopadhyay, 1962. See also Sirajul Islam, *The Permanent Settlement in Bengal: A Study of Its Operation, 1790–1890*, Dhaka: Bangla Academy, 1970, and *Ratnalekha Ray, Change in Bengal Agrarian Society, 1760–1850*, New Delhi: Manohar, 1979.
103. Guha, *Rule of Property*, pp. 104–107.
104. Ibid., pp. 171.
105. Sen, *Empire of Free Trade*, pp. 140–141.
106. Ibid., p. 141.
107. C. A. Bayly, *Rulers, Townsmen and Bazaars: North Indian Society in the Age of British Expansion, 1770–1870*, Cambridge, UK: Cambridge University Press, 1983, pp. 229–230.
108. James Rennell, *Memoir of a Map of Hindoostan, or the Mogul Empire*, London: W. Bulmer, 1793, p. 336.
109. Ibid., p. 337 (italics in the original).
110. See F. C. Hirst, *Notes on the Physical Geography of Bengal: From the Writings and Maps of Major James Rennell*, Calcutta: Bengal Secretariat Book Depot, 1925, pp. 6–7.
111. *W. H. Sleeman, Report on the Depredations Committed by the Thug Gangs of Upper and Central India*, Calcutta: O. H. Huttmann, 1840. See also Kim A. Wagner, *Thuggee: Banditry and the British in Early Nineteenth-Century India*, Basingstoke, UK: Palgrave Macmillan, 2007, p. 113.
112. Captain Thomas Hardwicke, "Narrative of a Journey to Srinagur," *Asiatick Researches*, vol. 6, 1799, pp. 313–314.
113. A. S. Altekar, *Banaras and Sarnath: Past and Present*, Varanasi, India: Benares Hindu University, 1947, p. 24.
114. For a brief discussion, see Rana P. B. Singh, *Banaras: Making of India's Heritage City*, Newcastle, UK: Cambridge Scholars, 2009, p. 82.
115. Mukerji, "Allahabad," pp. 663–664.
116. Nancy G. Cassels, *Religion and Pilgrim Tax under the Company Raj*, New Delhi: Manohar, 1988, pp. 50–51.

117. Partha Mitter, *Much Maligned Monsters: A History of European Reactions to Indian Art*, Chicago: University of Chicago Press, 1992, p. 106.
118. Christopher Hussey, *The Picturesque: Studies in a Point of View*, London: G. P. Putnam, 1927, p. 4.
119. William Hodges, *Travels in India: During the Years 1780, 1781, 1782, and 1783*, London: J. Edwards, 1793, p. 24.
120. Ibid., pp. 25–26.
121. Mildred Archer and William G. Archer, *Indian Painting for the British, 1770–1880*, London: Oxford University Press, 1955, p. 12.
122. Thomas Daniell and William Daniell, *A Picturesque Voyage to India*, London: Longmans, Hurst, Rees and Orme, 1810, p. ii.
123. Emma Roberts, *Scenes and Characteristics of Hindoostan with Sketches of Anglo-Indian Society*, London: W. H. Allen, 1835, p. 171.
124. Ibid., pp. 237, 239.
125. Hermione De Almeida and George H. Gilpin, *Indian Renaissance: British Romantic Art and the Prospect of India*, Burlington, VT: Ashgate, 2005.
126. W. S. Caine, *Picturesque India: A Handbook for European Travellers*, London: Routledge, 1891, p. 301.
127. See the *Journal of the Photographic Society of Bengal*, no. 3, 1857, p. 65; see also Sudeshna Guha, "Material Truths and Religious Identities: The Archeological and Photographic Making of Banaras," in Michael S. Dodson, ed., *Banaras: Urban Forms and Cultural Histories*, New Delhi: Routledge, 2012, pp. 47–48.
128. Fanny Parks, *Wanderings of a Pilgrim in Search of the Picturesque*, vol. 1, London: Pelham Richardson, 1850, p. 86.
129. Proby T. Cautley, *Report on the Ganges Canal Works*, London: Smith, Elder, 1860, p. 6.
130. Ibid., p. 19.
131. Ibid., p. 23.
132. Smith, *Short Account of the Ganges Canal*, p. 4.
133. Cautley, *Report on the Ganges Canal Works*, p. 52.
134. Charles Eliot Norton, *The Opening of the Ganges Canal*, Cambridge, UK: Metcalf, 1855, p. 7.
135. Ibid., p. 11.
136. J. S. Beresford, "Memorandum Dated August, 1875, on the Irrigation Duty of Water, and the Principles on Which Its Increase Depends," in *Punjab Irrigation Branch Papers*, no. 10, 1905, p. 2.
137. Ibid., p. 9.
138. Anthony Acciavatti, *Ganges Water Machine: Designing New India's Ancient River*, San Francisco: Applied Research and Design, 2015, p. 128.
139. Smith, *Short Account of the Ganges Canal*, p. 5.

140. *Railway Engineer*, vol. 9, 1888, p. 304.
141. *Railway Engineer*, vol. 13, 1892, p. 303.
142. See *The Dufferin Bridge: Description of the Oude and Rohilkund Railway Bridge at Benares*, Lucknow, India: London Printing Press, 1887, pp. 9–10.
143. Jawaharlal Nehru, quoted in Dorothy Norman, *Nehru, the First Sixty Years*, New York: John Day, 1965, p. 574.

Epilogue: The Two Bodies of the River

1. See the preface to Ernst Hartwig Kantorowicz, *The King's Two Bodies: A Study in Mediaeval Political Theology*, Princeton, NJ: Princeton University Press, 1997.
2. Burnouf, *Bhagavata Purana*, vol. 3, p. 220.
3. G. N. Das, *The Maxims of Kabir*, New Delhi: Abhinav, 1999, pp. 38–39.
4. Ibid., p. 39. English translations are by the author.
5. Linda Hess and Shukdev Singh, *The Bijak of Kabir*, New York: Oxford University Press, 2002, p. 14. For a discussion of *ulaṭbāṃsī* as a style of composition where words are endowed with hidden meaning, and riddles that can be understood only by the truly discerning (*gyān vicārī*), see Acharya Parshuram Chaturvedi, *Kabir Sahitya ki Parakh*, 2nd ed., Allahabad, India: Bharati Bhandar, 1963, pp. 154–155.
6. K. M. Sen, *Kabira*, vol. 1, Calcutta: Indian Publishing, 1911, p. 79.
7. Daya Ram Abrol, *The B40 Janam Sakhi*, ed. W. H. McLeod, Amritsar, India: Guru Nanak Dev University, 1980, pp. 84–85.
8. Akbar Allahabadi, *Gandhi Nama*, Allahabad, India: Kitabistan, 1948, p. 5.
9. Jawaharlal Nehru, *Discovery of India*, New Delhi: Oxford University Press, 1985, p. 51.
10. V. N. Chibber, *Jawaharlal Nehru: A Man of Letters*, Bangalore: Vikas, 1960, p. 192.
11. *Indian National Congress*, Allahabad, India: All India Congress Committee, 1926, p. 67.
12. *The Farakka Barrage*, New Delhi: Ministry of External Affairs, Government of India, 1976.
13. Punam Pandey, *India Bangladesh Domestic Politics: The River Ganges Water Issues*, Singapore: Springer, 2016, pp. 8–12.
14. S. K. Jain, Pushpendra K. Agarwal, and V. P. Singh, *Hydrology and Water Resources of India*, Dordrecht, The Netherlands: Springer, 2007, p. 377.
15. "Silt and Earthquakes Threaten New Dam," *New Scientist*, no. 1505, April 24, 1986.
16. Fred Pearce and Rob Butler, "The Dam That Should Not Be Built," *New Scientist*, no. 1753, January 26, 1991.
17. Sanjeev Khagram, *Dams and Development: Transnational Struggles for Water and Power*, Ithaca, NY: Cornell University Press, 2004, pp. 54–55.

18. David L. Haberman, *River of Love in an Age of Pollution: The Yamuna River of Northern India*, Berkeley: University of California Press, 2006, pp. 72–73.
19. Sunderlal Bahuguna and Tinzin Rigzin, *Fire in the Heart, Firewood on the Back*, Silyara, India: Parvatiya Navjeevan Mandal for Save Himalaya Movement, 1997, p. 83; see also Haripriya Rangan, *Of Myths and Movements: Rewriting Chipko into Himalayan History*, London: Verso, 2000, pp. 27–28.
20. Thomas Weber, *Hugging the Trees: The Story of the Chipko Movement*, New Delhi: Penguin, 1989, p. 75.
21. Bahuguna and Rigzin, *Fire in the Heart*, p. 183; S. Bahuguna, Vandana Shiva, and Mahesh N. Buch, *Environment Crisis and Sustainable Development*, Dehra Dun, India: Natraj, 1992, p. 312.
22. Richard White, *The Organic Machine*, p. 4.
23. Smith, *A Short Account of the Ganges Canal*, p. 4.
24. *The First Five Year Plan: A Summary*, New Delhi: Government of India, 1952, pp. 74–75.
25. R. Revelle and V. Lakshminarayana, "The Ganges Water Machine," *Science*, vol. 188, no. 4188, May 9, 1975, p. 611.
26. Leo Marx, *The Machine in the Garden* [1964], Oxford, UK: Oxford University Press, 2000, pp. 4–5.
27. Raghu Dayal, "Dirty Flows the Ganga: Why Plans to Clean the River Have Come a Cropper," *Economic and Political Weekly*, vol. 51, no. 21, June 18, 2016, p. 57.
28. See preface to Sivaramamurti, *Ganga*, Delhi: Orient Longman, 1976, p. v.
29. Personal communication with Dr. Sivabrata Chatterjee, former member, Expert Committee for Environmental Clearance of Development Projects, December 2015.
30. A. C. Shukla and V. Asthana, *Ganga: A Water Marvel*, New Delhi: Ashish, 1995, p. 228.
31. Mandal, R. B., *Water Resource Management*. New Delhi: Concept, 2006, p. 86. See also Guy C. Pegram, *River Basin Planning Principles: Procedures and Approaches for Strategic Basin Planning*, Manila, Philippines: Asian Development Bank, 2013, p. 111.
32. James C. Scott, *Seeing Like a State: How Certain Schemes to Improve the Human Condition Have Failed*, New Haven, CT: Yale University Press, 1998, pp. 4–6.

INDEX

The letter "f" after a page number indicates a figure.